Chemometric Methods in Forensic Science

Chemometric Methods in Forensic Science

Edited by

Vishal Sharma
Panjab University, India
Email: vsharma@pu.ac.in

Georgina Sauzier
Curtin University, Australia
Email: georgina.sauzier@curtin.edu.au

and

Simon W. Lewis
Curtin University, Australia
Email: s.lewis@curtin.edu.au

ROYAL SOCIETY
OF **CHEMISTRY**

Print ISBN: 978-1-83916-190-2
PDF ISBN: 978-1-83916-609-9
EPUB ISBN: 978-1-83916-610-5

A catalogue record for this book is available from the British Library

The Royal Society of Chemistry is a charity, registered in England and Wales, Number 207890, and a company incorporated in England by Royal Charter (Registered No. RC000524), registered office: Burlington House, Piccadilly, London W1J 0BA, UK, Telephone: +44 (0)20 7437 8656.

Visit our website at books.rsc.org

Dedications

In loving memory of my father S. N. Sharma, who inspired my love of science and taught me the value of hard work and dedication. This book is dedicated to you and your memory, with all my love.

Vishal Sharma

To Allan for always supporting my choices, however questionable – and to Maxwell, for reminding me that there's more to life than writing.

Georgina Sauzier

To the triumvirate of mentors who set me on this path – Wilhelm van Bronswijk, Max Houck and the late Jay Siegel. And to my father – my (legal) partner in crime for so many years.

Simon Lewis

Chemometric Methods in Forensic Science
Edited by Vishal Sharma, Georgina Sauzier and Simon W. Lewis

Published by the Royal Society of Chemistry, www.rsc.org

Foreword

Analytical chemists may think forensic chemistry is merely the application of their methods to samples of interest in criminal or civil investigations. Not true. Forensic chemistry is allied with but distinct from analytical chemistry in a number of ways. Beyond the ways people may normally think, the ones that put the non-forensic folks (and some forensic professionals) off their tea, forensic chemistry is a separate discipline that works in a defined domain. This foreword will briefly discuss how forensic chemistry is different from analytical chemistry in the perspective of this book, namely, chemometrics. It's not that analytical chemists don't use chemometrics; they do. Rather, it's that forensic chemists need to use chemometrics because of the differences between the two disciplines, outlined below.

First, forensic chemists do not get to choose their samples. Crime scenes, which most forensic chemists never visit, are difficult, confusing, complex, and a strenuous exercise in judgment sampling. From what the criminal(s) have left behind, what has survived any taphonomic (degradation) processes, what has been recognized, and what has been collected, arrives on the forensic chemist's bench to be considered for analysis. It may be old, dirty, compromised, decomposed, microscopic, or all of the above. The forensic chemist may have been dealt a lousy hand but they have to play to the best of their abilities, without bluffing. Also under the heading of samples is the fact that what gets submitted for analysis is at best suspected of being some type of substance, like "presumptive cocaine," "paint chips from the victim's trousers," or "hairs?" which in reality turn out to be nylon fibers. Thus, forensic chemists do not have the analytical luxury of knowing what they are analyzing, at least at first blush. The presumptive cocaine may turn out to be infant formula, for example. In this way, forensic chemists work in a Johari Window of known knowns, known unknowns, unknown knowns, and unknown unknowns. The known knowns are the standards used as

Chemometric Methods in Forensic Science
Edited by Vishal Sharma, Georgina Sauzier and Simon W. Lewis

Published by the Royal Society of Chemistry, www.rsc.org

references for the analysis. Known unknowns, being what the forensic chemist knows they do not know, like what the sample really is, are a mainstay of daily casework. Likewise, the unknown knowns, what the forensic chemist does not know but is nevertheless true about the sample, like whether it is cut with infant formula, aspirin, heroin, or scrubbing powder, are ever present. Finally, the worst, the unknown unknowns are what the chemist does not know and is not known about the sample. To stay with the drug example, while the analyst suspects the drug is cut with something (otherwise, the criminal economics are not profitable) and the drug dealer did cut it with something, the dealer may not know what they actually used. In the U.S Fentanyl may be cut with heroin, the latter's price being driven down by the opioid crisis, making heroin cheaper than other adulterating agents. Each case, each sample is a journey into some amount of uncertainty. All of the foregoing means that forensic chemists need to maximize their samples and any results or interpretations that come from their work.

Another distinction between analytical chemists and forensic chemists is that relevance is of the utmost importance in forensic analysis and research. A fascinating sample that does not help bring the case closer to being solved may be fascinating but it is not very useful. One might consider this to be a subset of the tired "basic *vs.* applied" argument, with analytical chemistry being "applied" and forensic chemistry being "too applied," but upon reflection, that is a poor perspective (see Donald Stokes' excellent book, Pasteur's Quadrant for why this dichotomy is passé). Relevance in any analysis is important but more so in forensic science. Crime scenes can produce enormous amounts of evidence, especially when one is interested in nanograms of possibly useful material, and sorting, categorizing, and deciding on what is most meaningful given the hypotheses (always plural) in the investigation. What would exculpate this person of interest? What would suggest a source for this material on the victim? What analyses would separate one set of things from another? A bad choice or a missed opportunity may give a criminal another chance to rob, rape, or murder. Relevance, therefore, is utmost in the mind of a forensic chemist.

A third factor that separates forensic chemistry from analytical chemistry is, well, the chemistry. Anything, literally anything, can be evidence. From the smallest pollen grains to a collapsed building, the category "evidence" contains all things natural and manufactured, in nearly any state (solid, liquid, gas) or status (fresh, degraded, deconstructed). In fact, except for impression evidence, like fingerprints or toolmarks, all forensic science is forensic chemistry. Modern forensic DNA analysis is really just chromatography, to be honest, to the chagrin of forensic biologists. And the patterned marks fingers leave behind are made up of, yes, chemicals, as are the residues of the discharge of firearms, the outsoles of shoes that make the prints in soils, the drugs the body metabolizes, and so on. Really, forensic chemistry sits at the core of the entire forensic discipline.

Finally, any analysis is meaningless unless it can be communicated. Here is possibly one of the more important ways that any forensic discipline differs from its siblings: Forensic professionals must communicate their results to non-scientists. Whether it is the police, the attorneys, the judges, or, most important, the jurors, forensic reports must be comprehensible to laypeople. Forensic professionals have struggled with how to convey the results of their analyses and comparisons to non-scientists. Bayes' theorem, likelihood ratios, and statistics of all sorts have been published and used to a greater or lesser extent. None have predominated, however. While likelihood ratios seem to be intuitive, they are difficult to calculate and, more importantly, are based on certain assumptions. In medicine and epidemiology, where these calculations are more often used, values like prevalence, true positives, false negatives, and so on are known. In forensic science, we often have no idea what these values are, making the use of Bayes' or likelihood ratios difficult or misleading for many types of evidence. Chemometrics on their own can convey forensic results and their meaningfulness to the average citizen or they could be used to provide more solid priors for likelihood ratios.

Chemometrics is a suite of statistical methods that maximize relevant chemical information, the exact four things outlined as distinguishing forensic chemistry from analytical chemistry. In this book lie numerous examples of the use of chemometrics on a variety of forensic samples. These examples make the case for chemometrics to become a routine and standard approach to interpreting forensic chemical results. It is data-driven, helping to discover patterns and clusters. This classification is at the heart of forensic chemistry, aiding in the sorting of those unknowns discussed earlier. By making sense of complicated methodologies and complex samples, chemometrics stands as the premier approach to communicating forensic meaning to investigators, the courts, and the public. This book is a valuable and significant milestone in supporting scientific understanding, critical thinking, and supporting the rule of law with informed analysis.

Max M. Houck, Ph.D., FRSC
Global Forensic & Justice Center
Florida International University

Preface

Recent years have seen remarkable advancements across the field of forensic chemistry and its related fields. New technologies and instrumental capabilities are continually being developed to enable more sensitive, more specific, or more informative analyses of physical evidence. However, these innovations also give rise to new challenges.

First, how do we ensure that the generated data is fit for purpose within the context of a criminal investigation? This encompasses not only conventional analytical chemistry metrics such as limits of detection, but also factors such as comprehensiveness in providing investigators with all necessary information; relevance to determining the facts of the offence; and timeliness in the light of ever-increasing caseloads. Subsequently, how can practitioners interpret this data and convey the resulting information in a way that is objective, reliable, and meaningful? Forensic experts face the challenging task of distilling vast amounts of information into a format that is accessible to non-experts. As the nature of this information becomes more sophisticated, additional tools are needed to effectively, yet accurately, translate it to lay and judicial audiences.

Chemometrics, or multivariate data analysis, has recently emerged as a means to address these challenges. The use of mathematical algorithms to model chemical data acquired from physical evidence can aid in optimising analytical parameters, performing multivariate quantification, and statistically assessing evidential strength. This in turn can enhance the credibility of practitioners, assist in the quality assurance of methods or their results, and support standardised evaluative reporting frameworks. Additionally, chemometric methods offer a highly efficient means of assessing large quantities of complex data, meeting the need for rapid and accurate analyses.

It is no surprise, then, that the forensic application of chemometrics is rapidly gaining popularity amongst researchers and practitioners alike.

Chemometric Methods in Forensic Science
Edited by Vishal Sharma, Georgina Sauzier and Simon W. Lewis

Published by the Royal Society of Chemistry, www.rsc.org

Multivariate statistical tools have the potential for use across a wide range of forensic disciplines involving chemical analysis, including illicit drug analysis, food forensics, questioned document examination, and environmental forensics. As a result, a thorough understanding of chemometric methodologies is needed for students, researchers, and practitioners of forensic science to remain up-to-date with contemporary practice. Working knowledge of what can or cannot be determined through chemometric means is also relevant to other forensic stakeholders such as laboratory managers, legal counsel, advisory groups, or even the general public.

In this book, we aim to provide a basic understanding of chemometric methods and their application across a range of forensic science disciplines that utilise chemical measurements. This volume will provide critical reviews of chemometric methods used in forensic research, their advantages, limitations, and emerging trends. Moreover, it will provide practitioners with a basic guide for integrating chemometrics within casework, illustrated through relevant case studies. Written in a flexible and accessible manner, this book is intended to serve as a centralised resource for undergraduate and postgraduate students, researchers, practitioners, and others who are interested in exploring the potential applications of chemometrics in forensic science.

We anticipate that the dissemination of knowledge included in this book will not only promote the application of chemometrics in forensic casework but will also be conducive to further research and critical discussion in the field. We hope this book will ultimately contribute to the advancement of forensic science through enhancing the validity and integrity of forensic evidence, thus strengthening the criminal justice system.

Vishal Sharma
Georgina Sauzier
Simon Lewis

Contents

Chemometric Methods in Forensic Science
Edited by Vishal Sharma, Georgina Sauzier and Simon W. Lewis

Published by the Royal Society of Chemistry, www.rsc.org

CHAPTER 1

Introduction

V. SHARMA,*[a] G. SAUZIER*[b] AND S. W. LEWIS*[b]

[a] Institute of Forensic Science and Criminology, Panjab University, India;
[b] School of Molecular and Life Sciences, Curtin University, Australia
*Emails: vsharma@pu.ac.in; georgina.sauzier@curtin.edu.au;
s.lewis@curtin.edu.au

1.1 Origins and Development of Chemometrics

The field of chemometrics originates from the emergence of chemical pattern recognition in the 1960s to 1980s, when scientific computing became readily accessible.[1] This advent in computing technology allowed mainstream scientists to apply multivariate statistics and applied mathematics to their work in the analytical sciences. For example, Jurs *et al.* reported the use of computerised learning machines to interpret combined data from infrared spectroscopy and mass spectrometry in 1969;[2] work that would now be recognised as chemometrics.

The actual term "chemometrics" is thought to have been coined in 1971, when organic chemist Svante Wold first referred to the Swedish equivalent "kemometri" in a grant proposal.[3] Wold is credited as an early pioneer of the field, later co-founding the Chemometrics Society (later renamed the International Chemometrics Society) with Bruce R. Kowalski in 1974. The society initially defined chemometrics as:

> "[t]he science of relating measurements made on a chemical system or process to the state of the system *via* application of mathematical or statistical methods."

Strictly speaking, this definition draws no formal distinction between chemometrics and statistical analytical chemistry. However, Kowalski asserted that

Chemometric Methods in Forensic Science
Edited by Vishal Sharma, Georgina Sauzier and Simon W. Lewis

Published by the Royal Society of Chemistry, www.rsc.org

chemometrics was underpinned by computerised methods to strengthen human pattern recognition capabilities.[4] In this volume, chemometrics is hence defined as the use of computationally intensive, multivariate statistical analyses applied to chemical systems and processes.

Following the establishment of the Chemometric Society, mainstream scientific journals such as Analytical Chemistry and Analytica Chimica Acta began publishing manuscripts featuring the term chemometrics from the mid-1970s. Many analytical chemists expressed scepticism of this early work, believing that complex data analysis was simply a way of compensating for poorly conducted experiments. It was not until the mid-1980s that chemometrics had become sufficiently well accepted that the first dedicated journals to the field, Journal of Chemometrics and Chemometrics and Intelligent Laboratory Systems, were introduced by the publishers Wiley and Elsevier, respectively.

Due to its roots in pattern recognition, early chemometrics was predominantly focussed on discrimination and classification algorithms such as partial least squares (PLS) modelling and principal component analysis (PCA). Contemporary chemometrics is recognised as incorporating a much wider range of machine learning methods, including multivariate regression and experimental design. Again, it is important to recognise that many of these methods were well-established prior to the recognition of chemometrics as a discipline. Simultaneously, chemometrics remains an evolving field with new or modified approaches continually being developed, an example being the introduction of random forest (RF) algorithms in the late 1990s. This evolution has been facilitated by increasing computational power allowing more sophisticated data analyses, along with open-source statistical packages such as R that allow scientists to develop and distribute customised algorithms.

1.1.1 Chemometrics as a Multidisciplinary Field

Although applied to chemical systems, the approaches used in chemometrics originate from the domains of computer science and mathematics. Additionally, chemical analyses are routinely used across the physical sciences, life sciences, and applied sciences. Chemometrics is hence not restricted to purely chemical problems, but instead interfaces with several fields including geology, biology, medicine, and engineering (see Figure 1.1). Within a forensic context, these methods can similarly be applied across diverse forensic disciplines including forensic chemistry, forensic anthropology, document examination, food forensics, pathology, forensic toxicology, environmental forensics, and criminalistics.

1.2 Chemometrics for the Forensic Enterprise

Interest in chemometrics applied to forensic disciplines is relatively recent but represents a rapidly growing field. Over the last 15 years, a steady rise

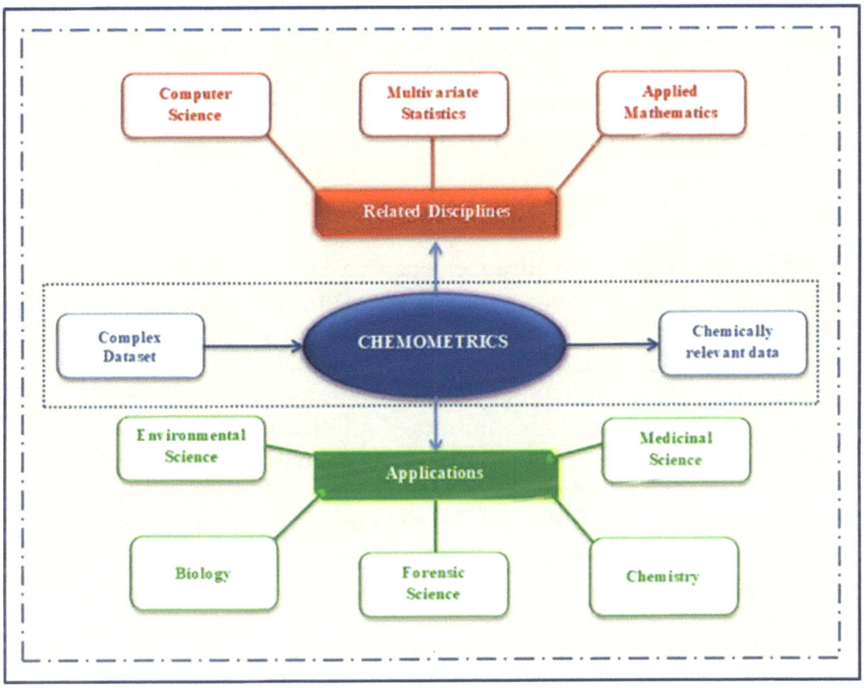

Figure 1.1 Schematic representation of the multidisciplinary nature of chemometrics.

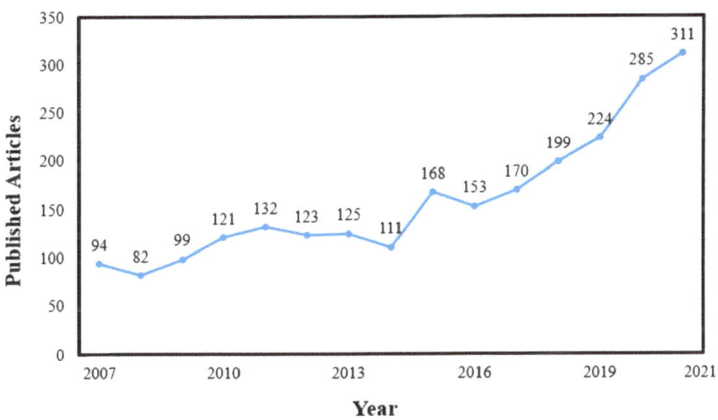

Figure 1.2 Total number of published research and review articles per year (from Elsevier, Wiley, Springer, and Royal Society of Chemistry journals) between 2007 and 2021, identified through database searches for both topic keywords "forensic + chemometrics".

can be observed in the number of published journal articles (research papers and reviews) featuring both topic keywords "forensic" and "chemometrics" from the publishing houses Elsevier, Wiley, Springer, and Royal Society of Chemistry (see Figure 1.2).

This can be attributed to several benefits that chemometric methods can provide in addressing contemporary issues faced by forensic scientists. It is essential to note that the aim of chemometric analysis in forensic science should be to provide information supporting, rather than replacing, the decision-making of the forensic expert.

1.2.1 Data Reduction

In recent decades, significant advancements have been made in the technology routinely available to forensic chemists. Emerging techniques such as direct analysis in real time-mass spectrometry, ultrafast chromatography, hand-held spectrometers, and hyperspectral imaging have become powerful tools for rapid qualitative and quantitative analysis. These methods allow examiners to extract increasing amounts of multivariate chemical data in the form of spectra, chromatograms, or similar output from physical evidence (see Figure 1.3). This raw data must then be decoded to produce meaningful information, such as identifying the composition of an unknown substance based on its chromatogram or establishing whether two samples potentially share a common origin based on the similarity or dissimilarity of their spectral profiles.

This data decoding has conventionally relied on visual assessments (sometimes referred to colloquially as "eyeballing") by the analyst. Put simply, the analyst will make estimates from which they can draw conclusions by looking at the data, either in its raw form or as expressed in spectra, chromatograms, *etc.* Even in disciplines where databases have been developed to assist with pattern matching or identification, these generally yield a list of potential results that must then be confirmed through visual analysis. The shift towards progressively more complex datasets increases the challenge of deciphering trends or patterns, which could result in key information being overlooked.

A potential way of overcoming this is to focus on a reduced portion of data, such as examining solely the fingerprint region of an infrared spectrum. However, this still requires the analyst to make an initial decision on which portions of the data contain useful information and should be retained (or *vice versa*). This is again conventionally based on a visual assessment and can result in useful but latent information being inadvertently discarded. Additionally, in casework scenarios where hundreds of exhibits may have been submitted for analysis, visually processing even a reduced portion of the generated data remains a time-consuming and tedious task.

Many chemometric methods are designed to simplify complex datasets through feature selection or combination. For example, RF algorithms inherently evaluate the relationship between inputs (raw data) and desired output (classifications), subsequently retaining only the input variables that are most useful for modelling. PCA on the other hand reduces data dimensionality by combining correlated variables into a reduced set of orthogonal "principal components". Chemometrics thus offers a means of

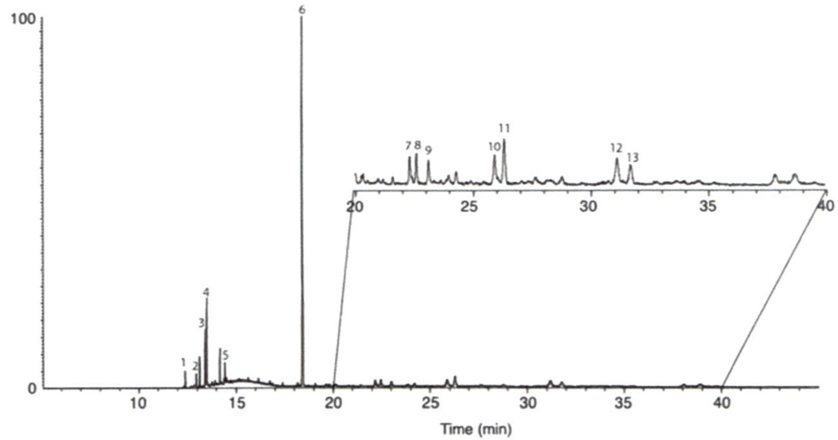

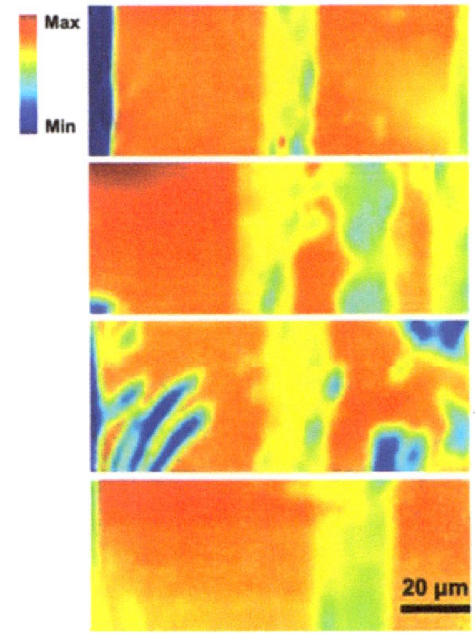

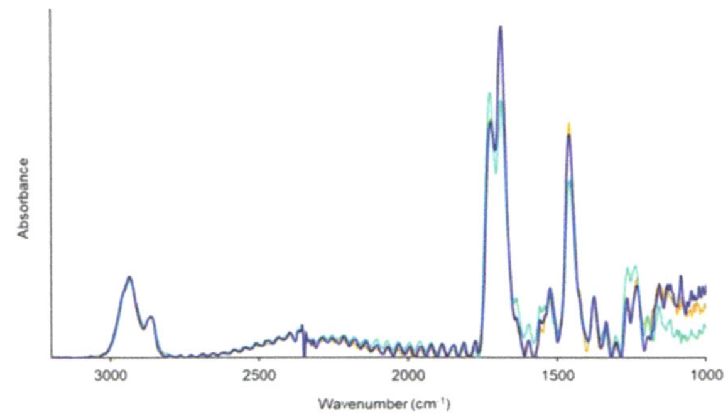

Sr	Sn	Sb	W
$7.41 \times 10^3 \pm 6.7\%$	$46.5 \pm 5.6\%$	$40.0 \pm 5.8\%$	$28.4 \pm 7.9\%$
$9.46 \times 10^3 \pm 3.9\%$	$59.1 \pm 16.0\%$	$36.2 \pm 5.9\%$	$45.2 \pm 4.2\%$
$6.64 \times 10^4 \pm 3.0\%$	$67.1 \pm 4.0\%$	$26.1 \pm 3.5\%$	$26.4 \pm 10.3\%$
$9.82 \times 10^3 \pm 5.6\%$	$73.3 \pm 3.3\%$	$45.1 \pm 6.9\%$	$24.0 \pm 14.0\%$
$1.64 \times 10^3 \pm 5.5\%$	$69.9 \pm 6.5\%$	$30.4 \pm 12.4\%$	$22.1 \pm 12.8\%$
$1.12 \times 10^3 \pm 6.4\%$	$33.9 \pm 10.5\%$	$13.9 \pm 6.0\%$	$20.1 \pm 5.6\%$
$1.72 \times 10^4 \pm 6.4\%$	$388 \pm 2.2\%$	$84.2 \pm 2.7\%$	$213 \pm 2.5\%$
$1.93 \times 10^3 \pm 7.0\%$	$75.5 \pm 21.0\%$	$33.4 \pm 10.6\%$	$24.1 \pm 10.3\%$

Figure 1.3 Examples of multivariate data likely to be generated through routine forensic examinations.

compressing key information into fewer variables, assisting the examiner in drawing conclusions. These methods can also be applied simultaneously to hundreds or thousands of samples, vastly improving the efficiency of forensic data analysis and potentially enabling large-scale data processing (with minimal loss of information) for intelligence purposes.

1.2.2 Improved Objectivity

A secondary issue with relying on visual data assessment for evidence interpretation is subjectivity. Forensic practitioners must depend on their expertise to make decisions, which will be influenced by their individual experience and training. Critical research and high-profile cases have also led to the growing recognition that forensic examiners can be influenced by cognitive or social biases. These may arise due to factors including initial expectations of results (confirmation or expectation bias), knowledge of investigative details (contextual bias), or a predisposition to assist the team retaining their services (role effects).[5] For these reasons, it is entirely possible that two examiners considering the same item of evidence could reach different conclusions. This inevitably creates opportunities for false positives or false negatives to occur.

The problem of subjectivity was highlighted in the broadly publicised National Academy of Sciences (NAS) report in 2009, which highlighted several issues regarding current interpretation protocols relating to forensic evidence. Although this report was commissioned to evaluate the state of forensic service provision specifically in the United States, it is generally accepted that many of the reported issues applied across international jurisdictions.

In particular, the report identified the need for more rigorous and objective means of evaluating forensic evidence, as well as fundamental research regarding accuracy and reliability in forensic disciplines:

> "A body of research is required to establish the limits and measures of performance and to address the impact of sources of variability and potential bias. Such research is sorely needed, but it seems to be lacking in most of the forensic disciplines that rely on subjective assessments of matching characteristics."[6]

This sentiment has been reiterated in other critical reports, including the 2016 report by the President's Council of Advisors on Science and Technology and a 2019 inquiry by the British House of Commons Science and Technology Committee.[7,8] These have resulted in strengthened efforts to establish statistically validated approaches for evidential assessment and interpretation.

Chemometric techniques address these concerns by providing well-established and validated statistical methods to provide objective measures of collected data. This output can include quantitative values for similarity or dissimilarity, variable contributions, or associated limits. Unlike human decision-making, a given chemometric algorithm will provide a consistent result when applied to the same set of data under the same set of conditions, and the process used to reach this result can be fully documented in mathematical terms. In addition to providing more objective measures of data, chemometric tools can also be used to evaluate contextual factors that could influence this data through experimental design. This information is essential in establishing "limits and measures of performance" considering potential sources of variability.

1.2.3 Evaluative Reporting

Issues of subjectivity are not limited to the forensic laboratory. As forensic experts are regularly called upon in legal proceedings to deliver their expert opinion, they must convey these opinions in a manner that will not be misconstrued by the courts. This is particularly challenging for comparative analyses where the expert is asked whether a questioned sample X could have originated from the known source Y. Non-standardised verbal descriptors along the lines of "consistent with" or "strongly supports" may convey various meanings that influence how triers of fact perceive and evaluate the evidence presented. The 2009 NAS report argued that forensic testimony should provide a clear rationale for any conclusions drawn, along with estimates of the scale of uncertainties associated with these conclusions, to allow a balanced interpretation of the evidence.[6] This has resulted in rising demand for statistical measures and formalised reasoning to be incorporated into forensic reporting.

Many jurisdictions are approaching this through evaluative reporting. Under this framework, the findings of an examination are evaluated given two alternative propositions.[9] Relative probability estimates for each proposition are determined and used to express the strength of the findings in the form of a likelihood ratio (LR). A challenge is that conventional feature-based LRs require data on the rarity of certain properties, which is not readily available in all forensic disciplines, as well as any statistical dependencies between variables measured in the analysis, making LRs highly challenging to determine in multivariate datasets.[10]

These issues can be at least partially overcome using score-based LRs in which a single metric (score) is determined to represent the similarity between two items under comparison, and this score converted to an LR using a univariate function. Chemometric data reduction methods such as PCA offer a way of rapidly deriving these scores in a way that ideally minimises the loss of information needed to produce reliable estimates. This hybrid approach has been of considerable interest in recent forensic research, although it is debated whether this provides a suitable proxy for the conventional Bayes factor in evaluative reporting.[11]

1.3 Purpose and Scope

Chemometrics demonstrates great potential as a supporting tool in forensic analysis and interpretation, but there is still a gap between existing research in this area and the deployment of chemometrics in operational casework. Forensic scientists must first understand how these methods should be appropriately applied to ensure relevant and scientifically sound outcomes. The acceptance of chemometric methods in forensic reporting also requires stakeholders such as legal counsel, court officers, and triers of fact to understand these methods and their limitations, preventing erroneous interpretations.

With these needs in mind, we present an introductory text for forensic students, practitioners, and other forensic stakeholders seeking a basic understanding of chemometric methods. This is not intended to provide an all-encompassing guide to chemometric methods, discipline areas or existing studies. Rather, the key purpose is to highlight the potential value of chemometrics across various forensic disciplines along with considerations for their proper use and interpretation, enabling more confident and transparent decision-making based on quantitative approaches. The remainder of this book will provide a tutorial on common chemometric methods, outline contemporary research applying these methods to forensic analyses, and discuss emerging trends and challenges.

Chapter 2 will provide a general introduction to common chemometric methods, aimed at those new to the field. This will include a discussion of key considerations that must be made when applying these methods and interpreting the resulting outputs, such as data pre-processing and appropriate model validation.

Chapter 3 reviews applications of chemometrics for the qualitative and quantitative profiling of illicit drugs or pharmaceuticals, including large-scale data analyses for forensic intelligence. Chapter 4 focusses on the discrimination and classification of fire debris and explosives, including optimisation studies into the sampling and storage of these evidence types. Chapter 5 examines how chemometric algorithms have been applied to provenance determination or adulterant analysis of food and beverage products.

In Chapter 6, the reader is introduced to chemometrics as a tool for characterising cosmetics and personal products, with a brief discussion on the challenges associated with "real" samples in a biological matrix. Chapter 7 provides a broad overview of chemometric characterisation of trace evidence including fibres, paints, glass, and soil. These applications include comparative analyses, provenance determination, and investigation of contextual factors that may affect analysis or interpretation of traces.

Chapter 8 focusses on the use of chemometrics within questioned document examination, with a focus on papers, toners, and inks. Chapter 9 considers chemometrics as a means of distinguishing or dating biological materials, particularly bodily fluids, while Chapter 10 focusses on the "chemical fingerprinting" of contaminants for environmental forensic investigations.

Finally, Chapter 11 discusses considerations for practitioners considering implementation of chemometric analyses in forensic workflows, including forensic reporting of chemometric outputs and the integration of chemometrics into existing interpretation frameworks.

Acknowledgements

Portions of this chapter were adapted from previous work,[12,13] reproduced with permission from the original authors and the Royal Society of Chemistry.

References

1. R. Brereton, *J. Chemom.*, 2014, **28**, 749.
2. P. C. Jurs, B. R. Kowalski, T. L. Isenhour and C. N. Reilley, *Anal. Chem.*, 1969, **41**, 1949.
3. B. Kowalski, S. Brown and B. Vandeginste, *J. Chemom.*, 1987, **1**, 1.
4. B. R. Kowalski, *Anal. Chem.*, 1975, **47**, 1152A.
5. Forensic Science Regulator, FSR-G-217: Cognitive Bias Effects Relevant to Forensic Science Examinations, 2020, The Forensic Science Regulator: Birmingham.
6. National Academy of Sciences, *Strengthening Forensic Science in the United States: A Path Forward*, National Research Council, Washington DC, 2009.
7. President's Council of Advisors on Science and Technology, *Forensic Science in Criminal Courts: Ensuring Scientific Validity of Feature-Comparison Methods*, Washington DC, 2016.
8. Science and Technology Select Committee, *Forensic Science and the Criminal Justice System: A Blueprint for Change*, London, 2011.
9. European Network of Forensic Science Institutes, *Guideline for Evaluative Reporting in Forensic Science*, 2015.
10. A. Martyna, G. Zadora and D. Ramos, *J. Anal. Appl. Pyrolysis*, 2018, **133**, 198.
11. C. Neumann and M. Ausdemore, *Law Probab. Risk*, 2020, **19**, 21.
12. G. Sauzier, PhD thesis, Curtin University, 2016.
13. G. Sauzier, W. van Bronswijk and S. W. Lewis, *Analyst*, 2021, **146**, 2415.

CHAPTER 2

Chemometric Methods

G. SAUZIER* AND S. W. LEWIS

School of Molecular and Life Sciences, Curtin University, Australia
*Email: Georgina.Sauzier@curtin.edu.au

2.1 Introduction

"Chemometrics" includes a range of methods including pattern detection, factor analysis, calibration, curve fitting, experimental design, and machine learning algorithms. These methods can generally be classified under three main categories according to the purpose of the analysis.

Pattern recognition techniques focus on extracting patterns from multivariate data. These techniques can be further categorised as unsupervised methods (also referred to as exploratory data analysis) that aim to identify trends, or supervised methods that model these trends for classification purposes. Regression methods provide quantitative prediction of sample properties and can also be applied for binary classification. Finally, experimental design techniques reveal information about chemical processes, enabling the design and optimisation of procedures.

Although chemometrics has very powerful tools, not all of these are necessarily suited to the forensic context. For example, limited use has been made of neural networks due to their lack of transparency (sometimes referred to as the "black box" effect), which conflicts with the need for reproducible and justifiable methods in forensic reporting. Chemometrics also does not compensate for insufficient (or poor quality) data and, like all statistical analyses, requires appropriate validation. The selection of samples, pre-processing, and validation approaches thus require careful consideration.

This chapter outlines the use and interpretation of selected chemometric methods, along with prior considerations that may affect their performance.

Chemometric Methods in Forensic Science
Edited by Vishal Sharma, Georgina Sauzier and Simon W. Lewis

Published by the Royal Society of Chemistry, www.rsc.org

Several references are made to open-access datasets that can be accessed for training purposes.

2.2 Initial Considerations

The quality of outputs from chemometric analysis is inherently limited by the quality of inputs, or to put it more commonly: "garbage in, garbage out". Before performing any analyses, it is vital to ensure an appropriate selection of data along with suitable pre-processing techniques to ensure its quality.

2.2.1 Data Selection

If analysis is carried out on data that does not truly represent the source, the results may be limited or inaccurate. This could cause probative information to be overlooked at best or produce misleading information at worst. This of course is not confined to analysis involving chemometric methods, and the purpose of this section is not to repeat existing literature or standards concerning sample and data selection. Rather, it is intended to outline the importance of these factors within the context of chemometric analysis.

The first consideration is whether the sample size is sufficient to support any conclusions drawn. This is largely dependent on the population under inquiry and the type of data analysis proposed. For example, take a feasibility study of the chemical variations between yellow-dyed polyester fibres. The target population is highly specific, high precision is not necessarily required, and manufactured fibres are expected to show far less variation than natural fibres. Under these circumstances, 10 samples could be considered appropriate. On the other hand, consider a validation study of an approach for profiling the elemental content of hair for race determination. We now have a broad population expected to show significant variation, requiring precise measures of error rates. In this context, 10 samples would be woefully insufficient.

This of course assumes that a large population of samples is available. In casework scenarios, there could be as few as two samples: a questioned and a known. This does not mean that chemometric techniques could not be applied; however, the analyst should note in their reporting that the results are specific to these samples and do not necessarily reflect broader trends.

The analysis of replicates is also essential for properly evaluating model performance. It is all too common for chemometric studies to declare successful differentiation between samples, each represented by a single data point. Without considering the impact of variability sources, "proper limits and measures of performance" cannot be established.[1] In some cases, the analysis makes use of replicates that have been averaged to a single data point prior to analysis. Even if statistical measures have determined no significant variation, all replicates should be included unless there is reason to believe they are outliers that are not a natural part of the population.

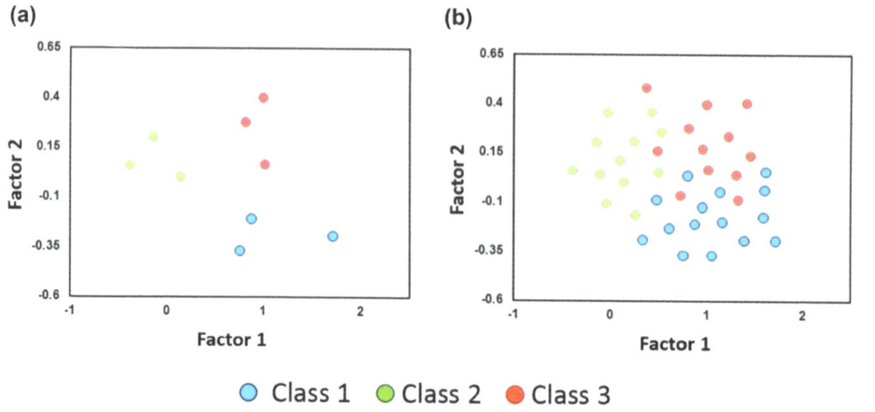

Figure 2.1 Theoretical cluster analysis performed (a) without and (b) with sample replicates, assuming three samples per class. The apparent differentiation between classes may change once comparative variation between replicates is examined.

The reason for this is to allow comparisons of inter-sample to intra-sample variation.

Take the following theoretical example of a cluster analysis carried out on product samples from different brands (see Figure 2.1). With a single measurement per sample, there appears to be differentiation between the brands. However, further replicates show that the variation between brands is of a similar magnitude to the variation within each brand. On that basis, they cannot be reliably distinguished. The replicates of each sample do not show statistically significant variation; rather, it is the fact that the natural variation causes the brands to display overlapping characteristics.

Three to five replicates per sample should generally be considered a minimum, although some standards may specify a greater number. Replicates should ideally consist of independent samples treated in the same manner, rather than repeated measures of a single sample. The latter fails to consider variability associated with sample collection, handling, and preparation. When taking replicates from a bulk material, consideration must also be given to the sampling approach to ensure that this is representative. In the case of mass manufactured products, this may include replicates across separate lot numbers to consider batch variation effects.

2.2.2 Pre-processing Techniques

Data pre-processing is an important stage in preparing data for chemometric analysis. Depending on the nature and quality of initial data, pre-processing may not be necessary. In fact, over-processing data can create just as poor outcomes as a lack of processing. The mathematical bases for these methods have been described elsewhere,[2–4] however the following section provides a basic understanding for the non-specialist reader. The optimal choice of pre-processing techniques for a given analysis depends on the dataset and is often best determined empirically.

2.2.2.1 *Baseline Correction*

Non-linear baselines or baseline drift can result from noise, changes in experimental conditions, or other interferences. Baseline corrections aim to compensate for this variation before data analysis. There are three main approaches that can be taken depending on the initial data.

In cases where the baseline has shifted from the "true" value, a baseline offset can be performed by subtracting the value of the lowest point from all variables, effectively setting the baseline value to zero. A linear baseline correction is used to subtract a sloped baseline, estimated from two variables, from the dataset. This sets the baseline value to zero at each of the selected points. Lastly, curvilinear baselines can be resolved through de-trending, in which a polynomial baseline is subtracted from the data. The degree of this polynomial is usually within 1–4 and estimated by a least-squares fit to the sample spectrum.

2.2.2.2 *Smoothing*

Smoothing is used to reduce noise in the data without reducing the number of variables. The basic premise is to estimate the "true" value of each point based on the surrounding values and/or fitting against a mathematical function.

The simplest smoothing method is a moving average, where the true value of a data point is calculated by averaging a chosen number of surrounding points (referred to as the segment size). These are referred to as an n-point moving average, where n is the segment size. However, this approach gives equal weight to all averaged points, which may not be desirable. In such instances, a Gaussian filter based on a normal distribution can be used as a weighting function, placing more weight on the immediately adjacent data points.

Alternatively, the Savitzky–Golay smoothing function fits a polynomial curve to a chosen segment size, then uses this polynomial to estimate the centre value. This is more computationally intensive but retains the shape and position of the original signal, which is particularly important for spectral and chromatographic data.

For all smoothing methods, increasing the segment size will result in better averaging but may result in a loss of features. Care should be taken not to over-smooth the data.

2.2.2.3 *Feature Scaling*

The purpose of feature scaling is to adjust all values to a common range or scale. This compensates for different units of measurement and minimises the effect of unintended instrumental or sample variations.

A common form of scaling is standardisation, in which each variable is divided by its standard deviation. This is ideal if planning to use chemometric methods that would heavily weight variables exhibiting large deviation, particularly if outliers are present. However, it will also increase the relative weighting of unimportant variables that describe random variation (noise).

An alternative set of approaches are those referred to as normalisation. In truth, these methods alter the range distribution of the data, and should not be confused with methods that fit data to a normal distribution. Typical modes of normalisation include:

- Min-max: rescales data to a fixed range, usually [0, 1]. While this ensures that variables are set to the same scale, it can be highly influenced by outliers containing extreme high or low values.
- Mean: scales data to a range of [−1, 1] by subtracting the mean from each value in a dataset. This is typically used prior to chemometric methods that assume zero-centred data.
- Vector: normalises each sample to a total vector length of 1 by mean-centring the data, then dividing by the square root of the sum of squared intensities. This is commonly used in pattern recognition, as it reduces the likelihood of variables with large values dominating distance-based similarity metrics.

2.2.2.4 Data Transformations

Applying a deterministic mathematical function to the values of a dataset is referred to as data transformation. These can reduce data skewness and kurtosis, which is important for chemometric methods that assume a uniform distribution of data. The two most common data transformations are the log and square root transformations, which respectively replace each data point with its logarithmic (to logarithm base 10, e, or 2) or square root value. Both transformations compress high values and spread low values, making them highly suited to right-skewed data. The inverse of these functions, the exponential and square transformations, can be used to correct for left skewness.

The log transformation has a powerful effect on the data distribution shape, although this may result in the skew becoming reversed. It also cannot process zero or negative values, which can be circumvented by adding a constant to the data beforehand. The square root transformation has a similar effect on data values, although it does so in a less aggressive manner. This transformation also has the advantage of being immediately applicable to zero values.

2.2.2.5 Scatter Corrections

Spectroscopic data collection is often affected by additive and multiplicative scatter effects caused by physical phenomena. Additive effects such as varying path length manifest as baseline displacement, while multiplicative effects such as particle size variations result in slope changes (particularly in low-absorbance spectral regions). Scatter corrections aim to eliminate variation caused by effects unrelated to the chemical nature of the sample, helping to estimate a more accurate spectrum. Some of these methods can also serve as generic signal processing tools.

The standard normal variate (SNV) transformation is routinely used to correct additive effects due to path length variations. SNV normalises spectra by subtracting the mean absorbance from each individual measurement, then dividing by the standard deviation. This is a highly popular scatter correction due to its relative simplicity, although it has a limited effect on multiplicative scatter.

Multiplicative scatter correction (MSC) instead regresses each spectrum onto a scatter-free reference (often an averaged spectrum for practicality), then corrects the spectrum using the reference slope and intercept. This is a relatively simple transformation that accounts for both additive and multiplicative effects. However, it cannot distinguish random scatter from that related to chemical variation, which can result in the loss of useful information. For this reason, an extended adaptation (EMSC) was developed to distinguish physical and chemical variation. EMSC considers the variation in relation to wavelength and enables spectral regions corresponding to known chemical constituents to be excluded.

A more recent approach is orthogonal signal correction (OSC). This method subtracts variation from the data matrix that is orthogonal to the response, on the assumption that this is irrelevant to modelling structure in the data. OSC is commonly used for near-infrared spectra.

2.2.2.6 *Derivatives*

Derivative transformations involve fitting data to a polynomial curve to determine the slope. This can correct for various baseline effects and accentuate hidden information in overlapping peaks, reducing interferences from broad band constituents. Some derivatives (such as Savitzky–Golay) also incorporate smoothing to compensate for the decreased scale and amplified noise of derivative functions.

First and second derivatives are the most frequently used. The first derivative measures the slope of the original curve, removing linear baseline shifts as the slope is zero. However, they can be less intuitive to interpret as the peaks and intercepts do not align with the original curve. The second derivative measures the rate of change in the slope, correcting linear baseline effects where the slope is constant. These derivatives may also provide simpler interpretation as peaks are produced in the same locations as the original data.

2.3 Methods

Forensic reporting requires that all methods used in the characterisation of evidence be documented and justifiable in a court of law. The use of chemometric methods within a forensic context thus requires a practical understanding of their parameters, interpretation, and potential limitations. This section outlines methods most often encountered in the current literature. Many institutions have developed in-house algorithms that modify these basic methods; however, these are beyond the scope of this chapter.

2.3.1 Unsupervised Pattern Recognition

Unsupervised learning aims to recognise patterns in datasets without setting any prior assumptions, thus providing minimal human intervention. Whilst these methods are not directly applicable to classification or regression, the information revealed can be used to inform subsequent data analysis. Some unsupervised methods also allow new samples to be projected onto a pre-existing dataset for comparative purposes. This should not strictly be considered as classification, as there is no prior assumption that specific classes exist.

2.3.1.1 Cluster Analysis

Cluster analysis is a group of unsupervised techniques that group objects (samples) based on their relative similarity, as determined using a chosen distance metric. The specific algorithm and parameters chosen will depend on the dataset and underlying purpose of the analysis.

A popular clustering approach is hierarchical cluster analysis (HCA), where objects are grouped according to a determined clustering pattern. This process usually follows a "bottom-up" approach, known as agglomerative HCA. Samples are initially treated as single-object clusters, and the distances between all possible pairs calculated. Those with the minimum distance are merged into a new cluster, and the process repeated until all samples have been recombined into a single dataset.

The developed clustering pattern will vary according to two main factors. The user must first select an appropriate metric that reflects the distance measure between samples. Some objects may be closely related according to one metric but not another. The second parameter is the linkage criteria, which describes how relationships between clusters are formed and thus influences the cluster shape. Common metrics and linkage criteria are described below (see Table 2.1).

The hierarchical relationship between clusters can be illustrated using a dendrogram, which plots objects according to their clustering pattern and relative distance. For example, Maric *et al.* used HCA to investigate over 700 automotive clear coats using attenuated total reflectance Fourier transform infrared (ATR–FTIR) spectroscopy (see Figure 2.2).[5] Using the squared Euclidean metric, Ward's linkage method, and the Calinski–Harabasz pseudo-F index as a stopping rule to determine the optimal number of clusters, eight classes were identified and associated with the country of vehicle manufacture.

Table 2.1 Common distance metrics and linkage criteria used for HCA. Reproduced from ref. 27 with permission from the Royal Society of Chemistry.

	Mode	Description
Metric	Euclidean	"Straight line" distance between objects in *n*-dimensional space
	Manhattan	Collective distance between objects along each dimension in *n*-dimensional space
	Mahalanobis	Distance between an object and a distribution determined from the covariance between dimensions
Linkage criterion	Single linkage	The shortest distance between objects of two clusters, resulting in long clusters.
	Complete linkage	The longest distance between objects of two clusters, resulting in compact clusters.
	Centroid linkage	The distance between central objects of two clusters, resulting in globular clusters
	Ward's linkage	Groups clusters so as to minimise the increase in variance, resulting in globular and fairly compact clusters.

In this instance, the horizontal axis represents the relative distance between samples, while vertical bars reflect the distance at which two samples or clusters are combined. Alternatively, samples are often plotted according to a calculated similarity measure, such as the ratio of distances between a selected pair to the maximum distance between any pair in the sample set.

The dendrogram provides a simple visual representation of similarity expressed in quantitative terms. In the example above, class 1 is the most dissimilar of the sample groupings, only merging with the remaining clusters at a relative distance of nearly 10. By contrast, classes 5 and 7 are merged at an early stage, being within a relative distance of approximately 0.4. These hierarchical relationships are consistent with the relative similarities of the initial data, as seen through inspection of the initial spectra. A typical class 1 spectrum lacks several of the absorbance bands seen in the spectra of other classes, likely explaining its greater separation distance.

2.3.1.2 Principal Component Analysis (PCA)

Identifying relationships within complex multivariate data such as spectra or chromatograms can be highly challenging. Additionally, it may be of interest to determine specific features leading to separation. Using cluster analysis, this information must be indirectly inferred through visual comparisons of the original data (seen above for HCA), which is not always practical for large datasets.

PCA reduces the dimensionality of multivariate data by extracting the orthogonal sources of variation, known as "principal components" or PCs.

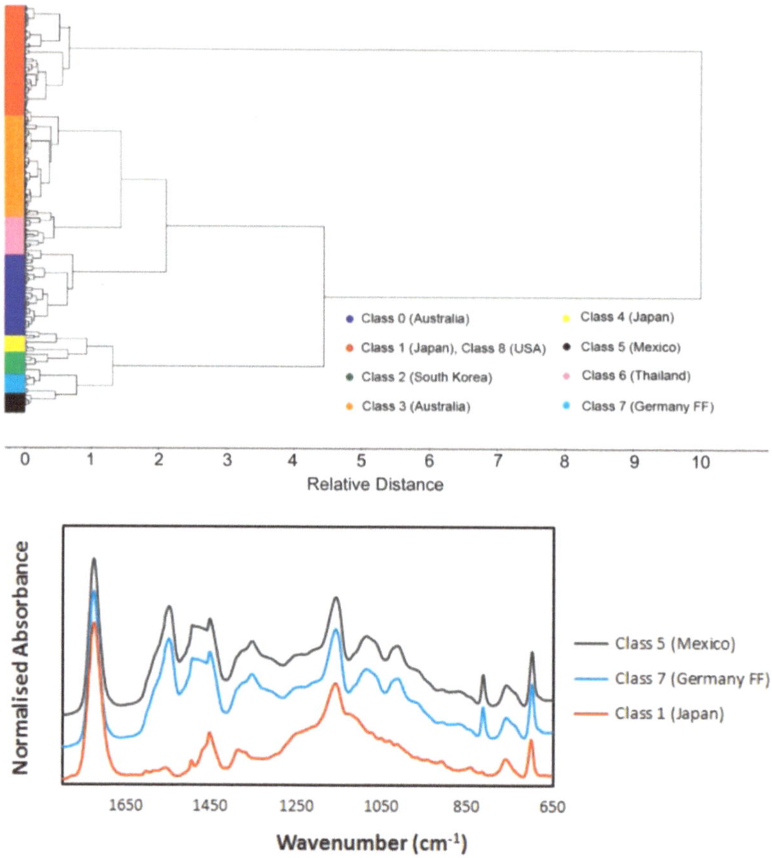

Figure 2.2 Dendrogram showing classes of automotive clear coats identified using HCA and representative spectra from selected classes typifying the relative similarity and dissimilarity. This dataset is available *via* Research Data Australia.[6] Reproduced from ref. 27 with permission from the Royal Society of Chemistry.

Each PC is a linear combination of the initial variables multiplied by a loading. Each PC is successively calculated to describe the greatest proportion of remaining variation, as shown for a simplified dataset below (see Figure 2.3). Here, samples are initially represented as data points in a multidimensional space described by the variables X, Y and Z. Each PC can be treated as a vector aligned with the direction of greatest dispersion. Projecting the samples onto the PCs then allows them to be described using the PC coordinates, known as scores, in place of the original variables.

The primary advantage of PCA is that fewer PCs are required to capture the majority of useful information in the dataset compared to the initial variables. Determination of which PCs are useful is typically done using a scree plot of the cumulative variance accounted for with each successive PC. A common rule is to only consider PCs accounting for at least 1% of total

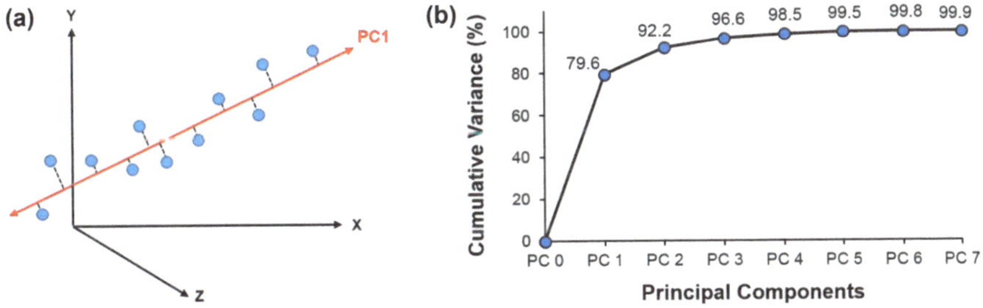

Figure 2.3 Diagram illustrating (a) generation of the first principal component (PC) for a simplified dataset described by variables *X*, *Y* and *Z*, with dashed lines indicating the projection of each sample onto the PC; and (b) example scree plot showing cumulative variance accounted for by each successive PC. Reproduced from ref. 27 with permission from the Royal Society of Chemistry.

variation, although it should be confirmed using scores plots that these components do not in fact contain information relevant to distinguishing specific sets of samples.

Mathematically, PCA transforms the original data matrix (*X*) into separate scores (*T*), loadings (*P*) and residuals (*E*) matrices according to the equation:

$$X = TP' + E.$$

This may be done through analysis of either the correlation or the covariance matrices, which are described by Otto.[7] In practical terms, the correlation method mean centres and scales the data to place equal weight on each variable, whereas the covariance method applies mean centring without scaling.

As most original variation is described by the first few PCs, multiplying the scores and loadings for these PCs can approximate the original data. The discrepancy between this approximation and the actual original data is described by the residuals matrix (*E*), which should ideally contain only random variation or noise. A useful way to evaluate the performance of the model is by plotting the residuals to assess their magnitude and pattern across each sample. Small and randomly dispersed residuals are ideal, as they indicate minimal loss of information.

By using the PC scores as a new coordinate system, a scores plot can be produced where similar samples are grouped together. Further information can be yielded from the PCA loadings, which describe the contribution of the original variables to each PC. Those with highly positive or negative loadings are significant in determining a sample's score along that component.

PCA plots provide an alternative way of visualising data that can uncover patterns not obvious from the raw information. As with HCA though, the initial data must still be examined to yield a complete understanding of these trends. This is illustrated in the example below (see Figure 2.4) where PCA was performed on ATR–FTIR spectra of 22 red-shaded lipsticks.[8]

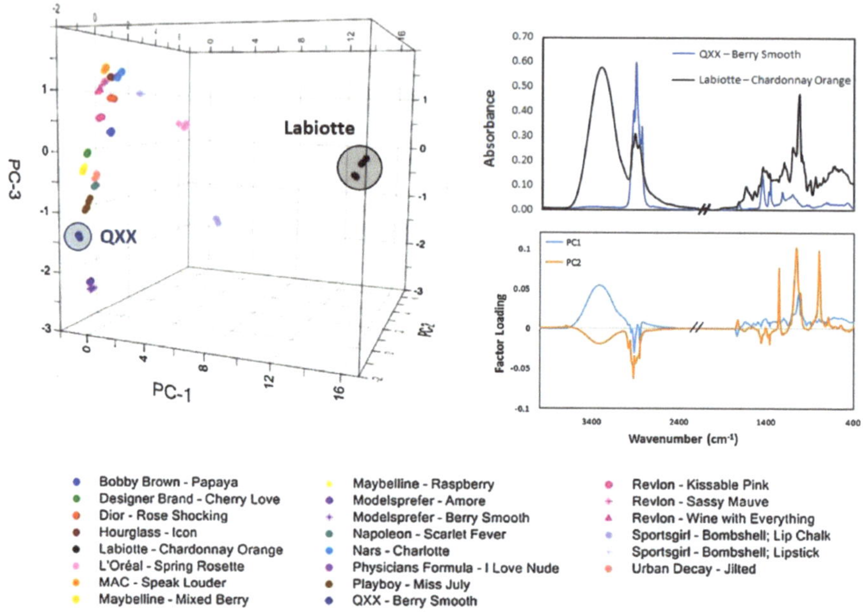

Figure 2.4 PCA scores and loading plots based on ATR–FTIR spectra of 22 red-shaded lipsticks. These plots revealed the Labiotte sample to be distinctive from other brands due to water and butylene glycol in its formulation. This dataset is available *via* Research Data Australia.[9] Reproduced from ref. 27 with permission from the Royal Society of Chemistry.

The Labiotte "Chardonnay Orange" lipstick obtained particularly high scores along PC1, which showed a strong positive correlation at *ca.* 3300 cm^{-1}. Initial spectra of this sample exhibited strong absorbance in this region, producing the more positive PC score. This infrared band was consistent with an OH stretching band, attributed to water and butylene glycol used as co-solvents in this formulation. The combination of PCA with examination of the initial data thus revealed the chemical basis for this sample's distinctiveness.

2.3.2 Supervised Pattern Recognition

If specific groupings of samples are known prior to data analysis, supervised learning methods can pair the input variables with the desired outputs and develop a function correlating the two. In other words, supervised algorithms use known data to create functions that can generate information about new samples. These methods are ideal for classification and regression as the model can be "trained" to detect and accurately model certain patterns. The model must then be validated to determine whether these patterns are generalisable to a broader dataset. The importance of rigorous validation cannot be overstated, as failure to establish accurate error rates has significant implications for criminal justice.

The simplest validation methods are based on resampling, where "test" samples simulating new inputs are taken from the original data. This is generally an iterative process, with the average predictive accuracy reported. A popular resampling approach is cross validation (CV), where a portion of the data is put aside as the test set. This portion may consist of individual samples (leave-one-out CV) or a block (leave-*p*-out CV and related variants). Another common mode of resampling is the bootstrap, where the same sample can be represented in the test set multiple times. Resampling validation methods are economical, but using the same data to train and test the model can over-estimate its performance. This is particularly the case for leave-one-out CV, which is not advisable unless samples are extremely limited.

A more robust approach is test set validation. Here, the dataset is divided into mutually exclusive training and test sets with the latter generally accounting for around 30% of the original data. To provide the most rigorous evaluation of the model, the test set should realistically represent the full sample population, including samples that are known to be similar or dissimilar.

2.3.2.1 k-*Nearest Neighbour (*k*NN)*

The *k*-nearest neighbour (*k*NN) algorithm classifies unknown samples according to their distance to a set of known classes. The distance metric between the unknown and each training sample is determined, with the unknown assigned to the class that is most common among its "*k*" nearest neighbours (see Figure 2.5). "*k*" is generally a small integer chosen through cross-validation. The selected value may alter the predictive result as shown below, where setting "*k*" to alternative values of 3 or 7 leads to the unknown sample being assigned to different classes. This example highlights how a degree of subjectivity will remain in chemometric analysis due to human selection of parameters.

*k*NN is simple to implement as the training phase is minimal and there is no need for classes to be linearly separated. The only assumption is that the data exists in a feature space where distance measures can be calculated. However, the *k*NN method is highly dependent on the selected "*k*" value, making it biased towards larger training classes reflecting a greater number of potential neighbours. Overly influential variables can also distort the results, as there is no reduction or screening of the data prior to classification.

2.3.2.2 *Linear Discriminant Analysis (LDA)*

Discriminant analysis (DA) aims to classify objects into known, mutually exclusive groups according to scores derived from a discriminant function. Like PCA, this function is a combination of original variables that can be used to revisualise the data. Unlike PCA however, the function aims at maximising the Fisher ratio of between-class to within-class variance.

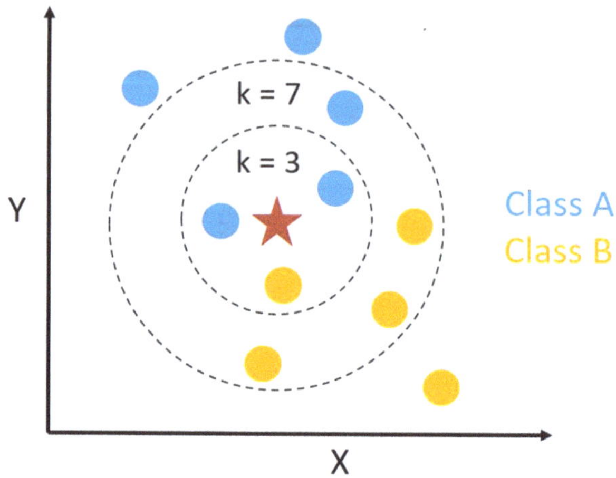

Figure 2.5 Representation of a *k*NN analysis where an unknown sample (red star) must be assigned to class A or class B, determined by the mode class value of the "*k*" closest training samples. The resulting classification may vary depending on the selection of "*k*".

If the separating function is assumed to be linear, this method is referred to as linear discriminant analysis (LDA). Because LDA requires a greater number of samples than variables, it is often carried out after a data reduction method (*e.g.* PCA) or variable selection.

The primary output of LDA is a confusion matrix, an example of which is shown below for analysis of ATR–FTIR spectra of three classes of food wrap (see Table 2.2).[10] Because each group must have a sample size exceeding the number of variables, the first two PCs from the PCA (retaining the majority of useful variance) were used in place of the original variables. Several other approaches to variable selection have also been described, such as the Wilks' lambda test statistic, Bayesian information criterion, or simply observing features common to (but variable amongst) each group.[12–14]

The confusion matrix compares the true class of each sample to their predicted classes using the LDA model. Values in the diagonal of the matrix (in bold in Table 2.2) reflect correctly assigned objects, which can be totalled to calculate the model's discrimination accuracy. This matrix can also assist in identifying classes that are not yet effectively separated, as these will produce misclassified samples.

As with any method, however, one should be wary of accepting these results at face value. LDA is a "hard" classification method that will assign samples – even those not belonging to any known classes – to the most similar class. These atypical samples can potentially be identified through their discriminant values (DVs), which act as distance measures between the sample and the centroid of each known class.

The DVs of a new spectrum classified by the food wrap model are shown below (see Table 2.3). This spectrum was actually generated from a lipstick

Table 2.2 LDA confusion matrix displaying results of a two-component LDA model constructed for three classes of polyethylene cling film based on their ATR–FTIR spectra. This dataset is available *via* Research Data Australia.[11] Reproduced from ref. 27 with permission from the Royal Society of Chemistry.

Actual	Predicted Class 1	Class 2	Class 3	% Correct
Class 1	**6**	0	0	100
Class 2	0	**30**	0	100
Class 3	0	0	**30**	100
Total	6	30	30	100

Table 2.3 Discriminant values of a lipstick spectrum against three known classes of polyethylene cling film, compared to training samples from each class. Bolded figures represent the lowest magnitude value and thus the predicted class. Reproduced from ref. 27 with permission from the Royal Society of Chemistry.

	Class 1	Class 2	Class 3
Lipstick spectrum	−13 417	−1735	**−150**
True class 1 sample average	**−2**	−130	−84
True class 2 sample average	−1324	**−2**	−27
True class 3 sample average	−5162	−198	**−2**

trace and would not be expected to belong to any of the current classes. Despite being assigned to class 3 as the closest match, the large magnitude DV compared to original samples from this class indicates a poor fit. Similar, small-magnitude DVs against multiple classes is indicative of similar groups that are not well separated. These values are hence an important metric in evaluating the quality of separation and prediction.

Samples may also be plotted according to their DVs against any two classes, producing a discriminant plot. This may be interpreted in a similar fashion to the PCA scores plot. However, the fundamental difference is that PCA is designed to explore the largest variation present regardless of whether this corresponds to particular sample features. LDA by contrast looks for variation that is useful in distinguishing pre-established groups. For this reason, the clustering pattern may differ slightly when examining the discriminant plot (see Figure 2.6). The discriminant plot may be used as a visual way of assessing the quality of separation.

2.3.2.3 *Soft Independent Modelling of Class Analogy (SIMCA)*

In some scenarios, classification needs to be more flexible than that offered by *k*NN or LDA. Soft independent modelling of class analogy (SIMCA) is a disjoint technique, meaning that it constructs individual models describing a classification boundary for each class. The SIMCA algorithm does this using PCA, with the ideal number of PCs for each model typically determined

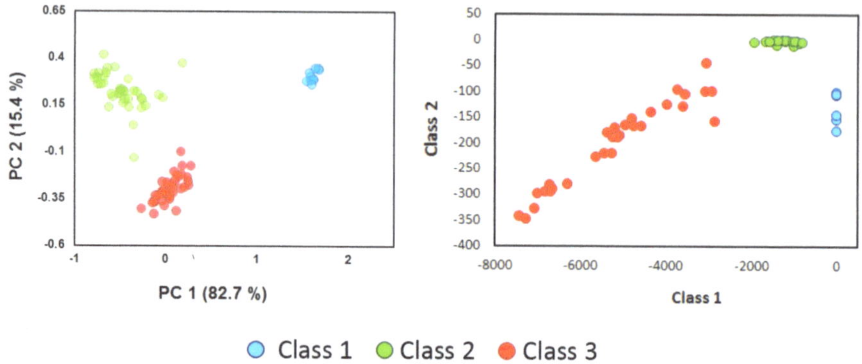

Figure 2.6 PCA scores plot (left) and LDA discriminant plot (right) based on ATR–FTIR spectra of nine polyethylene cling films. Varied shading has been used to distinguish the plots; however, each depicts the same spectra and three established classes. This dataset is available *via* Research Data Australia.[11] Reproduced from ref. 27 with permission from the Royal Society of Chemistry.

either through cross-validation or measures such as the Malinowski indicator function.[15,16] The number of retained PCs is highly influential on classification results, as too few could cause valuable information to be lost whilst too many will introduce noise.

Samples are classified according to their sample-to-model distance for each class as determined by two limits: residual variance and leverage. The residual variance of a sample is that which remains unexplained following projection onto a known class, determined as a residual sum of squares. The critical residual variance limit is generally determined using either a chi-square distribution or Jackson–Mudholkar approximation.[17] Leverage on the other hand represents the Mahalanobis distance between a sample and the centroid of a given class.

Unlike *k*NN or LDA methods that give a single classification based on the closest fit, SIMCA will assign unknown samples to all classes where the residual variance falls within the chosen thresholds. This also permits a "non-classification" result where the sample does not fall within the limits of any known classes, indicative of atypical samples. However, modelling each class separately means the algorithm is unable to compare within-class to between-class variance, potentially resulting in false negative classifications. This can be addressed by reducing the significance level (α), or the percentage of training samples accepted as outliers. By reducing α to expand the model boundaries, more samples may be accepted for membership of a given class, although at the risk of increased false positive classifications.

In the following example, SIMCA was used to model three series of driver's licence cards issued in Western Australia based on their cross-sectional ATR–FTIR spectra (see Table 2.4).[18] The resulting models were used to assign spectra obtained from one licence of each series along with a Swiss licence not belonging to any of the initial series. The 2001 test sample was assigned

Table 2.4 SIMCA classification results for three Western Australian driver's licences and one Swiss driver's licence against constructed models for 2001, 2011 and 2014 series Western Australian Driver's licences. Numbers indicate the number of replicate spectra accepted for classification. This dataset is available *via* Research Data Australia.[18] Reproduced from ref. 27 with permission from the Royal Society of Chemistry.

	2001 series	2011 series	2014 series
2001 test sample	4	0	4
2011 test sample	0	4	0
2014 test sample	0	0	2
Swiss test sample	0	0	0

to both the 2001 and 2014 series, which had been established to contain similar polymer blends, whereas the Swiss licence was correctly identified as atypical. Half of the replicate spectra from the 2014 test sample were incorrectly rejected, demonstrating the high sensitivity of SIMCA to within-class variance.

SIMCA classifications are often visualised using a Cooman plot. This maps the sample-to-model distance of an object against any two groups, similar to an LDA discriminant plot. As with PCA, residuals plots can also be used as an additional tool to evaluate and optimise the SIMCA model.

Two other important metrics in the SIMCA output are the modelling and discriminatory power. Modelling power is analogous to a factor loading with PCA, showing the weighting of an initial variable against a particular class. A high modelling power (approaching 1) indicates more relevant variables. By contrast, discriminatory power describes the importance of a variable in distinguishing two class models. Variables with low modelling and low discriminatory power are more likely to contain noise and can potentially be omitted from the model. Another important indicator is the model distance, which reflects how well two classes are separated based on residual variance. Models with an inter-class distance above three can typically be considered well-separated.

2.3.2.4 Support Vector Machines (SVM)

In situations where known classes cannot be linearly separated, support vector machines (SVMs) are considered by many as an ideal classifier. In SVM analysis, samples are considered as points in a multi-dimensional space. Boundaries between different classes are established in the form of a separating gap or hyperplane, mapped to be as wide as possible (see Figure 2.7a). The hyperplane (classifier) can be expressed in terms of the samples lying closest to its boundaries, which are known as support vectors. Classification occurs by mapping new samples into the same space as the training set and assigning them to a known class based on which side of the boundary they fall on.

To establish the classifier between non-linearly separable classes, the data is transformed through a mathematical function (referred to as a kernel).

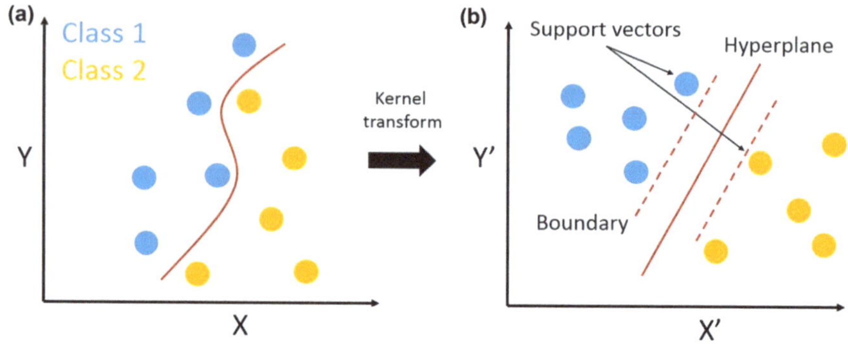

Figure 2.7 Representation of an SVM analysis with (a) two hypothetical sample classes that cannot be linearly separated; (b) the same classes mapped onto a higher dimensional space in which a linear hyperplane can be constructed and converted back into the lower dimensional space as a non-linear classifier. Reproduced from ref. 27 with permission from the Royal Society of Chemistry.

This transforms the data into a higher-dimensional space where linear separation is achievable (see Figure 2.7b). The mathematical basis behind such kernel transformations (the most common being linear, polynomial or Gaussian radial basis function) or computation of the classifier function are beyond the scope of this chapter but have been covered in other literature.[19]

In the example below, SVM was applied to the driver's licence series previously shown for SIMCA (see Table 2.5).[20] Because each individual model can only perform binary classification, multi-class problems require separate models to distinguish each possible pair of classes. In this case, all three Western Australian licences were correctly classified, with no incorrect rejections. However, the "hard" classification mode has also resulted in the Swiss licence being assigned to the most similar (2011) series despite having a different chemical composition.

SVM models are useful for high dimensionality data because they are a relatively simple way of allowing non-linear separation. They also work with fewer training samples and can handle datasets with more variables than samples, unlike LDA. On the other hand, there are no immediate metrics to indicate the reliability of discrimination. Results are also highly dependent on the kernel function and parameter selection, making optimisation potentially more complicated.

2.3.2.5 *Decision Trees and Random Forest (RF)*

Decision trees are a popular tool in machine learning for constructing and visualising classification rules. The tree is made up of several nodes, each representing a test on a particular sample attribute. Depending on the outcome of the test, the sample is directed along the corresponding branch to the next node, where the process repeats until a class label (leaf) is reached.

Table 2.5 SVM classification results for three Western Australian driver's licences and one Swiss driver's licence against constructed models for 2001, 2011 and 2014 series Western Australian Driver's licences. Numbers indicate the number of replicate spectra accepted for classification. This dataset is available *via* Research Data Australia.[18] Reproduced from ref. 27 with permission from the Royal Society of Chemistry.

	2001 series	2011 series	2014 series
2001 test sample	4	0	0
2011 test sample	0	4	0
2014 test sample	0	0	4
Swiss test sample	0	4	0

Whilst decision trees have the advantages of both simplicity and transparency, they are inherently unstable. Small changes in the dataset can lead to significant changes in the tree structure, which in turn can lead to errors. This can be remedied through use of a random forest (RF), an ensemble of independent decision trees (see Figure 2.8). These trees are built *via* a "bagging" process, wherein a random subset of samples is selected for each tree and a random subset of features used to build each tree's structure. The trees thus have low correlation to each other, compensating for individual errors. Input from an unknown sample is simultaneously processed by each tree, and the majority result given as the overall classification.

While generally applied to classification, RFs can also be used for exploratory analysis as they are inherently designed to probe sample dissimilarities. They can also be used for regression, with the mean predicted value given as the overall output. RFs are hence a versatile tool that are relatively robust to noise and outliers. Unlike many other classifiers, they can also handle missing data values. On the other hand, the complexity of the forest network can be challenging to interpret and is more computationally intensive to build and train. As of yet, RFs represent an emerging area of interest within the forensic context.

2.3.2.6 *Artificial Neural Networks (ANN)*

Artificial neural networks (ANNs) are computational models intended to mimic the design and function of biological neural structures. ANNs consist of interconnected nodes or "neurons" usually organised into an input layer, one or more hidden layers, and an output layer. An example of a fully connected network is shown below (see Figure 2.9a), where each neuron is linked to all other neurons in the adjacent layers. Other architectures include pooled (feed-forward) networks, where each neuron connects to a single neuron in the subsequent layer, or recurrent networks where links can be formed to neurons in the same or previous layers. The more complex the network, the more powerful it becomes. However, the degree of training required to establish relationships also increases.

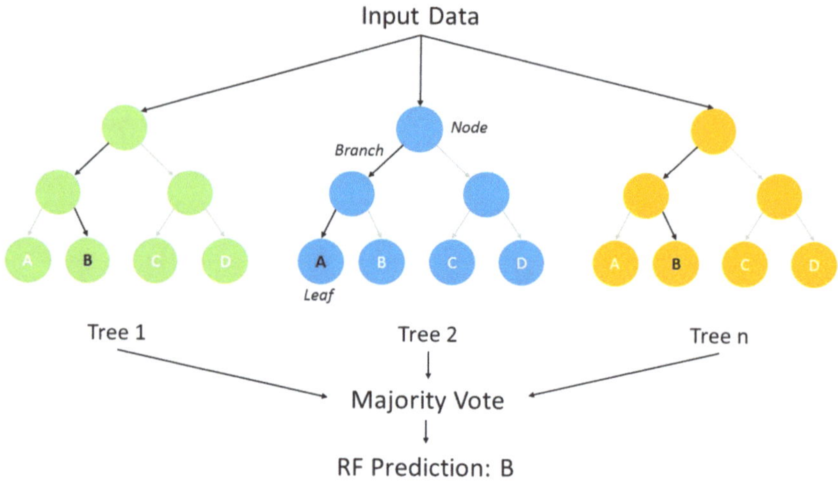

Figure 2.8 Basic architecture of a random forest model. Reproduced from ref. 27 with permission from the Royal Society of Chemistry.

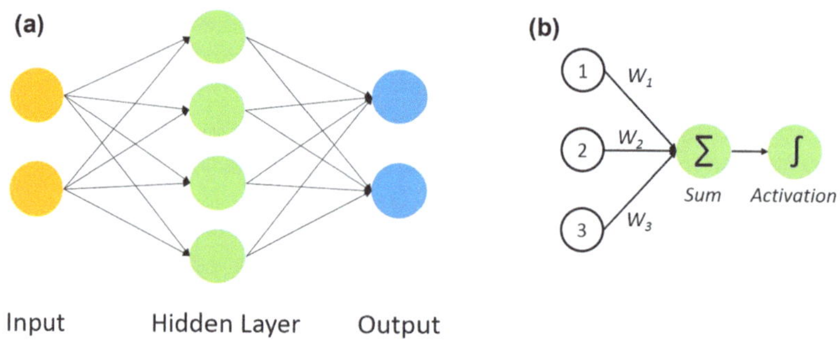

Figure 2.9 General schematic representing (a) the architecture of a fully connected ANN with a single hidden layer; and (b) the processing functions within a single neuron. Reproduced from ref. 27 with permission from the Royal Society of Chemistry.

Inputs received by a given neuron are multiplied by a weighting through various mathematical operations (see Figure 2.9b), effectively reflecting the strength of the neural connection. The resulting products are summed and processed through a non-linear activation function, producing an output that must exceed a certain threshold to be passed on to subsequent neurons. An appropriate response is then generated based on the information reaching the final output layer.

The key advantage of ANNs is that they learn from example in an analogous way to biological neurons. As with other supervised methods, the network is trained by generating predictions for a training set and evaluating the number of correct results. Any errors are used to adjust the hidden layers until the desired accuracy is achieved, a process known as back-propagation. The architecture of ANNs is highly flexible as there is no initial relationship assumed between inputs and outputs, and the functions within the neurons can be non-linear. This is ideal for analysing complex data, although the lack of an assumed model may require more training data to effectively identify relationships. The lack of transparency in the hidden layers also means there is no direct information on how these relationships have been derived. This has traditionally limited the use of ANNs in forensic analysis, where transparency around forensic decision-making and the underlying data is paramount.

2.3.3 Multivariate Regression

Multivariate regression methods establish the quantitative relationships between several predictor (independent) variables and a dependent response. These models can then predict responses for new samples based on their predictor data. Such approaches can also be adapted as binary classifiers by coding numeric values (usually zero and one) for each class.

2.3.3.1 Multiple Linear Regression (MLR)

The simplest multivariate regression technique is multiple linear regression (MLR), which constructs a linear combination of variables to describe the dataset in the form:

$$y_i = \beta_0 + \beta_1 x_{i1} + \beta_2 x_{i2} + \ldots + \beta_p x_{ip} + \varepsilon$$

where y_i is the response variable, β_0 is an intercept, β_p is the coefficient loading applied to the corresponding predictor variable x_{ip}, and ε is the residual error. This regression equation is constructed to give the closest approximation of the initial data points, *i.e.* to minimise the residuals.

p-values can be generated for the coefficients to identify variables with low importance that could be omitted from the model, in this scenario the null hypothesis that a variable's coefficient is not significantly different from zero, or the variable's influence is negligible. Low p-values (generally <0.05, representing a 5% probability) disprove this hypothesis and indicate variables important to measuring the response. This is important because MLR is another method requiring more samples than predictor variables. Another limitation of MLR is the assumption that there is no major correlation or multicollinearity between predictors, which is an incorrect assumption for most spectral and chromatographic data.

2.3.3.2 Principal Component Regression (PCR)

A possible means to overcome the limitations of MLR is to first remove multicollinearity and simplify the variable set through a dimensionality reduction technique. Principal component regression (PCR) does this by applying PCA to the original variables, retaining selected PCs as new

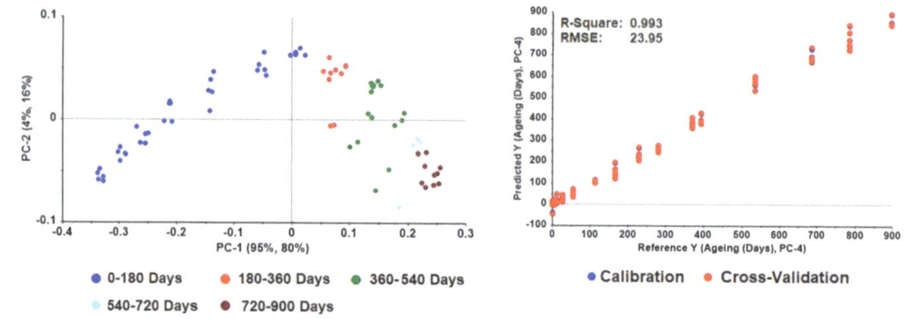

Figure 2.10 PCR results showing (left) scores plot for the first two factors and (right) regression plot of predicted *versus* actual ages for blue ballpoint ink exposed to light over 32 months. This dataset is available *via* Research Data Australia.[21] Reproduced from ref. 27 with permission from the Royal Society of Chemistry.

orthogonal predictors. PCs are generally retained based on either their eigenvalues or a scree plot.

Above is the output of PCR used to model the ageing of writing ink analysed by diffuse reflectance visible spectroscopy for 32 months (see Figure 2.10). PCA is first carried out to produce a scores plot identifying systematic changes that may be linked to the response. In this instance, more positive scores along PC 1 were linked to increased ageing. A regression plot based on the first four PCs was produced to visualise the predicted *vs.* actual responses for the dataset based on re-substitution of the calibration data and leave-*p*-out cross-validation. Poor overlap between the calibration and validation points suggests that the dataset contains outlier points that substantially alter the model structure.

The regression model initially appeared highly accurate based on cross-validation, exhibiting a strong linear correlation ($R^2 = 0.993$) and a low root mean square error (RMSE) of approximately 24 days between actual and predicted age. This suggested the model could predict the age of an unknown ink deposit to within one month. However, prediction of a separate validation set gave several negative age estimations and a high relative standard deviation (RSD) amongst replicate spectra. Predictive performance was particularly poor for samples within the first six months of ageing, when chemical changes appeared most rapid (Table 2.6). It was concluded that the changes taking place this ageing period were too variable to be precisely modelled. This example highlights the importance of appropriate test set validation for chemometric models.

2.3.3.3 *Partial Least Squares Regression (PLSR)*

A drawback to PCR is that it only considers variation amongst the predictors, not the response. That is, heavier weighting is applied to predictors that vary the most, even if this variation doesn't appear to alter the response.

Table 2.6 Actual *versus* estimated age of a validation blue ballpoint ink deposit exposed to light over 24 months using four-factor PCR and PLSR models. Predictions for each time interval were averaged over five replicate spectra. This dataset is available *via* Research Data Australia.[22] Reproduced from ref. 27 with permission from the Royal Society of Chemistry.

	PCR		PLSR	
Actual age (days)	Predicted age	% RSD	Prediction age	% RSD
1	−52	18	−47	17
6	−41	14	−35	15
21	−18	56	−12	82
42	−13	61	−4	194
98	28	57	39	40
147	85	21	93	18
504	452	10	453	10
596	430	50	432	47
617	587	6	588	6
672	632	4	633	4

Partial least squares regression (PLSR) instead aims to maximise the covariance between X and Y, applying greater weight to predictors that vary with the response. The assumption is that such predictors will allow more accurate relationships to be established, although this of course does not guarantee higher predictive performance. In the case of the inks referred to above, age estimation was no better using a PLSR model than PCR (see Table 2.6).[22] This indicates that in this scenario, the predictors showing the greatest overall variation were also those most strongly correlated with the response.

Another common application of the partial least squares algorithm is binary classification. Partial least-squares-discriminant analysis (PLS-DA) uses dummy variables to represent the two classes to be distinguished. Regression of new samples onto the model results in a classification based on whether the estimated value falls close to zero (class A) or one (class B). This process is illustrated below, where PLS-DA was used to differentiate authentic and counterfeit perfumes analysed using paper spray mass spectrometry (see Figure 2.11).[23]

As with SVM however, several separate models are required to apply a binary PLS classifier to multi-class problems. PLS-DA is also sensitive to within-class variability and can be difficult to interpret if predicted values lie between zero and one. Nonetheless, the weights and loadings generated through PLS-DA can provide valuable insights into the basis of discrimination.

2.3.4 Experimental Design

The methods and parameters (including sample preparation, storage, or instrumental settings) used in forensic analysis must be demonstrably fit for purpose. Experimental design methods are ideal tools in identifying and modelling these factors to yield optimal analytical results. This chapter focusses on multivariate designs that concurrently investigate several factors,

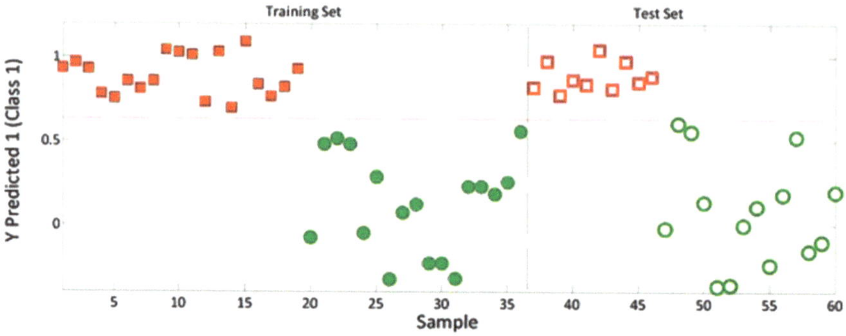

Figure 2.11 PLS-DA predictions of authentic (red; coded as 1) and counterfeit (green; coded as 0) perfume samples. Full symbols represent training samples, empty symbols represent validation samples, and the horizontal dashed line indicates the estimated decision threshold. Reproduced from ref. 23 with permission from the Royal Society of Chemistry.

rather than one-factor-at-a-time (OFAT) approaches. Although simpler, OFAT designs fail to consider potential interactions between factors. For example, the optimal inlet temperature for a gas chromatographic analysis may vary according to injection volume. The multivariate approach is more efficient in such cases as it directly assesses interactions between them, whereas an OFAT approach requires re-optimisation with each factor.

2.3.4.1 *Factorial Designs*

Statistical optimisation usually begins with a screening design to identify key factors impacting the target response. By far the most common screening approach is a factorial design, which estimates the main effects and interactions by switching each variable (factor) between discrete levels. The simplest of these is a two-level (2^k) factorial design, where "k" describes the number of continuous factors under investigation, each with two levels ("low" and "high"). A 2^3 factorial design is graphically represented below (see Figure 2.12a) as a cubic design region, where the three factors form the dimensions of the cube and the levels of each factor form the vertices. Experiments are run using parameters defined at each vertex, and the resulting data used to produce a linear plane of the response according to each factor. As such, the minimum number of experiments required for a two-level factorial design is equal to 2^k. Additional replicates will increase the model's ability to detect significant effects at the cost of additional time and resources required.

Analysis of variance (ANOVA) is used to assess the model fit and to identify any significant factors or factor interactions. Table 2.7 shows the ANOVA for a 2^3 factorial design maximising cocaine extraction from hair using enzymatic hydrolysis, with two replicates of each point producing 16 total experiments.[24] The factors investigated were buffer solution volume, pronase E concentration ([P-E]) and dispersing mass to hair mass ratio (D-S);

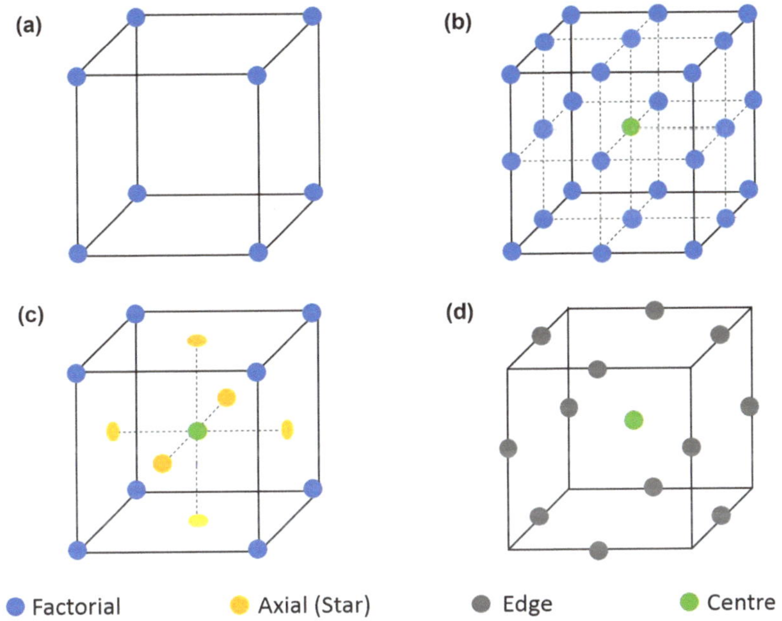

Figure 2.12 Graphical representation of minimum experimental runs required for (a) a two-level full factorial, (b) a three-level full factorial, (c) a face-centred central composite, and (d) Box–Behnken designs investigating three continuous variables. Reproduced from ref. 27 with permission from the Royal Society of Chemistry.

Table 2.7 ANOVA for data from a 2^3 factorial design optimising cocaine extraction from hair samples based on enzymatic hydrolysis. Note that the three-way interaction ABC can also be assessed but has been omitted for clarity. This dataset is available as supplementary information associated with the corresponding research article.[24] Reproduced from ref. 27 with permission from the Royal Society of Chemistry.

Parameter	*p*-value	Parameter[a]	Value
Model significance	<0.001	R^2	0.922
A: volume (mL)	<0.005	Adjusted R^2	0.854
B: [P-E] (mg mL^{-1})	<0.001	Predicted R^2	0.689
C: D-S	0.487		
AB:	0.888		
AC	0.095		
BC	0.337		

[a] Adjusted R^2 is used when comparing models with differing numbers of factors.

coded respectively as variables *A*, *B*, and *C*. A high R^2 value indicates a good fit of the model, whilst a predicted R^2 of 0.689 suggests reasonable predictive capability. The predicted R^2 is akin to cross-validation accuracy, where each data point is removed and re-estimated from the remaining data. If the predicted R^2 is substantially lower than the overall R^2, the model is likely "over-fitted" to the initial data, rendering it unable to adapt to new information.

ANOVA was performed on each factor to identify those with a statistically significant effect on the recovered amount of cocaine. This process takes the null hypothesis that the effect of a given factor on the response is not statistically significant, with the resulting *p*-value reflecting the probability of observing the response obtained if this assumption is true. Low *p*-values thus indicate parameters that do significantly affect the response. In this instance, the [P-E] and buffer volume were found to significantly affect cocaine recovery using enzymatic hydrolysis extraction.

This was reinforced by examination of the Pareto chart (see Figure 2.13a), which shows the ranked standardised effects of each factor based on *t*-statistics.

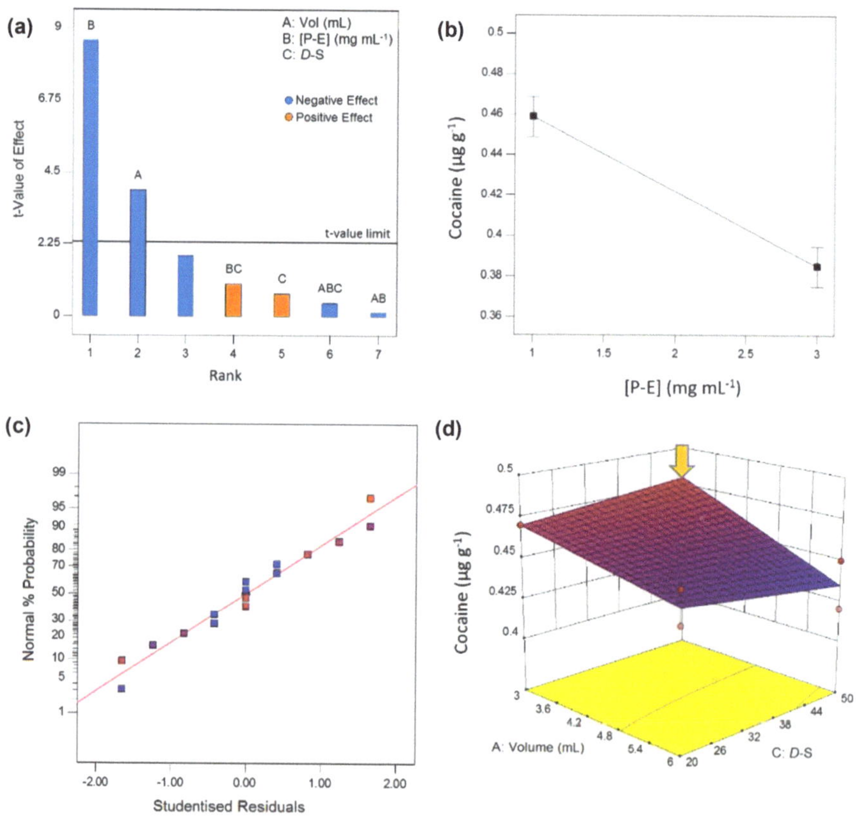

Figure 2.13 Outputs from a 2^3 factorial design optimising cocaine extraction from hair samples consisting of (a) a Pareto chart; (b) a one-factor plot for pronase-E concentration; (c) a normal plot of residuals; and (d) a response surface for the concentration of cocaine as a function of buffer volume and dispersing mass to hair mass ratio. The arrow indicates optimum parameters on the response surface. This dataset is available as supplementary information associated with the corresponding research article.[24] Reproduced from ref. 27 with permission from the Royal Society of Chemistry.

Both factors fell above the *t*-value limit, indicating a statistically significant effect. The Pareto chart and one-factor plots (see Figure 2.13b) showed these effects to be negative, meaning that lower buffer volumes and [P-E] led to a greater extraction efficiency of cocaine from hair. A normal plot of residuals (see Figure 2.13c) was used to check the distribution of residuals for each of the 16 experimental runs (two replicates of each factorial point). Non-linearity of the residuals plot could indicate the presence of outlying samples, or an undetected factor not accounted for within the design.

We can then examine the response surface for the concentration of extracted cocaine according to buffer volume and D-S, assuming that [P-E] is minimal (see Figure 2.13d). This surface is modelled as a linear plane with the average response at each experimental run acting as anchor points. The highest cocaine recovery was obtained using a minimal [P-E] and buffer volume (as expected) and a D-S of 50. However, these may not in fact be the absolute optimal extraction parameters. This is because 2^k factorials assume that the response changes linearly with each factor. Detecting non-linearity (curvature) in the response requires experimental data obtained at values between the low and high levels. One approach to this would be a three-level factorial (3^k) design, but this can quickly lead to an unfeasible number of experimental runs. Studying three factors at three levels (a 3^3 full factorial design) with no replicates already requires 27 experiments (see Figure 2.12c)!

Each additional factor in a 2^k design also doubles the number of experiments required. In scenarios where this becomes impractical, some design points can be omitted to create a fractional factorial design. This design is more economical, although some factors or interactions will not be estimated.[25] Another alternative is a Plackett Burman design, which generates designs with a number of experiments between those of a 2^k factorial. On the other hand, simplifying the design in this manner can result in confounding main effects with factor interactions.[26]

2.3.4.2 Central Composite Designs (CCD)

A central composite design (CCD) augments a two-level factorial design with replicated centre points where all factors are set to a median level, plus a set of axial (star) points where a single factor takes on levels above and below the median (see Figure 2.12c). These additional design points allow the detection and modelling of curvature in fewer runs than a three-level factorial design. The axial and centre points can also be run without the embedded factorial, producing a Box–Behnken design (Figure 2.12d) requiring even fewer experiments.[25] Although highly cost-efficient, this can result in less precise modelling of the response.

Different CCD models can be produced based on whether the axial points fall beyond, within, or on the factorial points. Circumscribed designs have axial values outside of the factorial points, allowing a larger investigative range to be covered. In some cases, however, such extreme values may not be practically or safely achievable. Inscribed designs that have axial points within the factorial space are ideal in these scenarios. Circumscribed and inscribed designs are both

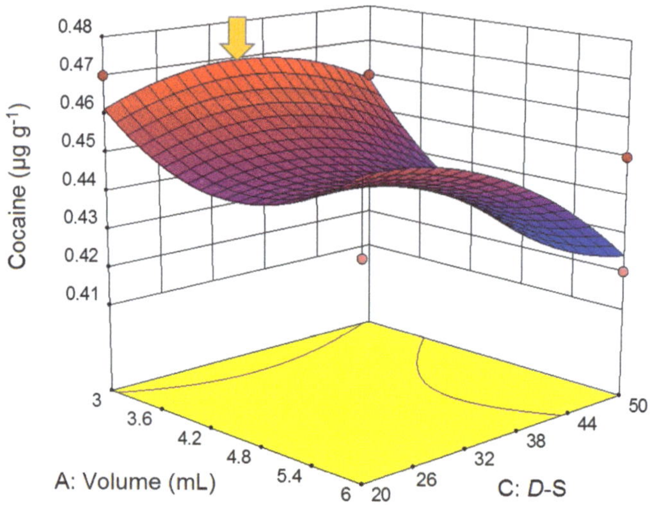

Figure 2.14 Response surface for the concentration of cocaine as a function of buffer volume and dispersing mass to hair mass ratio using a circumscribed CCD. The arrow indicates optimum parameters on the response surface. This dataset is available as supplementary information associated with the corresponding research article.[24] Reproduced from ref. 27 with permission from the Royal Society of Chemistry.

rotatable, meaning the prediction error is consistent across the design area. Face-centred CCDs re-use the factorial points as the axial points. Although non-rotatable, face-centred designs generally require fewer experiments, and may be necessary if values between or beyond the factorial levels are not practicable.

Returning to the example of cocaine extraction through enzymatic hydrolysis, a non-linear "saddle" response surface produced by a circumscribed CCD is shown above (see Figure 2.14). This design was created by collecting further data at axial and centre points added (in duplicate) to the original factorial design. ANOVA again found the pronase E concentration and buffer volume to be negatively associated with cocaine concentration. However, the optimal dispersing mass to hair mass ratio was found to be 35 rather than 50, due to curvature in the response. This "true" optimum was undetected in the factorial design because no data was collected at median values of each factor. This example shows how CCDs are powerful approaches for optimising evidential protocols.

2.4 Conclusion

This chapter has demonstrated a variety of chemometric tools that offer powerful means of extracting, comparing, and optimising data for forensic purposes. However, as with any forensic method, these tools should not be used blindly. Chemometric procedures require careful data selection, preparation, and validation to ensure proper measures of performance.

Above all, a thorough understanding is needed of the current limitations and how these may be resolved through future research.

Acknowledgements

Portions of this chapter were adapted from the author's previous work,[27] reproduced with permission of the Royal Society of Chemistry. Professor Wilhelm van Bronswijk (Curtin University) is thanked for assistance with reviewing this material.

References

1. National Academy of Sciences, Strengthening Forensic Science in the United States: A Path Forward. Committee on Identifying the Needs of the Forensic Sciences Community, National Research Council, Washington DC, 2009.
2. Å. Rinnan, F. van den Berg and S. B. Engelsen, *TrAC, Trends Anal. Chem.*, 2009, **28**, 1201.
3. P. Lasch, *Chemom. Intell. Lab. Syst.*, 2012, **117**, 100.
4. J. Engel, J. Gerretzen, E. Szymańska, J. J. Jansen, G. Downey, L. Blanchet and L. M. C. Buydens, *TrAC, Trends Anal. Chem.*, 2013, **50**, 96.
5. M. Maric, W. van Bronswijk, S. W. Lewis and K. Pitts, *Anal. Methods*, 2012, **4**, 2687.
6. K. van der Pal, G. Sauzier, M. Maric, S. W. Lewis and W. van Bronswijk, Published Dataset, 2020, DOI: 10.25917/5f51c4681c11e.
7. M. Otto, *Chemometrics: Statistics and computer application in analytical chemistry*, Wiley-VCH, Weinheim, Germany, 2nd edn, 2007.
8. J. X. W. Wong, G. Sauzier and S. W. Lewis, *Forensic Sci. Int.*, 2019, **298**, 88.
9. J. X. W. Wong, G. Sauzier and S. W. Lewis, Published Dataset, 2019, DOI: 10.25917/5bee60501fdf0.
10. C. J. Telford, B. A. Burrows, G. Sauzier, W. van Bronswijk, M. M. Houck, M. Maric and S. W. Lewis, *Anal. Methods*, 2017, **9**, 192.
11. C. J. Telford, B. A. Burrows, G. Sauzier, W. van Bronswijk, M. M. Houck, M. Maric and S. W. Lewis, Published Dataset, 2018, DOI: 10.25917/5b692f7c1f8e3.
12. J. O. Berger, L. R. Pericchi, J. K. Ghosh, T. Samanta, F. De Santis, J. O. Berger and L. R. Pericchi, *Lect. Notes-Monograph Ser.*, 2001, **38**, 135.
13. V. Todorov, *Stat. Methods Appl.*, 2007, **15**, 395.
14. A. L. Setser and R. Waddell Smith, *Forensic Chem.*, 2018, **11**, 77.
15. J. B. M. Dröge and H. A. Van 'T Klooster, *J. Chemom.*, 1987, **1**, 221.
16. J. B. M. Dröge, W. J. Rinsma, H. A. Van 'T Klooster, A. C. Tas and J. Van Der Greef, *J. Chemom.*, 1987, **1**, 231.
17. A. T. Hermane, S. Pierre-Yves, L. Pierre, H. Philippe and Z. Eric, *Anal. Chim. Acta*, 2021, **1144**, 130.
18. J. McGann, M. Willans, G. Sauzier, M. J. Hackett, S. W. Lewis, J. McGinn, T. Trubshoe and W. van Bronswijk, Published Dataset, 2020, DOI: 10.25917/5f3a028d3d388.

19. A. Tharwat, *Knowl. Inf. Syst.*, 2019, **61**, 1269.
20. J. McGann, M. Willans, G. Sauzier, M. J. Hackett, S. W. Lewis, J. McGinn, T. Trubshoe and W. van Bronswijk, *Forensic Sci. Int. Rep.*, 2020, **2**, 100149.
21. G. Sauzier, J. McGann, S. W. Lewis and W. van Bronswijk, Published Dataset, 2018, DOI: 10.4225/06/5b35e1f822c13.
22. G. Sauzier, J. McGann, S. W. Lewis and W. van Bronswijk, *Anal. Methods*, 2018, **10**, 5613.
23. J. A. R. Teodoro, H. V Pereira, D. N. Correia, M. M. Sena, E. Piccin and R. Augusti, *Anal. Methods*, 2017, **9**, 4979.
24. M. Míguez-Framil, P. Cabarcos, M. J. Tabernero, A. M. Bermejo, P. Bermejo-Barrera and A. Moreda-Piñeiro, *J. Chromatogr. A*, 2013, **1316**, 15.
25. A. I. Khuri and S. Mukhopadhyay, *Wiley Interdiscip. Rev. Comput. Stat.*, 2010, **2**, 128.
26. M. J. Anderson and P. J. Whitcomb, *RSM Simplified: Optimising Processes Using Response Surface Methods for Design of Experiments*, CRC Press, Boca Raton, Florida, 2005.
27. G. Sauzier, W. van Bronswijk and S. W. Lewis, *Analyst*, 2021, **146**, 2415.

CHAPTER 3

Illicit Drugs and Pharmaceuticals Analysis

A. POPOVIC, C. ROUX AND M. MORELATO*

Centre for Forensic Science, University of Technology Sydney, PO Box 123, Broadway, NSW 2007, Australia
*Email: Marie.Morelato@uts.edu.au

3.1 Introduction

Drug consumption and trafficking is one of the biggest challenges for health and law enforcement organisations worldwide.[1] As drug markets are continually evolving, it is essential to understand them to disrupt and reduce their impact.[2] Given the large number of analytical techniques used to analyse drugs and the complex datasets they generate, chemometric techniques are commonly and increasingly used to "design or select optimal experimental procedures; to provide maximum relevant chemical information by analysing chemical data and to obtain knowledge about chemical systems" (ref. 3, p. 1). This chapter will summarise the common chemometric techniques that are applied in illicit drugs and pharmaceutical analysis, with a particular emphasis on drug profiling, outlining existing research to date and emerging trends in the field.

3.2 Drug Problem

In 2019, an estimated 275 million people had used drugs (*i.e.*, substance controlled under the international drug control conventions) at least once in the previous year and approximately 36 million people suffered from

Chemometric Methods in Forensic Science
Edited by Vishal Sharma, Georgina Sauzier and Simon W. Lewis

Published by the Royal Society of Chemistry, www.rsc.org

drug use disorders.[1] The illicit use of drugs can encompass a range of substances:

- **Illicit drugs:** "In the context of international drug control, 'drug' means any of the substances in Schedule I and II of the 1961 Convention, whether natural or synthetic" (ref. 4, p. 63). (*e.g.*, cannabis, cocaine and amphetamine-type stimulants (ATS)).
- **Authentic pharmaceuticals:** "Drugs that are available from a pharmacy, over the counter or by prescription, which may be subject to misuse, *i.e.*, when used for purposes, or in quantities, other than medical purposes for which they were prescribed" (ref. 5, p. 154) (*e.g.*, benzodiazepines, over-the-counter codeine and doping agents).
- **Counterfeit pharmaceuticals:** "Medicines which are deliberately and fraudulently mislabelled with respect to identity or source" (ref. 6, p. 168).

During the past decade, the types of substances available on the drug market have considerably expanded.[7] As well as traditional plant-based substances (cannabis, cocaine, heroin, *etc.*), the market for synthetic illicit drugs and pharmaceuticals has seen a considerable diversification.[7] The diversification and growth of these markets is primarily driven by business considerations, globalisation, technological developments and changes in law enforcement activities, as illustrated by the recent rise of online anonymous marketplaces (situated on the Dark Web).[8] In addition to the increased prominence of illicit drugs, roughly 10% of the world's medicines are counterfeits, with new cases being detected every week.[9]

Over the years, forensic science has primarily positioned itself as a service provider for the criminal justice system.[10] Unfortunately, this focus on the criminal justice system has limited its capacity to provide knowledge about crime systems. This limitation is a major impediment in the area of illicit drugs. Traditionally, each case is treated individually, and information related to illicit drugs is owned by different organisations (*e.g.* law enforcement agencies, health department and laboratory), failing to highlight criminal networks or an overarching overview of the problem. Information conveyed by traces (*e.g.* drug specimens) exist beyond the evaluation imposed by the criminal justice system and should be integrated into proactive styles of policing (such as intelligence-led policing) and security studies.[11–14] A significant opportunity exists to transform the contribution of forensic science through the systematic integration of forensic case data (*e.g.* drug specimens) to provide knowledge about criminal activities before they occur and thus disrupt and prevent crimes (*i.e.*, forensic intelligence).[15,16] In the drug area, forensic intelligence can be considered as "a systematic process in which each new specimen is compared with existing data organised in a memory† built upon earlier seizures of known origin references. It can then be interpreted in the context of the current understanding of the crime situation in order to obtain intelligence" (ref. 12, p. 5).

Although the detection and confirmation of illicit drugs and counterfeit pharmaceuticals is important (relatively simple and fast methods available), more information can be gained from the physical characteristics and impurities present in the drug. By generating a memory of chemical and physical profiles and systematically analysing it, it is possible to provide not only tactical and operational intelligence, but it can also provide strategic level trends‡ (*e.g.*, overview of drug markets).[6,19]

3.3 Drug Profiling

Drug profiling is defined as "the extraction of physical and chemical profiles (*i.e.*, characteristics of a specimen specifically chosen with respect to the purpose of the process) to be used in the application of policies against the illegal use of drugs" (ref. 20, p. 248). In an illicit drug context, profiling can be used to link drug specimens, identify synthetic routes, and identify common origins between seizures and trafficking routes.[12,21–24] In a pharmaceutical context, profiling can be used to differentiate between authentic and counterfeit specimens, or to link counterfeit specimens, similar to illicit drugs.[25,26]

With respect to physical characteristics, they include dimensions (diameter and height), form (*i.e.*, powder, liquid or tablet), shape, colour, logo, break line or score (*i.e.*, middle line on a tablet assisting in splitting a tablet into two), granularity and packaging.[27] These characteristics are easily extracted and can form a physical profile (see Table 3.1).

In contrast to physical characteristics, chemical characteristics are more difficult to extract and require more advanced analytical instruments. For illicit drugs, chemical characteristics include impurities, precursors, byproducts, solvents and cutting agents (adulterants and diluents). With regards to pharmaceuticals, the main chemical characteristics that are targeted are active pharmaceutical ingredients (APIs), excipients and impurities.[28] There is a large variety of analytical techniques that are used

---

†The memory is a broad concept that can be defined as an organised structure representing the knowledge the organisation operating the process has, at a certain time, about criminal activities (*e.g.* trends, linked cases, patterns, serial crimes, *etc.*). In contrast to the database, the memory contains information that is systematically structured to help form meaningful outcomes.[16]

‡Tactical intelligence supports front-line officers in taking case-specific action and is relevant to specific investigations.[17,18] It lacks a greater understanding of long-term or wider geographical problems.[18] Operational intelligence assists in planning crime reduction activities and relates to crime series. It is considered as a "meso-level" and supports decision-makers that are responsible for geographical areas or who command teams.[18] Finally, strategic intelligence provides an understanding of patterns and functioning of criminal behaviour and environments. It is future-oriented, proactive and used by top-level managers. It will influence not just the police activities but also non-law enforcement agencies, such as health and policy.[18]

Table 3.1 Current articles focusing on physical profiling for drug intelligence purposes.

Sample type	Analytical method
MDMA	Visual[22,32,45]
Heroin	Visual[60]
Heroin packaging	General measurements[60]

Table 3.2 Common analytical methods used in illicit drug profiling.

Analytical method	ATS profile	Cocaine profile	Heroin profile
GC–MS	By-products[44,48,49,91,116–123]	Alkaloids[30,124,125]	By-products[30,57,72,126–128]
GC–FID	By-products[42,129]		By-products[51,52,56,60,71,130–132]
CE–DAD	Adulterants[120,121]		
LC–ELSD	Diluents[120,121]		
ICP–MS	Elements[120,121]	Elements[63]	Elements[38,53,59]
CE	Enantiomers[133]		Alkaloids[127]
IRMS	Isotope ratios[133]	Isotope ratios[134,135]	
ATR–FTIR		Adulterants & diluents[65,67]	
HS GC–MS		Solvents[62,136]	Solvents[127,136]

to chemically profile drugs; most can be categorised into either chromatographic-based (*e.g.* high-performance liquid chromatography (HPLC) and gas chromatography (GC)) or spectroscopic-based techniques (*e.g.*, near infrared spectroscopy (NIR), IR and Raman spectroscopy).[25] These analytical techniques can be applied to drugs to extract different profiles, *e.g.* GC can be applied to illicit drugs to extract a "by-product" profile. In contrast, GC does not have as many applications for pharmaceuticals as very few contain volatile substances.[25] NIR tends to be the more favoured analytical techniques when it comes to pharmaceuticals as it is rapid and non-destructive.[29] The selection of characteristics and analytical technique will inevitably influence the results obtained from the chemometric methods applied.[30] Table 3.2 includes references to some recent studies focusing on chemical profiling of illicit drug specimens.

Drug profiling is based on the assumption that a great variation exists in physical and chemical properties between specimens coming from different sources (*e.g.*, clandestine laboratory), while variations in profile of specimens synthesised by the same source (*e.g.*, the same laboratory) are assumed to be relatively small.[31,32] As a result, chemometric approaches may be used to determine the degree of similarity between specimens, allowing them to be grouped into clusters and assigned a link value (*i.e.*, similarity score).[33,34]

Drug profiling is not just used to confirm or disprove a link between two cases. Its biggest potential lies within a systematic procedure in which each new specimen is compared to existing data stored in a memory based on previous specimen profiles and interpreted in the light of current knowledge of the situation.[20,35,36] Specimens can be pre-organised into groups having similar profiles. The physical/chemical profile of a new specimen is acquired and compared to the pre-determined groups to obtain an indication of a possible link between the new specimen and the existing data. For example, recently detected counterfeit pharmaceuticals can be checked against the database to see if they are similar to specimens detected in the past.[37] As drug profiling data is usually high-dimensional, chemometric methods are used to aid the grouping of specimens or evaluation of similarity.[25]

3.4 Chemometrics Techniques

As forensic scientists use a multitude of analytical methods for the analysis of illicit drugs and pharmaceuticals, large and complex datasets are generated. These datasets should thus be subjected to various mathematical and statistical techniques. In the case of drug profiles, it is common to pre-treat or pre-process the data.[2,38] Analyses obtained from two different instruments may not exhibit the same sensitivity and data scaling is often required. Pre-treatments are employed to reduce the instrumental influence, and therefore represent an essential part of the overall profiling process.[25,26,39] Pre-treatments are also used to minimise the effect of significant differences in the concentration of compounds present in a specimen.[8] The most common pre-treatments are outlined in Table 3.3.

Table 3.3 Common pre-treatments used in illicit drug profiling.

Pre-treatment	Formula
Logarithm (L)	$\log(x_i)$
Normalisation (N)	$N_i = \dfrac{x_i}{\sum_{i=1}^{n} x_i}$
Standardisation (S)	$S_i = \dfrac{x_i}{S.D._i}$
Fourth square root ($4R$)	$\sqrt[4]{x_i}$
Normalisation + logarithm ($N+L$)	$\log(N_i)$
Normalisation + standardisation ($N+S$)	$\dfrac{N_i}{S.D._i}$
Normalisation + fourth square root ($N+4R$)	$\sqrt[4]{N_i}$
Normalisation + standardisation + logarithm ($N+S+L$)	$\log(S+N)_i$
Normalisation + standardisation + fourth square root ($N+S+4R$)	$\sqrt[4]{(S+N)_i}$

In addition to pre-treatments, drug profiles are commonly subjected to pattern recognition methods to extract information about the supplied specimens and classify them into desired groups. These pattern recognition methods can be broken down into two groups: unsupervised pattern recognition (UPR) and supervised pattern recognition (SPR) methods. The differentiating factor of SPR methods is that they perform modelling using specimens with a known class, whereas UPR methods do not require this known class. Figure 3.1 outlines the current body of research focusing on unsupervised and supervised pattern recognition methods for illicit drug and counterfeit pharmaceuticals. Specifically, the left pane of Figure 3.1 visualises the studies explored in this chapter and trend of publication time. A Sankey chart (right pane) stems from the left pane in Figure 3.1, which depicts the count of and relationship between analytical technique groups, chemometric method group and goals within each article. Additionally, the analytical technique group and article aim are subgrouped by drug type (illicit drug (ID) at the top) or counterfeit pharmaceutical (CP) at the bottom) based on the study.

3.5 Unsupervised Pattern Recognition

There are two main types of UPR methods (see Figure 3.2): projection methods[40] (*e.g.*, principal component analysis (PCA) and projection pursuit (PP)) and clustering methods[41] (*e.g.*, hierarchical clustering analysis (HCA) and *k*-means clustering (KMC)). These methods aim to uncover any hidden patterns in multi-dimensional data, such as similarities between specimens or the explanatory variables used.[25]

3.5.1 Illicit Drugs

In the literature, chemometrics techniques are usually applied according to a specific goal. UPR methods can be used in a variety of different ways depending on the goal of the analysis. The articles in Table 3.4 highlight some of the different goals encountered in the literature. The most common goal is testing the feasibility of UPR to group specimens into their geographic origin. Other goals include testing the viability of UPR to group specimens into their original seizure or batches. These goals will be explored in more detail in the following sections, highlighting the specific methods used and their feasibility. Considering the literature explored, PCA was the most frequently documented method, closely followed by HCA (see Table 3.4).

With respect to PCA, previous studies delivered mixed results; no distinct pattern could be observed for the specimens in several articles,[42–45] while in other studies[39,40] PCA was used to successfully infer production batches of cocaine specimens.[46,47] Looking at clustering methods, HCA generally worked well for seizure level clustering.[42,48] Furthermore, HCA was also used to successfully differentiate between specimens of different synthetic routes.[49]

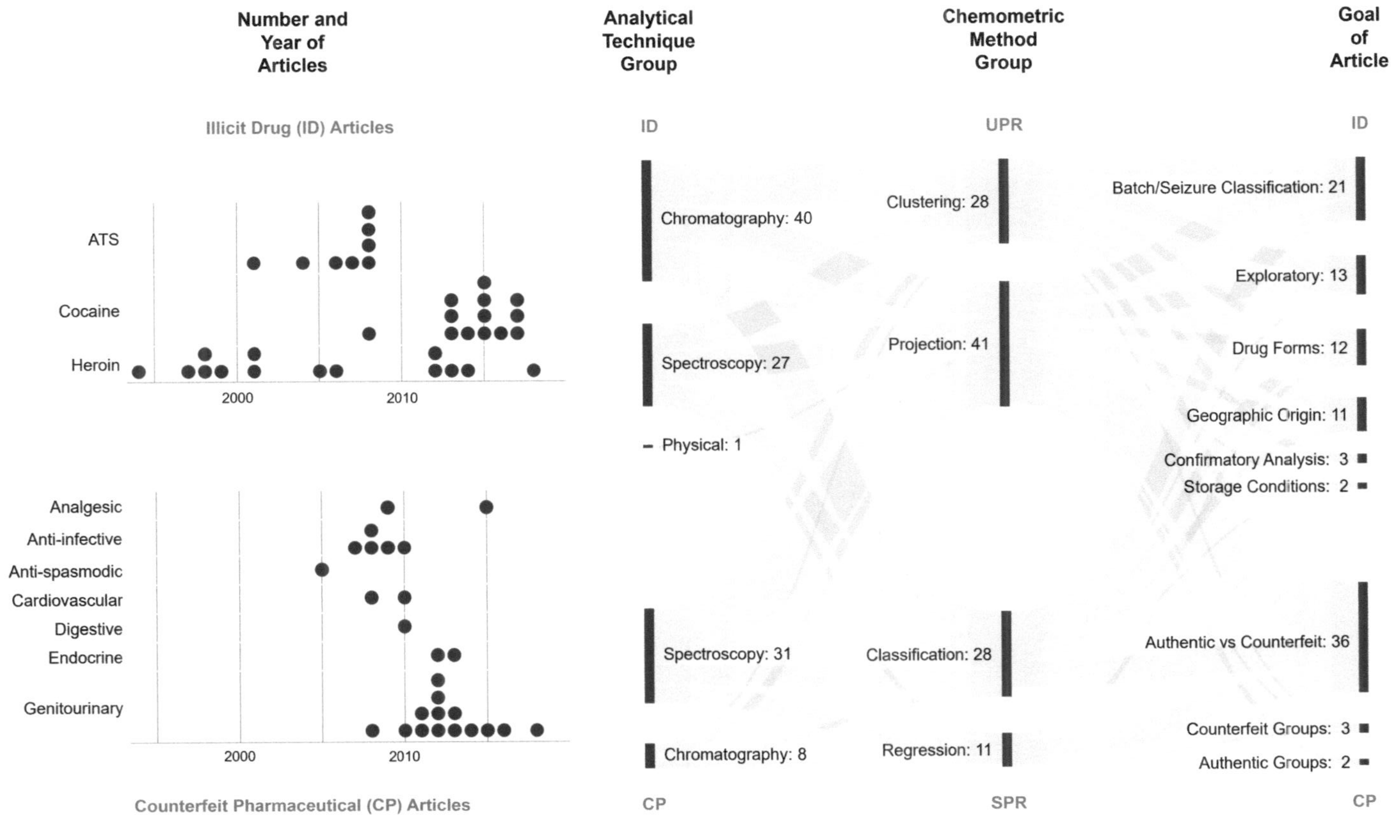

Figure 3.1 Overview of research focused on using chemometric methods for illicit drugs and counterfeit pharmaceuticals.

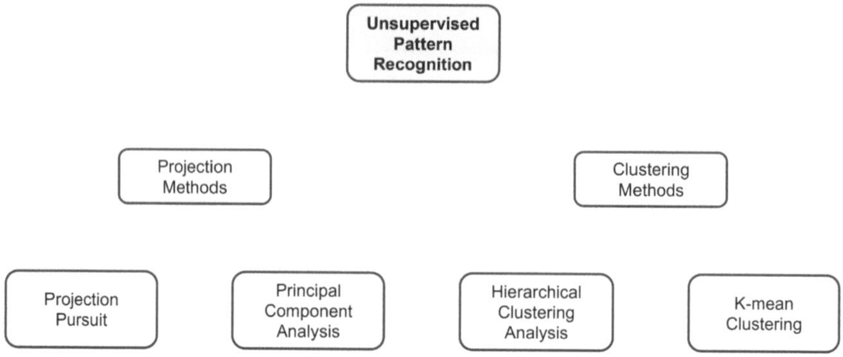

Figure 3.2 Common UPR methods using illicit drug and counterfeit pharmaceutical analysis.

Table 3.4 Unsupervised pattern recognition methods and specific goals featured in illicit drug analysis.

Specific goal	PCA	HCA	KMC
Exploratory analysis	Heroin[38,50–53]	ATS[118] Cocaine[54] Heroin[53,55]	Heroin[71]
Batch/seizure classification	ATS[42–45] Cocaine[46,47] Heroin[56,57]	ATS[42,43,48,49] Heroin[56,57]	Heroin[56]
Geographic origin	Heroin[59,60]	Cocaine[61–64]	Heroin[53]
Drug forms	Cocaine[65–67]	Cocaine[65,67,68]	
Confirmatory	Cocaine[64]	Heroin[52,72]	Heroin[50]
Dimensionality reduction	Heroin[71]		Heroin[72]

It is also common for these methods to be used sequentially; HCA worked well for evaluating PCA generated clusters.[50]

3.5.1.1 *Exploratory Analysis*

This subsection looks at reported studies that did not have a specific goal set. In these studies, the aim was to conduct an exploratory analysis and understand whether there were any observable patterns in the data. Overall, PCA was demonstrated to be useful as an exploratory tool (for heroin specimens in particular).[38,50–53] For example, Chan *et al.* used PCA to explore which pre-treatment method was the most discriminative for heroin specimen classification.[53] Furthermore, HCA was also used as a preliminary clustering technique to explore natural clusters in the data.[53–55] This use of the technique does not feature greatly in the literature but is useful when the specimen to seizure relationship is not clear or known. For example,

HCA and KMC were used as an exploratory technique where specimens were subsequently grouped into two clusters.[53] These clusters were subsequently found to be correlated to the geographic origin of the specimens, indicating that the profiles used in the study would be useful for future specimen geographic origin determination.

3.5.1.2 Batch/Seizure Classification

Several studies focused on using chemometric techniques to group specimens by seizure or batch. This goal looks at the feasibility of chemometric methods in differentiating linked and unlinked specimens, an important feature in forensic intelligence.[2] Focusing on PCA, suboptimal results were observed when trying to group ATS specimens into their original seizure,[42–45] however, PCA worked well for grouping cocaine and heroin specimens into their original batches.[46,47,56,57]

HCA also produced suboptimal results for ATS specimens. Although indicating potential grouping, not one original seizure was correctly clustered by HCA.[43] However, in other studies HCA was shown to be an adequate technique to evaluate class membership (seizure[42,48,57] and batch[49,56] classification) for ATS, cocaine and heroin specimens, with the exception of a few misclassifications. For example, Dujourdy *et al.* were able to identify batch level trends for methylamphetamine (MA) specimens synthesised *via* ephedrine (EPH).[49] As MA synthesised *via* EPH can be achieved in a variety of ways (*e.g.* through Birch reduction, Nagai and Moscow routes[58]) it is important to be able to differentiate between batches of specimens. Using HCA, Dujourdy *et al.* identified the presence of six clusters, indicative of different synthesis routes.[49]

In another study, KMC was able to differentiate each of the samples according to the specimen they were derived from (NB: specimens were cut into five samples differing by purity).[56] KMC can be used as a tool for clustering specimens regardless of whether the specimen origin is known or not. If the origin is not known, a chosen specimen or sample can be used to represent the typical profile of a cluster.[56] Chan *et al.* experimented using high, medium and low purity samples as starting points for KMC to optimise the clustering capabilities of the method.[56] The results showed that the samples with medium level purities produced the best separation instead of over training the model by using only low or high purity specimens as the training set.[56] The biggest take away from these findings is that the initial choice of the training set will drive how well KMC performs.[56]

3.5.1.3 Geographic Origin

Knowing the geographic origin of a given specimen is an important piece of information, especially in investigations that aim at establishing trafficking routes. PCA is a useful tool for inferring the geographic origin of specimens by identifying clusters in the data.[59,60] For example, PCA clustered heroin specimens into two broad groups based on the presence of trace elements identified in specimens.[59] In previous research, trace elements (specifically Lanthanides) were found to be significantly useful for classifying heroin specimens based on geographic location.[53] It should be noted that the presence of the trace elements at the time of plant cultivation is not guaranteed; they may have been introduced at any stage throughout the manufacturing process.

Similarly, HCA performed well for splitting cocaine specimens into their geographic origins, with the exception of some misclassifications.[61–64] Leveraging seizure information, Dujourdy *et al.* labelled specimens as either "Colombian", "Bolivian" or "Peruvian".[62] The authors noted that the purchase country may not necessarily be the origin country; however, these countries were used as the tentative geographic origins of specimens.[62] Upon solvent analysis, each presumed geographic origin group was found to contain differing profiles: "Bolivian" specimens contained high quantities of acetone and diethyl ether, "Colombian" specimens contained high concentrations of methyl ethyl ketone, hexane or chloroform or ethyl acetate, *N*-propyl acetate, and toluene, "Peruvian" specimens also frequently contained acetone, however, in lower concentrations to "Bolivian" specimens.[62] HCA was used to confirm if a separation between specimens of differing geographic origins was possible. Three distinct clusters were observed: the "Columbian" specimens were well separated; however, some "Peruvian" and "Bolivian" specimens were mixed across the two groups. Although the method showed minor classification errors, it is possible that some specimens did not originate from their purchase location. Noting this limitation, the method still shows promising results for the use of HCA and geographic origin determination.

3.5.1.4 *Different Forms of the Drug*

As a specimen's degree of purity, adulterant profiling, solvents and form (*e.g.*, salt and base) depend on the specific refining laboratory, it is possible to track the composition of specimens. Specifically, PCA was used to differentiate cocaine specimens based on form (HCl and base) and/or purity.[65–67] Rodrigues *et al.* and Materazzi *et al.* demonstrated that cocaine specimens of various purities could be differentiated *via* PCA.[65,66] Additionally, using loading plots, the authors were able to infer which impurity peaks were responsible for the discrimination between specimens.[65,67]

The HCAs seen in ref. 65, 67 and 68 show that specimens were clustered by chemical form and purity level. For example, Marcelo *et al.* found that HCA correctly clustered salt and base forms of cocaine. It is important to note that the separation between the two clusters may not be due to form alone; it is possible that different levels of purity and adulterants used in the trafficking process influence clustering.[67] At a high-level, Marcelo *et al.* demonstrated that 513 Brazilian cocaine specimens could be correctly clustered into salt and base groups by HCA.[67] In the same HCA model, the cocaine specimens were clustered into smaller groups, each having a different major adulterant present. These subgroups offer further insight into the precise origin of specimens, as the groups may indicate that specimens are from the same refining laboratory or batch.[69]

3.5.1.5 Dimensionality Reduction

Not all variables contribute equally to the specimen profiles and should be removed if they are deemed to be redundant. This can be achieved through applying the following criteria (amongst others): optimal target variables (1) should be present in most specimens at a sufficient concentration; (2) should have a small intra-variability for specimens within seizures and large inter-variability for specimens between seizures; (3) should be uncorrelated to each other.[70]

To negate the effect of compounds with linear relationships to each other, Ratle *et al.* used PCA to remove correlated variables.[71] This method successfully reduced dimensionality in the data, which resulted in the first two principal components being used for subsequent analyses in the study.[71] In a similar manner, Myors *et al.* opted to use KMC to remove compounds with redundant information.[72] In this instance KMC was used to cluster compounds (parameters), whereby larger clusters were further investigated to establish one compound as an appropriate alias. The resultant parameters were supplied as inputs into predictive logistic models. Accuracy of the model is generally improved without redundant information.

3.5.1.6 Application of Several UPR Methods – Confirmatory Analysis

Occasionally, UPR methods may be used in a semi-supervised way, *i.e.*, one UPR method followed by another. This method of subsequently applying UPR methods may be used for a variety of reasons and can include a different combination of methods. For example, PCA results were confirmed by HCA[52] and KMC[50] and showed that the visual interpretation of specimens clustered in PCA score plots was robust. On the other hand, PCA can be used to further explore clusters formed by HCA; by leveraging PCA loading plots it is possible to identify which variables were drivers of the HCA clustering.[64]

Another instance of semi-supervised pattern recognition saw HCA used to evaluate the clustering of parameters by KMC.[72] Myors *et al.* initially used KMC to reduce the number of parameters (compounds) in the raw data; the clustering of parameters was closely replicated by HCA. Additionally, HCA found that many of the clusters contained parameters that were highly correlated to each other, indicating that the number of parameters could be dropped even further.[72]

3.5.2 Counterfeit Pharmaceuticals

Counterfeit, falsified and substandard pharmaceuticals result in socioeconomic cost and damage to health. They not only increase disease prevalence, antimicrobial resistance and produce adverse health effects, they also lead to socioeconomic loss.[73] As counterfeit medicines pose a significant threat to public health, it is important to develop models that can differentiate between authentic and counterfeit specimens (between-group variations), as well as within groups (*i.e.*, within authentic or counterfeit specimens). Several studies have investigated the potential of using UPR to differentiate between authentic and counterfeit specimens.[69,74–86] Similarly, several articles also looked at differentiating specimens within groups of counterfeit[69,78,87,88] and authentic[81,89] pharmaceuticals (see Table 3.5).

As shown in Table 3.5, PCA tends to be the most frequently used technique for exploratory analysis involving the comparison of pharmaceuticals; however, most articles present mixed results for the method. Within the same articles, PCA performance varied based on the drug type evaluated,[78] the analytical technique used[77,87] and whether or not pharmaceutical coating was retained.[88] Similarly, PP performed differently depending on the drug analysed,[78] while PP was outperformed by PCA in another study.[79] The clustering methods (HCA and KMC) explored in this chapter proved to be more suitable than projection methods for differentiating pharmaceuticals, whether that be between-group[78] or within-group comparisons.[87]

3.5.2.1 Comparison Between Groups of Authentic and Counterfeit Pharmaceuticals

This section dives deeper into the performance of techniques used to differentiate authentic from counterfeit pharmaceuticals. As mentioned in the previous section, PCA has been shown to work successfully for differentiating authentic from counterfeit pharmaceuticals,[74,79,80,82,83,87] whereas several articles[69,76–78,84,85] have presented mixed results for PCA. For example, the latter study conducted by Deconinck *et al.* compared counterfeit Viagra and Cialis specimens to their authentic counterparts.[78] In contrast to PCA and HCA, PP was not able to differentiate Viagra counterfeits from authentic specimens. This was also the case for Cialis specimens.[78] However,

Table 3.5 Unsupervised pattern recognition methods and specific goals featured in counterfeit pharmaceutical analysis.

Specific goal	PCA	PP	HCA	KMC
Between-group comparison	Anti-infective[74,87] Cardiovascular[77,80] Digestive[81] Genito-urinary[69,76,78,79,82–85]	Genitourinary[78,79,82]	Genitourinary[78,82]	
Within-group comparison: authentic	Analgesic[89] Digestive[81]			
Within-group comparison: counterfeit	Analgesic[88] Anti-infective[87] Endocrine[90] Genitourinary[78]	Genitourinary[78]	Genitourinary[69,78]	Anti-infective[87]

the use of HCA was able to successfully differentiate Cialis specimens in the same study.[78]

Additionally, PCA had mixed results depending on the analytical technique used to extract the data. One article by Custers *et al.* (2016) utilised photodiode-array (PDA) detection as well as mass spectrometry (MS).[76] When using PCA, data acquired from MS was more discriminatory than the data produced by PDA.[76] Similarly, mixed results were obtained for PCA when using data from NIR spectroscopy (NIRS) and NIR chemical imaging (NIR-CI).[77] In this instance, the data obtained *via* NIR-CI was more discriminatory than that of NIRS. Similarly, Lopes *et al.* used NIR-CI data to successfully differentiate authentic and counterfeit specimens of Heptodin (*i.e.*, an anti-infective drug) *via* PCA.[87] These articles suggest that the performance of the chemometric techniques are driven by the drug type as well as the analytical technique used to extract the raw data.

3.5.2.2 *Comparison Within Groups of Authentic and Counterfeit Pharmaceuticals*

Although manufacturing standards are stringent for authentic pharmaceuticals, variation can still exist between batches.[74] For comparison purposes, it is important to know the extent of this deviation. Silvestre *et al.* found that using Nuclear Magnetic Resonance (NMR) data, authentic aspirin and paracetamol specimens from the same batch were grouped together and specimens from different batches (same supplier) were distinctly separated using PCA.[89] In a similar manner, other articles have explored the use of PCA, PP, HCA and KMC for differentiating groups of counterfeit specimens.[78,87,88,90] This exercise is just as important as between-group comparisons, as even the same batch of counterfeit pharmaceuticals can exhibit a large variation in chemical makeup.[78]

Depending on the drug type used, PCA had varying results; within-group differentiation of Viagra and Cialis counterfeits was not possible,[78] while PCA was able to show some separation within groups of Heptodin[87] and modelled counterfeits.[88] In contrast, PP was able to differentiate between counterfeit pharmaceuticals to an extent for Viagra, however this was not the case for Cialis.[78]

While Lopes *et al.* (2009) demonstrated that PCA was a sufficient technique to differentiate authentic and counterfeit pharmaceuticals, the authors used KMC to establish clusters of counterfeit specimens alone.[87] In this instance, the clusters that resulted from KMC did not add any additional value to what was observed with PCA.[87] The danger of over-classification, or even under-classification, was addressed while determining the appropriate number of groups in the data. Over-classification is likely to occur, as counterfeiters' primary purpose is to make the product seem as similar as possible while avoiding quality inspection. As some counterfeiting activities are classified as "back-yard", the variation in counterfeits from the same source is likely to be greater than the variation in genuine items from the same source, genuine products that are made under tight regulatory and quality controls.[87]

Expanding on the difficulty of differentiating groups of counterfeit specimens, one article demonstrated the value of retaining the coating of specimens.[88] That is, PCA demonstrated that it is possible to differentiate coated and uncoated specimens as well as coated specimens from each other. However, it was difficult to cluster the different uncoated specimens, leading to the conclusion that the coating contains an important discriminatory feature that adds to the analysis.[88]

3.6 Supervised Pattern Recognition

In contrast to UPR, when using SPR, the number of clusters is known before any methods are applied to the data. SPR aims to assign unclassified objects to an existing cluster, rather than looking for natural clusters in the data. SPR can be used to create logical rules that help determine which group a specimen belongs to, *e.g.*, a specimen's geographic origin or to distinguish authentic and counterfeit specimens. Depending on how they are used, most SPR methods can perform discrimination, classification and regression.[25] The main difference between discrimination and classification methods is the algorithm that determines how specimens are grouped.[25] Classification methods allow a specimen to be placed into any number of groups, while discrimination methods force a specimen to be placed into one group.[25] The most common SPR methods with respect to drugs are summarised in Figure 3.3.

3.6.1 Illicit Drugs

As with the UPR section, the following sections will breakdown the use of SPR methods for specific goals. The articles in Table 3.6 outline some goals

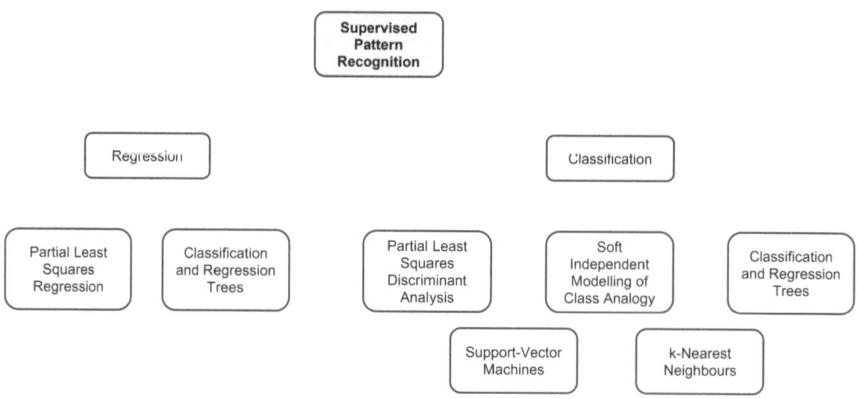

Figure 3.3 Common SPR methods using illicit drug and counterfeit pharmaceutical analysis.

Table 3.6 Supervised pattern recognition methods and specific goals featured in illicit drug analysis.

Specific goal	PLS-DA & PLSR	LDA	ANN
Drug form	ATS[49,91] Cocaine[65–68,92]	Cocaine[46]	
Batch/seizure classification		Heroin[94]	ATS[43] Heroin[57]
Geographic origin		Heroin[56,59,93]	Heroin[51,71]
Storage conditions		Cocaine[46,95]	
Variable reduction		ATS[44]	

that have been highlighted in the literature. Articles looking at differentiating specimens based on "drug form" tend to be the most prominent; this goal includes sub goals which relate to differentiating/classifying specimens based on routes of synthesis, purity levels and base *vs.* "salt form". These goals and others will be explored in the sections below.

As seen in Table 3.6, partial least squares (PLS) is one of the most frequently used methods; it was used for both discrimination (PLS-DA) and regression (PLS-R) purposes. In addition, the use of LDA was prominent for differentiating illicit drug specimens. The prominence of PLS and LDA may come down to their ease of application and interpretability, compared to other methods.

3.6.1.1 Drug Form

The classification of illicit drug specimens according to their form (*e.g.*, salt *vs.* base) is important from an investigative point of view due to different trafficking routes and distribution networks; the profile of each form is different due to adulterants and solvents used in the refining process. If we take the example of cocaine, it can be consumed in its base form or its salt form. For example, in Brazil, cocaine is mostly taken in its base form, known as crack, and salt cocaine is the most usually trafficked form to Europe. Furthermore, the treatment differs between salt and base cocaine in terms of drug addiction.[67] This section looks at articles that differentiated specimens based on "drug form" which groups together several sub-goals; synthesis routes, purity content and base *vs.* "salt form".

To optimise the separation between specimens, PLS-DA and PLSR can be used. As a drug's chemical form tend to be discrete variables (*e.g.* base or salt form), PLS-DA is a great method to use to differentiate them.[49,65,67,91,92] Articles looking at ATS tend to focus on differentiating specimens based on route of synthesis, while articles using cocaine data focus on classifying specimens based on purity and base/salt form. Similarly, articles looking at differentiating specimens based on purity (continuous variable) have used PLS-R successfully.[66,68]

PLS-DA was successful in classifying and optimising models that differentiated specimens based on drug form.[49,65,67,91,92] For example, Dujourdy *et al.*

elucidated that their model produced a very high correlation coefficient for both calibration and validation, demonstrating the applicability of their model in discriminating between MA synthesised *via* the two most prominent precursors (benzyl methyl ketone and ephedrine).[49] The authors also used their validated model to make predictions with new unclassified specimens, which resulted in a 5% misclassification rate. Additionally, the models developed by Marcelo *et al.* (2015)[67] and Grobério *et al.* (2015)[92] each correctly identified 100% of specimens as being either cocaine HCl or base. In the case of ref. 92 by Groberio *et al.* (2015), the lack of error was achieved after outliers (4.4% of the dataset) were removed.

In addition to PLS-DA, LDA was modelled on pure cocaine specimens.[46] Test "cut" specimens (containing cutting agents) were classified according to clusters observed through PCA. In both cases, prediction error only started occurring when specimens contained more than one cutting agent and a cocaine percentage of less than 60%.

3.6.1.2 Geographic Origin

Several articles have used SPR to determine the geographical origin of a specimen.[51,56,59,71,93] For heroin profiling, ANN architectures such as RBF and MLP have been featured several times in the literature,[51,71] with Ratle *et al.*[71] evaluating alternative ANN such as PNN and an additional non-parametric method *k*-ANN. PNN and *k*-ANN outperformed the MLP and RBF methods, but this outcome may be attributable to the application of a dimensionality reduction technique such as PCA. Another SPR technique applied to heroin profiling data is DA. The technique can be used to discriminate specimens based on geographic origin.[56,59,93] DA models produced some misclassification. Amongst other things, the error could be due to the small specimen size or uncertainty of the country of origin. DA learns from available specimens and a small dataset could potentially overtrain the model and misclassify specimens.[56,94] Furthermore, if there is a high inaccuracy in the original assumption of the country of origin, this could increase the misclassification rate during the training of a model.[93]

3.6.1.3 Batch/Seizure Classification

Some articles used ANN to classify specimens into their original units at different stages of manufacture. That is, Waddell *et al.*[43] classified ecstasy tablets into their original seizures using various MLP models and retained the six optimum ANN configurations. These configurations produced an accurate classification rate of 96–99%. Misclassification can become a major issue if seizure intra-variability is high. However, this type of variation will always be present in clandestinely produced illicit substances.[43] Conversely, when discriminating between heroin batches, Klemenc[57] found that PCA and HCA were less erroneous than *k*-ANN. It should be noted that this misclassification by the SPR technique could be attributed to the small specimen size. Another SPR technique applied to heroin profiling data is DA. The technique was used to discriminate specimens based on batch trends; the model resulted in a good separation of specimen batches, with some misclassification (91.9% accuracy).[94] As with the findings in the geographic origin Section (3.6.1.2), it is possible to overtrain a model and increase error rates with a small sample size.[56,94]

3.6.1.4 *Storage Conditions*

As a specimen's profile may degrade overtime, it is important to note how specific storage conditions may affect this process.[46,95] For example, cocaine specimens were subjected to different storage conditions and their alkaloid and residual solvent profiles were generated over time.[95] LDA predicted the level of association between specimens. The authors suggested improving the statistical comparison of cocaine profiling by combining the two profiles. The stability of residual solvent profiles over time, as well as the specificity of alkaloid profiles, proved to produce lower prediction error rates.

3.6.1.5 *Dimensionality Reduction*

As mentioned in Section 3.5.1.5, if any variables are deemed redundant, they should be eliminated. For example, LDA was applied to MDMA data to ascertain whether a reduction of target variables offered the same level of discrimination.[44] It was found that the model employing the reduced set of variables presented minor misclassification. Furthermore, the model was also able to identify the variables mainly responsible for specimen discrimination.

3.6.2 Counterfeit Pharmaceuticals

Similar to Section 3.5.2, this section will highlight the use of SPR methods for counterfeit pharmaceuticals. Table 3.7 outlines the specific methods used to differentiate between authentic and counterfeit drugs,[74,76–81,84,85,96–101] as well as groups of counterfeit[78] drugs.

As seen in Table 3.7, partial least squares (PLS) (including regression (PLSR) and discriminant analysis (PLS-DA)) tends to be the most common method used for differentiating authentic pharmaceuticals from counterfeit versions, authentic pharmaceuticals from each other as well as different groups of counterfeit pharmaceuticals. Regardless of drug type, PLSR[77,78,84,97,98] and PLS-DA[76,80,85,96,99] performed well for differentiating authentic and counterfeit pharmaceuticals. In addition, articles using soft independent modelling by class analogy (SIMCA) and support vector machines (SVM) reported 100% accuracy rates for their models,[74,78,79,85,99,101] except for some models using Cialis data where one counterfeit sample was misclassified as authentic.[76,79,85]

Table 3.7 Supervised pattern recognition methods and specific goals featured in counterfeit pharmaceutical analysis.

Specific goal	PLS-DA & PLSR	SIMCA	*k*-NN	CART	SVM
Between-group	Anti-infective[96–98] Genitourinary[76,78,84,85] Cardiovascular[77,80] Endocrine[99]	Anti-infective[81] Antispasmodic[74] Genitourinary[76,78,79,83,85,101] Endocrine[99]	Genitourinary[76,78,83,85,100,101,137]	Genitourinary[78,79,100,101]	Genitourinary[78]

The performance of classification and regression tree (CART) greatly varied based on the drug type being used and its sample size. Two articles report successful differentiation of authentic and counterfeit Viagra specimens using CART.[78,100] However, some articles also show that CART results in unacceptable errors where counterfeit specimens are categorised as genuine.[78,79,101] This error most likely stems from small sample sizes in the classes being analysed. The recommendation for improving accuracy rates is to either gather more specimens for the class question or alter the class so that it contains more specimens.

As with CART, articles using *k*-nearest neighbors (*k*-NN) reported conflicting results for the methods used in differentiating authentic and counterfeit specimens. When using Cialis data, the method results in many misclassifications concerning counterfeit specimens considered to be genuine.[76,85] One study combined Viagra and Cialis specimens which resulted in poor classification between authentic and counterfeit specimens.[101] Conversely, individual Viagra and Cialis data had 100% correct classification of authentic and counterfeit specimens.[78] These results indicated that *k*-NN is not a suitable method for an expanded model which combines different drug types, rather it performed more accurately when datasets of drugs are analysed separately.

All mentioned methods work well to differentiate authentic and counterfeit specimens, except for Cialis data. Some models were not able to differentiate authentic and counterfeit Cialis; profiles extracted from Cialis tend to have certain nuances that make it more difficult to differentiate batches compared to other types of drugs. This shows again that the strategy necessary to discriminate between genuine and counterfeit medicines will be different depending on the group of products considered. This can be due to the nature of the active ingredients and excipients, but also due to the diversity of producers of a type of counterfeit medicines.[78] Cialis is less counterfeited than Viagra due to the popularity of the brand name of Viagra.

3.7 Emerging Trends: The Move to Decentralised Drug Analyses

The analysis of drugs (illicit and counterfeit medicine) is challenging, not only to obtain timely and reliable results but also to produce added value from the generated data.[102] In addition, laboratories are being challenged by a decentralisation of forensic capabilities. This decentralisation movement is mainly driven by an increasing demand for timely and cheaper support to investigative and intelligence activities.[103] As a result, there has been an increase in the development of field-deployable analytical technologies.

NIR and Raman technologies have been intensively used in the pharmaceutical industry for quality control,[104–106] to analyse counterfeit pharmaceuticals[81,90,107–109] and to identify and quantify illicit drugs.[68,110–114]

Chemometrics methods are also commonly used. The following examples are not an exhaustive list but show the benefit and advances of the decentralisation of analyses, bringing the lab to the scene. For example, in a recent article,[102] various combinations of algorithms and parameters were applied to the calibration dataset of cocaine, heroin and cannabis street specimens. The best results for cocaine and cannabis were obtained using randomized tree methods while for heroin, the best algorithms were a group of gradient-boosting classifier and regressor methods. The authors mentioned that the difference in algorithms is certainly due to the diversity of cutting agents in the calibration datasets. In another article, Kranenburg *et al.* (2020) used an NIR spectrometer on a large set of cocaine street specimens and applied multiple chemometric models to characterize and determine the purity of street cocaine specimens (*i.e.*, ANN and bagging tree regression methods).[115] Lu *et al.* presented a unique identification approach for differentiating counterfeit pharmaceuticals using portable Raman technology.[90] This approach incorporated a modified local straight-line screening algorithm (manipulation of the Raman spectra) combined with PCA. The model was successful in differentiating 40 batches of counterfeit hypoglycemic drugs, having an accuracy of 96.35%.[90]

In the future, the appropriate use of such technologies as well as the use of chemometrics can free up the laboratory from high-volume/low importance cases (*i.e.*, backlog), so they can focus on more complex or more important cases.[103] This is essential given the number of drugs seized and the data they generate.

3.8 Conclusion

A variety of analyses (both qualitative and quantitative) are performed on illicit drugs and pharmaceuticals, leading to complex and often multidimensional datasets. The decentralisation of forensic capabilities due to an increasing demand for timely and cheaper support to investigative and intelligence activities has complexified the possibilities and data generated. There is thus a need to use chemometrics to deal with this huge amount of data. This chapter discussed the most common chemometric techniques applied in the area of illicit drugs and pharmaceuticals. These techniques, if applied correctly, with a clear purpose and combined with the correct interpretation of the results, can provide additional information to aid decisions regarding crime disruption, prevention and reduction. They may be a valuable tool to support intelligence-led policing. However, they are only a tool and the type of chemometrics techniques used and the results generated will always depend on the type of analytical technique used, the type of drug (whether it is illicit or pharmaceutical), and most importantly the aim of the application and the research questions.

References

1. World Drug Report 2021, United Nations Office on Drugs and Crime, New York, 2021.
2. A. Popovic, M. Morelato, C. Roux and A. Beavis, *Forensic Sci. Int.*, 2019, **302**, 109911.

3. D. L. Massart, B. G. Vandeginste, L. Buydens, S. De Jong, P. J. Lewi and J. Smeyers-Verbeke, *Handbook of chemometrics and qualimetrics: Part A*, Elsevier, 1998.
4. United Nations Office on Drugs and Crime, *Terminology and Information on Drugs*, United Nations Publications, 3rd edn, 2016.
5. Australia's health 2016, Australian Institute of Health and Welfare, Canberra, 2016.
6. K. Dégardin, Y. Roggo and P. Margot, *J. Pharm. Biomed. Anal.*, 2014, **87**, 167–175.
7. *Cross-cutting issues: Evolving trends and new challenges*, United Nations Office on Drugs and Crime, New York, 2020.
8. European Monitoring Centre for Drugs and Drug Addiction and Europol, *EU Drug Markets Report: In-Depth Analysis*, Publications Office of the European Union, Luxembourg, 2016.
9. W. IMPACT, *World Health Organization International Medical Products Anti-Counterfeiting Taskforce*, 2008, www.who.int/entity/impact/FinalBrochureWHA2008a.pdf.
10. O. Ribaux and S. Caneppele, in *The Routledge International Handbook of Forensic Intelligence and Criminology*, ed. Q. Rossy, D. Décary-Hétu, O. Delémont and M. Mulone, Routledge Taylor & Francis Group, London and New York, 2018, pp. 137–148.
11. D. Barclay, in *Handbook of Forensic Science*, ed. R. W. J. Fraser, Willan, Collumpton, 2009, pp. 337–358.
12. M. Morelato, A. Beavis, M. Tahtouh, O. Ribaux, P. Kirkbride and C. Roux, *Forensic Sci. Int.*, 2013, **226**, 1–9.
13. S. Bitzer, N. Albertini, E. Lock, O. Ribaux and O. Delémont, *Sci. Justice*, 2015, **55**, 509–513.
14. S. Kind, *J. Forensic Sci. Soc.*, 1994, **34**, 155–164.
15. M. Morelato, S. Baechler, O. Ribaux, A. Beavis, M. Tahtouh, P. Kirkbride, C. Roux and P. Margot, *Forensic Sci. Int.*, 2014, **236**, 181–190.
16. O. Ribaux, A. Girod, S. J. Walsh, P. Margot, S. Mizrahi and V. Clivaz, *Law Probab. Risk*, 2003, **2**, 47–60.
17. M. Peterson, *Intelligence-led Policing: The New Intelligence Architecture*, US Department of Justice, Washington, DC, 2005.
18. J. H. Ratcliffe, *Integrated Intelligence and Crime Analysis: Enhanced Information Management for Law Enforcement Leaders*, Washington, 2007.
19. A. Popovic, M. Morelato, S. Baechler, A. De Grazia, M. Tahtouh, C. Roux and A. Beavis, *Drug Test. Anal.*, 2022, **14**, 481–495.
20. P. Esseiva, S. Ioset, F. Anglada, L. Gasté, O. Ribaux, P. Margot, A. Gallusser, A. Biedermann, Y. Specht and E. Ottinger, *Forensic Sci. Int.*, 2007, **167**, 247–254.
21. H. Inoue, Y. T. Iwata and K. Kuwayama, *J. Health Sci.*, 2008, **54**, 615–622.
22. Q. Milliet, C. Weyermann and P. Esseiva, *Forensic Sci. Int.*, 2009, **187**, 58–65.

23. M. Collins, J. Huttunen, I. Evans and J. Robertson, *Aust. J. Forensic Sci.*, 2007, **39**, 25–32.
24. M. Morelato, D. Franscella, P. Esseiva and J. Broséus, *Int. J. Drug Policy*, 2019, **73**, 7–15.
25. B. Krakowska, D. Custers, E. Deconinck and M. Daszykowski, *J. Pharm. Biomed. Anal.*, 2016, **127**, 112–122.
26. P. D. D. O. Sardela, V. F. Sardela, A. M. D. S. da Silva, H. M. G. Pereira and F. R. de Aquino Neto, *Forensic Toxicol.*, 2019, **37**, 465–473.
27. M. Lopatka and M. Vallat, *Forensic Sci. Int.*, 2011, **210**, 188–194.
28. D. Custers, P. Courselle, S. Apers and E. Deconinck, *Rev. Anal. Chem.*, 2016, **35**, 145–168.
29. A. El-Gindy and G. M. Hadad, *J. AOAC Int.*, 2012, **95**, 609–623.
30. P. Esseiva, L. Gaste, D. Alvarez and F. Anglada, *Forensic Sci. Int.*, 2011, **207**, 27–34.
31. B. Remberg and A. H. Stead, *Bull. Narc.*, 1999, **51**, 2.
32. A. Bolck, C. Weyermann, L. Dujourdy, P. Esseiva and J. van den Berg, *Forensic Sci. Int.*, 2009, **191**, 42–51.
33. O. Guéniat and P. Esseiva, *Le profilage de l'héroïne et de la cocaïne*, in *Une Méthodologie Moderne de Lutte Contre le Trafic Illicite*, Presses polytechniques et universitaires romandes, Lausanne, 2005.
34. I. Baer and P. Margot, *Forensic Sci. Int.*, 2007, **167**, 229–233.
35. O. Ribaux, A. Baylon, C. Roux, O. Delémont, E. Lock, C. Zingg and P. Margot, *Forensic Sci. Int.*, 2010, **195**, 10–16.
36. O. Ribaux, A. Baylon, E. Lock, O. Delemont, C. Roux, C. Zingg and P. Margot, *Forensic Sci. Int.*, 2010, **199**, 63–71.
37. K. Dégardin, Y. Roggo and P. Margot, *Forensic Sci. Int.*, 2015, **248**, 15–32.
38. M. Z. Yusoff, K. H. Chang and A. F. L. Abdullah, *Aust. J. Forensic Sci.*, 2018, **0618**, 1–11.
39. M. Vosough, H. Mohamedian, A. Salemi and T. Baheri, *J. Chromatogr. Sci.*, 2014, **53**, 233–239.
40. M. Daszykowski, B. Walczak and D. Massart, *Chemom. Intell. Lab. Syst.*, 2003, **65**, 97–112.
41. D. Massart and L. Kaufman, *The interpretation of analytical chemical data by the use of cluster analysis*, Wiley, New York, 1983.
42. W. Krawczyk and A. Parczewski, *Anal. Chim. Acta*, 2001, **446**, 107–114.
43. R. J. Waddell, N. NicDaeid and D. Littlejohn, *Analyst*, 2004, **129**, 235–240.
44. C. Weyermann, R. Marquis, C. Delaporte, P. Esseiva, E. Lock, L. Aalberg, J. S. Bozenko Jr., S. Dieckmann, L. Dujourdy and F. Zrcek, *Forensic Sci. Int.*, 2008, **177**, 11–16.
45. R. Marquis, C. Weyermann, C. Delaporte, P. Esseiva, L. Aalberg, F. Besacier, J. S. Bozenko Jr., R. Dahlenburg, C. Kopper and F. Zrcek, *Forensic Sci. Int.*, 2008, **178**, 34–39.
46. M. Monfreda, F. Varani, F. Cattaruzza, S. Ciambrone and A. Proposito, *Sci. Justice*, 2015, **55**, 456–466.

47. L. Stride Nielsen, P. Villesen and C. Lindholst, *Forensic Sci. Int.*, 2017, **280**, 194–199.
48. K. Kuwayama, K. Tsujikawa, H. Miyaguchi, T. Kanamori, Y. Iwata, H. Inoue, S. Saitoh and T. Kishi, *Forensic Sci. Int.*, 2006, **160**, 44–52.
49. L. Dujourdy, V. Dufey, F. Besacier, N. Miano, R. Marquis, E. Lock, L. Aalberg, S. Dieckmann, F. Zrcek and J. S. Bozenko Jr., *Forensic Sci. Int.*, 2008, **177**, 153–161.
50. F. Besacier, H. Chaudron-Thozet, M. Rousseau-Tsangaris, J. Girard and A. Lamotte, *Forensic Sci. Int.*, 1997, **85**, 113–125.
51. P. Esseiva, F. Anglada, L. Dujourdy, F. Taroni, P. Margot, E. D. Pasquier, M. Dawson, C. Roux and P. Doble, *Talanta*, 2005, **67**, 360–367.
52. K. W. Chan, G. H. Tan and R. C. S. Wong, *Aust. J. Forensic Sci.*, 2013, **45**, 153–163.
53. R. Myors, R. J. Wells, S. V. Skopec, R. Iavetz, Z. Skopec, P. Crisp, A. Ekangaki and J. Robertson, *Anal. Commun.*, 1998, **35**, 403–410.
54. J. J. Zacca, T. S. Groberio, A. O. Maldaner, M. L. Vieira and J. W. Braga, *Anal. Chem.*, 2013, **85**, 2457–2464.
55. H. Neumann, *Forensic Sci. Int.*, 1994, **69**, 7–16.
56. K. W. Chan, G. H. Tan and R. C. S. Wong, *Aust. J. Forensic Sci.*, 2012, **44**, 299–309.
57. S. Klemenc, *Forensic Sci. Int.*, 2001, **115**, 43–52.
58. N. Stojanovska, S. Fu, M. Tahtouh, T. Kelly, A. Beavis and K. P. Kirkbride, *Forensic Sci. Int.*, 2013, **224**, 8–26.
59. C. Liu, Z. Hua, Y. Bai and Y. Liu, *Forensic Sci. Int.*, 2014, **239**, 37–43.
60. K.-W. Chan, G.-H. Tan and R. C. S. Wong, *Aust. J. Forensic Sci.*, 2012, **44**, 353–369.
61. E. J. Magalhaes, C. C. Nascentes, L. S. Pereira, M. L. Guedes, R. A. Lordeiro, L. M. Auler, R. Augusti and M. E. de Queiroz, *Sci. Justice*, 2013, **53**, 425–432.
62. L. Dujourdy and F. Besacier, *Forensic Sci. Int.*, 2008, **179**, 111–122.
63. C. Liu, Z. Hua and X. Meng, *Forensic Sci. Int.*, 2017, **276**, 77–84.
64. B. Pagano, I. Lauri, S. De Tito, G. Persico, M. G. Chini, A. Malmendal, E. Novellino and A. Randazzo, *Forensic Sci. Int.*, 2013, **231**, 120–124.
65. N. V. S. Rodrigues, E. M. Cardoso, M. V. O. Andrade, C. L. Donnici and M. M. Sena, *J. Braz. Chem. Soc.*, 2013, **24**, 507–517.
66. S. Materazzi, A. Gregori, L. Ripani, A. Apriceno and R. Risoluti, *Talanta*, 2017, **166**, 328–335.
67. M. C. Marcelo, K. C. Mariotti, M. F. Ferrao and R. S. Ortiz, *Forensic Sci. Int.*, 2015, **246**, 65–71.
68. C. Perez-Alfonso, N. Galipienso, S. Garrigues and M. de la Guardia, *Forensic Sci. Int.*, 2014, **237**, 70–77.
69. R. S. Ortiz, K. de Cássia Mariotti, M. H. Holzschuh, W. Romão, R. P. Limberger and P. Mayorga, *Forensic Sci. Int.*, 2013, **229**, 13–20.
70. M. Morelato, Forensic drug profiling: a tool for intelligence-led policing, PhD thesis, University of Technology Sydney, 2015.

71. F. Ratle, A.-L. Terrettaz, M. Kanevski, P. Esseiva and O. Ribaux, Pattern analysis in illicit heroin seizures: a novel application of machine learning algorithms, *European Symposium on Artificial Neural Networks*, Bruges, 2006, pp. 665–670.
72. R. B. Myors, S. V. Skopec, R. J. Wells and P. T. Crisp, *Analyst*, 2001, **126**, 679–689.
73. The WHO Member State Mechanism on Substandard and Falsified Medical Products, World Health Organisation, 2020.
74. O. Y. Rodionova, L. P. Houmøller, A. L. Pomerantsev, P. Geladi, J. Burger, V. L. Dorofeyev and A. P. Arzamastsev, *Anal. Chim. Acta*, 2005, **549**, 151–158.
75. K. Kwok and L. S. Taylor, *J. Pharm. Biomed. Anal.*, 2012, **66**, 126–135.
76. D. Custers, B. Krakowska, J. De Beer, P. Courselle, M. Daszykowski, S. Apers and E. Deconinck, *Talanta*, 2016, **146**, 540–548.
77. T. Puchert, D. Lochmann, J. Menezes and G. Reich, *J. Pharm. Biomed. Anal.*, 2010, **51**, 138–145.
78. E. Deconinck, P.-Y. Sacre, P. Courselle and J. D. Beer, *Talanta*, 2012, **100**, 123–133.
79. D. Custers, M. Canfyn, P. Courselle, J. O. De Beer, S. Apers and E. Deconinck, *Talanta*, 2014, **123**, 78–88.
80. P. De Peinder, M. Vredenbregt, T. Visser and D. D. Kaste, *J. Pharm. Biomed. Anal.*, 2008, **47**, 688–694.
81. O. Y. Rodionova and A. Pomerantsev, *TrAC, Trends Anal. Chem.*, 2010, **29**, 795–803.
82. M. de Veij, A. Deneckere, P. Vandenabeele, D. de Kaste and L. Moens, *J. Pharm. Biomed. Anal.*, 2008, **46**, 303–309.
83. P.-Y. Sacré, E. Deconinck, L. Saerens, T. D. Beer, P. Courselle, R. Vancauwenberghe, P. Chiap, J. Crommen and J. O. D. Beer, *J. Pharm. Biomed. Anal.*, 2011, **56**, 454–461.
84. P.-Y. Sacré, E. Deconinck, T. D. Beer, P. Courselle, R. Vancauwenberghe, P. Chiap, J. Crommen and J. O. D. Beer, *J. Pharm. Biomed. Anal.*, 2010, **53**, 445–453.
85. P.-Y. Sacré, E. Deconinck, M. Daszykowski, P. Courselle, R. Vancauwenberghe, P. Chiap, J. Crommen and J. O. D. Beer, *Anal. Chim. Acta*, 2011, **701**, 224–231.
86. R. S. Ortiz, K. C. Mariotti, N. V. Schwab, G. P. Sabin, W. F. Rocha, E. V. de Castro, R. P. Limberger, P. Mayorga, M. I. M. Bueno and W. Romão, *J. Pharm. Biomed. Anal.*, 2012, **58**, 7–11.
87. M. B. Lopes and J.-C. Wolff, *Anal. Chim. Acta*, 2009, **633**, 149–155.
88. S. Neuberger and C. Neusüß, *J. Pharm. Biomed. Anal.*, 2015, **112**, 70–78.
89. V. Silvestre, V. M. Mboula, C. Jouitteau, S. Akoka, R. J. Robins and G. S. Remaud, *J. Pharm. Biomed. Anal.*, 2009, **50**, 336–341.
90. F. Lu, X. Weng, Y. Chai, Y. Yang, Y. Yu and G. Duan, *Chemom. Intell. Lab. Syst.*, 2013, **127**, 63–69.
91. K. Andersson, E. Lock, K. Jalava, H. Huizer, S. Jonson, E. Kaa, A. Lopes, A. Poortman-van der Meer, E. Sippola, L. Dujourdy and J. Dahlén, *Forensic Sci. Int.*, 2007, **169**, 86–99.

92. T. S. Groberio, J. J. Zacca, E. D. Botelho, M. Talhavini and J. W. B. Braga, *Forensic Sci. Int.*, 2015, **257**, 297–306.
93. A. Johnston and L. A. King, *Forensic Sci. Int.*, 1998, **95**, 47–55.
94. K. Janhunen and M. D. Cole, *Forensic Sci. Int.*, 1999, **102**, 1–11.
95. L. Stride Nielsen, C. Lindholst and P. Villesen, *Forensic Sci. Int.*, 2016, **269**, 42–49.
96. F. E. Dowell, E. B. Maghirang, F. M. Fernandez, P. N. Newton and M. D. Green, *J. Pharm. Biomed. Anal.*, 2008, **48**, 1011–1014.
97. J. Dubois, J.-C. Wolff, J. K. Warrack, J. Schoppelrei and E. N. Lewis, *Spectroscopy*, 2007, **22**, 40–50.
98. J. Wolff, J. Warrack, L. Kidder and E. Lewis, *Pharmaceutical Manufacturing*, 2008, vol. 7.
99. R. da Silva Fernandes, F. S. L. da Costa, P. Valderrama, P. H. Março and K. M. G. de Lima, *J. Pharm. Biomed. Anal.*, 2012, **66**, 85–90.
100. E. Deconinck, P.-Y. Sacré, D. Coomans and J. D. Beer, *J. Pharm. Biomed. Anal.*, 2012, **57**, 68–75.
101. D. Custers, T. Cauwenbergh, J. Bothy, P. Courselle, J. De Beer, S. Apers and E. Deconinck, *J. Pharm. Biomed. Anal.*, 2015, **112**, 181–189.
102. F. Coppey, A. Bécue, P.-Y. Sacré, E. M. Ziemons, P. Hubert and P. Esseiva, *Forensic Sci. Int.*, 2020, **317**, 110498.
103. E. Casey, O. Ribaux and C. Roux, *J. Forensic Sci.*, 2019, **64**, 127–136.
104. H. Yan and H. W. Siesler, *J. Pharm. Biomed. Anal.*, 2018, **160**, 179–186.
105. M. Alcalà, M. Blanco, D. Moyano, N. W. Broad, N. O'Brien, D. Friedrich, F. Pfeifer and H. W. Siesler, *J. Near Infrared Spectrosc.*, 2013, **21**, 445–457.
106. L. Harper, J. Powell and E. M. Pijl, *Harm Reduct. J.*, 2017, **14**, 1–13.
107. F. Been, Y. Roggo, K. Degardin, P. Esseiva and P. Margot, *Forensic Sci. Int.*, 2011, **211**, 83–100.
108. O. Y. Rodionova, A. Titova, N. Demkin, K. Balyklova and A. Pomerantsev, *Talanta*, 2019, **195**, 662–667.
109. O. Y. Rodionova, K. Balyklova, A. Titova and A. Pomerantsev, *Forensic Chem.*, 2018, **8**, 82–89.
110. A. F. da Silva, T. S. Grobério, J. J. Zacca, A. O. Maldaner and J. W. B. Braga, *Forensic Sci. Int.*, 2018, **290**, 169–177.
111. C. A. F. de Oliveira Penido, M. T. T. Pacheco, E. H. Novotny, I. K. Lednev and L. Silveira Jr., *J. Raman Spectrosc.*, 2017, **48**, 1732–1743.
112. M. C. Hespanhol, C. Pasquini and A. O. Maldaner, *Talanta*, 2019, **200**, 553–561.
113. C.-M. Liu, Y. Han, S.-G. Min, W. Jia, X. Meng and P.-P. Liu, *Forensic Sci. Int.*, 2018, **290**, 162–168.
114. C. G. Pederson, D. M. Friedrich, C. Hsiung, M. von Gunten, N. A. O'Brien, H.-J. Ramaker, E. van Sprang and M. Dreischor, Pocket-size near-infrared spectrometer for narcotic materials identification, *Proc. SPIE, Next-Generation Spectroscopic Technologies VII*, 2014, vol. 9101, p. 91010O.
115. R. F. Kranenburg, J. Verduin, Y. Weesepoel, M. Alewijn, M. Heerschop, G. Koomen, P. Keizers, F. Bakker, F. Wallace and A. van Esch, *Drug Test. Anal.*, 2020, **12**, 1404–1418.

116. F. Bonadio, P. Margot, O. Delémont and P. Esseiva, *Forensic Sci. Int.*, 2008, **182**, 52–56.
117. F. Bonadio, P. Margot, O. Delémont and P. Esseiva, *Forensic Sci. Int.*, 2009, **187**, 73–80.
118. K. Kuwayama, H. Inoue, J. Phorachata, K. Kongpatnitiroj, V. Puthaviriyakorn, K. Tsujikawa, H. Miyaguchi, T. Kanamori, Y. T. Iwata and N. Kamo, *Forensic Sci. Int.*, 2008, **175**, 85–92.
119. P. Gimeno, F. Besacier, H. Chaudron-Thozet, J. Girard and A. Lamotte, *Forensic Sci. Int.*, 2002, **127**, 1–44.
120. M. Morelato, A. Beavis, M. Tahtouh, O. Ribaux, K. P. Kirkbride and C. Roux, *Forensic Sci. Int.*, 2015, **246**, 55–64.
121. M. Morelato, A. Beavis, M. Tahtouh, O. Ribaux, P. Kirkbride and C. Roux, *Sci. Justice*, 2014, **54**, 32–41.
122. K. Tanaka, S. Seta, T. Inoue and T. Ohmori, Impurity Profiling Analysis of Illicit Methamphetamine by Capillary Gas Chromatography, *J. Forensic Sci.*, 1994, **39**, 500–511.
123. M. M. van Deursen, E. R. Lock and A. J. Poortman-van der Meere, *Sci. Justice*, 2006, **46**, 135–152.
124. J. Broseus, S. Huhtala and P. Esseiva, *Forensic Sci. Int.*, 2015, **251**, 87–94.
125. S. Lociciro, P. Hayoz, P. Esseiva, L. Dujourdy, F. Besacier and P. Margot, *Forensic Sci. Int.*, 2007, **167**, 220–228.
126. J. Broséus, B. Debrus, O. Delémont, S. Rudaz and P. Esseiva, *Forensic Sci. Int.*, 2013, **230**, 16–28.
127. M. Collins, E. Casale, D. B. Hibbert, S. Panicker, J. Robertson and S. Vujic, *J. Forensic Sci.*, 2006, **51**, 597–602.
128. A. C. Allen, D. A. Cooper, J. M. Moore, M. Gloger and H. Neumann, *Anal. Chem.*, 1984, **56**, 2940–2947.
129. H. Inoue, T. Kanamori, Y. T. Iwata, Y. Ohmae, K. Tsujikawa, S. Saitoh and T. Kishi, *Forensic Sci. Int.*, 2003, **135**, 42–47.
130. V. Dufey, L. Dujourdy, F. Besacier and H. Chaudron, *Forensic Sci. Int.*, 2007, **169**, 108–117.
131. P. Esseiva, L. Dujourdy, F. Anglada, F. Taroni and P. Margot, *Forensic Sci. Int.*, 2003, **132**, 139–152.
132. L. Strömberg, L. Lundberg, H. Neumann, B. Bobon, H. Huizer and N. W. van der Stelt, *Forensic Sci. Int.*, 2000, **114**, 67–88.
133. M. Collins, *Aust. J. Forensic Sci.*, 2017, **49**, 591–604.
134. J. R. Ehleringer, D. A. Cooper, M. J. Lott and C. S. Cook, *Forensic Sci. Int.*, 1999, **106**, 27–35.
135. J. R. Ehleringer, J. F. Casale, M. J. Lott and V. L. Ford, *Nature*, 2000, **408**, 311–312.
136. D. R. Morello and R. P. Meyers, *J. Forensic Sci.*, 1995, **40**, 957–963.
137. M. J. Anzanello, R. S. Ortiz, R. P. Limbergerb and P. Mayorga, *J. Pharm. Biomed. Anal.*, 2013, **83**, 209–214.

CHAPTER 4

Fire and Explosions Investigation

A. J. HORROCKS,[a] K. PITTS,*[b] D. DETATA[b] AND R. DUNSMORE[b]

[a] School of Molecular and Life Sciences, Curtin University, GPO Box U1987, Perth, Western Australia 6845, Australia; [b] Physical Evidence, Forensic Science Laboratory, ChemCentre, PO Box 1250, Bentley DC, Western Australia 6983, Australia
*Email: kpitts@chemcentre.wa.gov.au

4.1 Introduction

Deliberately set fires and explosions are ongoing criminal activities that can cause severe damage to people, property, and the environment.[1,2] Ignitable liquid residues (ILRs) and explosive residues (ERs) are residual amounts of material that are found after the event of a fire or explosion that originate from the source of the fire or explosion. ILRs and ERs are important sources of evidence throughout the proceedings of a forensic investigation into arson and explosive cases. Examples of explosive materials that can be used to create improvised explosive devices have been provided (see Figure 4.1). Successful analysis of ILRs and ERs may provide information about the materials used to initiate the fire/explosion, even being able to go as far as to determine specific brands of material used to construct the event. A challenge when analysing ILRs and ERs is that a large volume of data is produced from many different analytical techniques such as infrared (IR) and Raman spectroscopy, gas chromatography (GC), and mass spectrometry (MS). The sheer volume of data can become an issue when analysing the data to determine the source materials used. A solution to this issue is through

Chemometric Methods in Forensic Science
Edited by Vishal Sharma, Georgina Sauzier and Simon W. Lewis

Published by the Royal Society of Chemistry, www.rsc.org

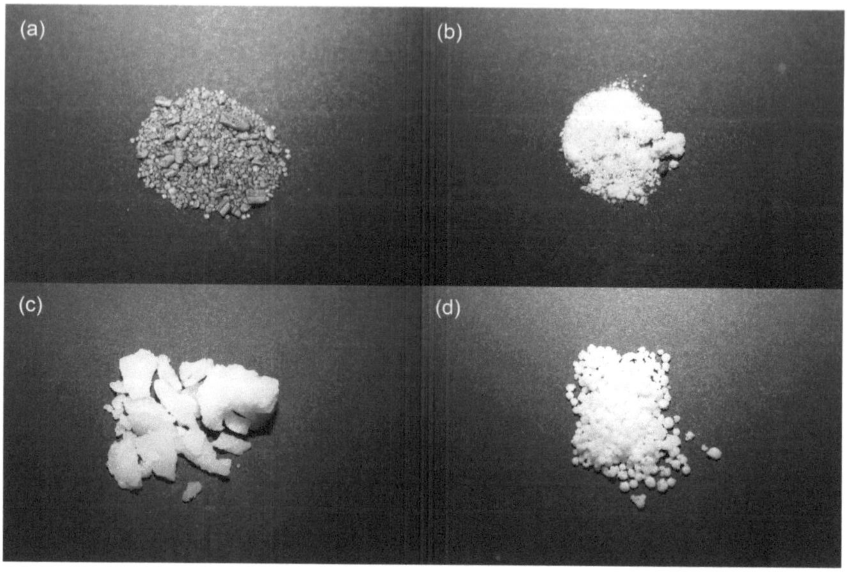

Figure 4.1 Explosive materials that can be used in the formation of IEDs: (a) sparkler material mixed with granulated sugar, (b) homemade potassium perchlorate, (c) homemade ammonium nitrate, and (d) laboratory grade urea nitrate. Photos courtesy of ChemCentre.

the use of chemometrics. Chemometric methods allow for multivariable analysis to be conducted on the data collected to process the data faster and at a level that would otherwise be impractical to complete manually. In recent years, there has been increasing research into the application of chemometrics in arson and explosives forensic analysis, which has improved sampling methods, detection, and discrimination of many different ILRs and ERs. The application of chemometrics is imperative for the continual improvement of forensic processes in this field, highlighting the importance of further research in this area.

4.2 Chemometrics

Chemometrics, otherwise known as multivariable analysis, is a discipline that utilises well-established statistical methods to model chemical data.[2] Many chemometric approaches are used in the forensic chemistry discipline for pattern recognition, calibration, and experimental design, such as principal component analysis (PCA), linear discriminant analysis (LDA), partial least squares regression (PLSR), and central composite design (CCD), to name a few.[2] Throughout forensic investigations of ILRs and ERs, a large amount of chemical data can be generated through the implementation of different analytical processes. Chemometric techniques can reveal trends in this data that may not otherwise be evident. For example, an IR spectrum of explosive material will exhibit peaks at many different wavelengths but may appear visually similar to others. The use of chemometric methods can then highlight the maximum variance and/or correlations between samples. Evaluation of these correlations allows for the potential discrimination of sources.

One of the most common analytical techniques used for the analysis of ILs and explosive materials is GC–MS, which produces data in the form of retention times and fragmentation patterns of mass-to-charge ratios.[3] Chemometric methods can be used to examine these datasets to find key points of variation or correlation to distinguished classes so that when an unknown substance is analysed, it may be compared and matched to a known source or classification. The classification and discrimination of sources using chemometric methods is a well-known process and has been used in many areas such as drug analysis,[4] DNA analysis,[5] and food analysis.[6] This works the same for materials found throughout arson and explosion investigations. The chemometric examination of large datasets first begins with the process also known as unsupervised learning or exploratory analysis.[2] There are several chemometric techniques, such as PCA or hierarchical cluster analysis (HCA), that can achieve this. PCA is an unsupervised variance examination technique used for handling multivariate data without prior knowledge about the samples under investigation. PCA aims to reduce the dimensionality of a dataset consisting of large numbers of variables whilst keeping maximum variation in the dataset.[2] PCA achieves this by transforming the original set of correlated variables into orthogonal variables that are referred to as principal components (PCs).[2] By examining these PCs, trends and potential classes can be established within the data.

Other chemometric methods can then be applied after PCA is completed, such as discriminant analysis techniques (*e.g.* LDA), for the classification of unknown samples. These classifications commonly reference a predefined PCA model. Classification models such as LDA are referred to as supervised learning, as they use datasets with pre-specified groupings.[2] The use of chemometrics is of valuable interest in the proceedings of arson and explosion investigations, as it can be used to create classification models for the determination and classification of ILRs and ERs found in criminal investigations. In some cases, the classification models can associate samples to specific brands of materials used. For example, PCA has been previously used to construct a model to characterise and discriminate homemade urea nitrate samples prepared using different starting materials (see Figure 4.2).

Another important application of chemometrics is through using the available statistical analysis methods, such as CCD, to improve experimental design. The design of an experiment is one of the most important steps when creating or trying to improve analytical methods. Any analytical method aims to gain as much information from the sample as you possibly can. Therefore, the process of analysis in question must be optimised in the recovery of information from samples. For example, there are many steps involved in gathering a sample of an ILR or ER for the analysis of the said sample. Each step throughout this process has many variables that could affect the recovery of information from the sample. For example, many types

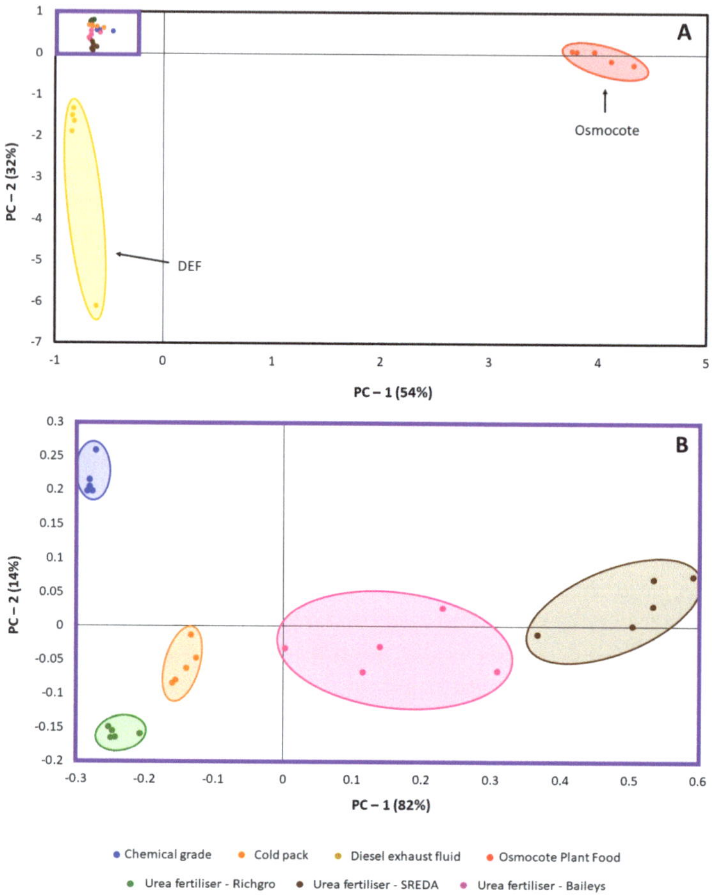

Figure 4.2 PCA scores of homemade urea nitrates prepared using different starting materials through the use of trace elemental data from ICP-MS: (A) entire PCA scores, (B) magnified upper-left section of A. Data courtesy of Joshua D'Uva.

of sampling and analytical methods are available for collecting and analysing ILRs and ERs from a crime scene, some of which may be more efficient than others. The optimal methods must be chosen for each step to recover as much information as possible. Therefore, during the experimental design, these factors would be compared against each other, creating a large amount of data that would be difficult to process manually, especially if you are comparing processes by changing only one variable at a time. The use of chemometrics in the design of an experiment allows for statistical analysis of the data collected, which would otherwise have taken much longer than most research scientists have the time for. Therefore, through using chemometric methods, optimal sample, extraction, and analysis methods for the analysis of ILRs and ERs can be determined.

4.3 Arson

Arson is the act of deliberately setting fire to property or objects and can cause serious damage to people, property, and the environment. Fire is a chemical process involving three main elements: fuel, oxidiser, and an initiator.[1] Accelerants are volatile liquids such as petroleum products that are commonly used in arson cases as they can spread fire easily across any surface.[1] Ignitable liquids (ILs) are accelerants that have been used for arson, and it is often found that ILRs are found throughout forensic investigations of arson cases. The most common ILs encountered are products such as gasoline, diesel, kerosene, and other types of alcohols as they are easily obtainable and ignitable.[7,8] ILRs are important in forensic investigations as they can be used to determine the cause of a fire and possibly the source of the IL used. Fire debris is another important form of evidence as the ILRs can be encapsulated within debris as carbon that is produced during the fire can absorb and contain the ILRs.[1,8] Proper identification of ILRs can be a challenge as there are many hindering factors such as interfering pyrolytic products and fire suppression agents used by firefighters.[8] An example of the aftermath of an arson attack has been provided (see Figure 4.3), where fire debris and the aforementioned hindering factors can be seen throughout, demonstrating the challenging environment that forensic practitioners deal with.

4.3.1 Analytical Techniques

Many techniques are available for extracting and analysing ILRs from fire debris found throughout arson investigations. Common methods of separation of ILRs from fire debris samples for headspace analysis include using adsorbent activated carbon strips, solid-phase microextraction,

Figure 4.3 Aftermath of a controlled burn scene in a bedroom. Photo courtesy of ChemCentre.

dynamic headspace adsorption with Tenax, and headspace sorptive extraction.[7] Many different analytical techniques can be used to identify ILs, some of which include GC–FID, GC–MS, Raman spectroscopy, and IR spectroscopy. GC–MS is one of the most common techniques for identifying and quantifying ILs through the analysis of the recovered ILRs. A recurring issue throughout the analysis of ILRs, however, is that the samples collected from crime scenes often contain interfering residues. These interfering residues within these samples can be introduced from varying sources. Examples include pyrolytic products produced throughout the event of the fire, fire suppression agents used by firefighters, environmental conditions, and biological substances.

4.3.2 Experimental Design

The detection and identification of ILRs throughout forensic investigations rely heavily on effective sampling and storage of the samples before technical analysis. The storage of items that may contain ILRs is of particular importance as many ILRs are composed of volatile hydrocarbons, which, due to their high vapour pressure, will evaporate, decreasing the chance of a successful analysis. The container in which ILRs are stored must enable the retention of the volatile organic compounds associated with ILs without loss to the environment or absorption into the container itself. It is also important that the container does not release any volatiles into the sample, as this contamination would affect the resulting analysis.[9] The recommended containers for fire debris evidence consist of metal cans suited with compression lids, glass jars with Teflon lids, and special polymer bags.[10,11] An example of the metal can containers used by practising crime scene investigators in Australia for both ILRs and ERs has been provided (see Figure 4.4). Previous research investigating the performance

Figure 4.4 Metal cans suited with compression lid, used for the collection and storage of ILRs and ERs. Photos courtesy of ChemCentre.

of fire debris evidence storage has focused on comparing different sampling containers whilst using the same extraction method. For example, Williams *et al.* investigated the performance of different commercial containers for the collection and storage of fire debris evidence by comparing the recovery rates of a set of standard volatile hydrocarbons using passive headspace concentration of ILs onto activated charcoal as the extraction method.[10] With the aid of chemometric methods, the performance of the storage containers could be assessed with a higher degree of optimisation. For example, comparing different storage containers whilst using the same extraction method will provide useful information on the recovery of ILRs whilst using a specific container, much as Williams *et al.* showed. However, if the same experiment was conducted with a different extraction method, such as through the use of solid-phase microextraction, which uses a specifically coated fibre for the absorption of the ILRs from the headspace, the results may vary. Chemometric methods could be applied here to compare the performance of the storage containers in reference to varying extraction methods to find an optimal process of storing and extracting ILRs from a crime scene. This would be important as different extraction methods may interact with different materials from the storage containers to affect recovery. It is also possible that more sensitive extraction methods will pick up interferences that may otherwise have been missed.

The sampling and extraction of ILRs from fire debris are extremely important in the analytical process. For a successful analysis, the extraction of the ILRs from fire debris samples must be optimal for the best recovery of evidence. Chemometric methods can be used to assess the performance and allow for the optimisation of extraction methods used for ILRs. For example, a common method for the extraction of ILRs from fire debris is through passive headspace adsorption onto activated charcoal strips.[12,13] Many factors affect the recovery rate from this extraction, such as time, temperature, strip size, and sample concentration. In 1996, Newman *et al.* investigated the effect of these variables on the recovery of ILRs on activated charcoal strips.[13] As this was investigated before chemometric techniques were commonly used in research, the comparison of the variables was split into time and temperature, strip size, using only a single time and temperature variable, and then sample concentration, using a standardised strip size, time, and temperature. If the variables were to all be compared against each other, changing one variable at a time, the volume of data collected would be too large to process manually. Using a chemometric method such as CCDs allows for the processing of data that would be collected if each of these variables were to be compared against each other. Using CCD may have possibly revealed a more optimal pathway for the extraction of ILRs using activated charcoal strips that would otherwise not be known if certain variables work together differently. For example, a sample collected with a larger strip size may benefit from the implementation of using a higher temperature for desorption.

4.3.3 Pattern Recognition

Developments in chemometric methods over the years have resulted in an increased amount of research into implementing chemometrics in forensic analysis for the detection and classification of ILRs. Chemometric methods allow for the classification of target analytes such as ILRs through pattern recognition. Chemometric pattern recognition can be achieved using unsupervised multivariate analysis methods to first reduce the dimensionality of collected data before a discriminant analysis can be applied for the classification of unknowns to be made. For example, in the analysis of ILRs, through the use of chemometric techniques such as PCA and LDA, a pattern recognition model can be created using standard compounds of common ILRs, so that the classification of an unknown ILR may be made, therefore identifying the ILR. There is a large amount of research on the use of chemometric approaches for the determination and classification of ILRs.

4.3.4 Gasoline

Gasoline is one of the most common sources of ILRs found throughout criminal investigations as it is easily accessible and highly flammable.[7,8] There has been much research completed on the determination and classification of gasoline for forensic purposes. Doble *et al.*, for example, have investigated the classification of premium and regular gasoline through the use of GC–MS, PCA, LDA, and artificial neural networks (ANNs).[14] Throughout this study, 88 samples of gasoline were analysed by GC–MS, for which PCA was then performed on the dataset. It was found that most of the variability in the data was explained in the first four principal components. LDA was then performed on the data, resulting in the correct classification of 93.2% of the gasoline samples as either premium or regular.[14] However, when attempting to further classify the samples into their seasonal formulation, the model performed poorly. ANNs were then trained to recognise structures within the data to correctly classify all the gasoline samples as premium or regular, as well as correctly classifying 97% of the samples to their correct seasonal formulation.[14] These results are strong evidence of the usefulness of chemometric approaches in the analysis of ILRs.

Going one step further in the discrimination of gasoline, Barnett *et al.* have investigated the use of direct analysis in real-time mass spectrometry (DART–MS) and chemometric methods in the discrimination of brands of gasoline.[15] Throughout this research, 87 gasoline samples were analysed using DART–MS before the collected data was processed using analysis of variance (ANOVA) followed by PCA for both the full MS and selective ion spectra.[15] PLS-DA was then chosen as the classification method of the PCA model, and it was found to provide a classification rate of 100%.[15] This is another example of a great pattern recognition model being created for the classification of gasoline, which has important implications for possible forensic applications. Another study that investigated the differentiation of gasoline samples according to their brands was a study conducted by Monfreda *et al.* In this study, SPME-GC–MS was used to gather data from 50 samples of gasoline.[16] The chemometric methods used in this study were PCA and DA. Similarly, this investigation resulted in a classification ability of 100%.

Research by Ardila *et al.* shows the use of Raman spectroscopy to characterise gasoline with chemometric analysis.[17] In this study, 19 samples of gasoline were analysed using Raman spectroscopy before the chemometric methods of PCA, PLS and PLS-DA were used to classify the gasoline samples by distributor. PCA was performed on the data collected from the Raman spectroscopy, with 97.47% of the total variance of the dataset being provided in the first four principal components. Classification of the samples was then performed using PLS-DA, which resulted in no misclassifications of all 19 gasoline samples. The implementation of chemometric approaches to Raman analysis of gasoline samples showed promising results in the use of Raman spectroscopy as a tool for the classification of gasoline.

Sinkov *et al.* conducted an investigation into the application of chemometric classification on actual casework arson samples to determine if gasoline was present.[18] In this study, 156 casework samples provided by the Royal Canadian Mounted Police (RCMP) were analysed following RCMP protocols, which follow ASTM methods E1618 and E1412. These procedures involved passive headspace extraction onto activated carbon strips, followed by elution with carbon disulfide and analysis with GC–MS. Each sample was analysed in duplicate, with one of the duplicates containing a perdeuterated alkane ladder. A deuterated alkane ladder-based alignment and a CR-guided automated approach for variable selection were applied to generate PLS-DA and SIMCA models to classify the arson casework samples based on gasoline content. The PLS-DA and SIMCA models produced in this study were able to correctly classify each sample as either gasoline-containing or gasoline-free with no false positives or negatives.[18] This study is among a few that demonstrate the application of chemometric techniques to real casework data, therefore, highlighting the importance of research in this field. Other examples of research into the use of chemometric approaches to aid in the classification of gasoline include work by Rudnev *et al.*, de Figeuiredo *et al.*, Silva *et al.*, and Petraco *et al.*[18–21]

4.3.5 Diesel

Diesel is an alternative IL commonly encountered throughout forensic investigations, hence it is important to be able to correctly identify and classify diesel to aid in forensic investigations. Novak *et al.* studied the use of combined cluster and discriminant analysis as a chemometric approach in the characterisation of diesel fuels.[22] Twenty-five samples of diesel were collected, representing 20 different brands, before being analysed using GC-isotope ratio MS (GC-IRMS). Combined cluster and discriminant analysis (CCDA), which combines the two multivariate grouping methods, HCA and

LDA, was performed on the data so that classifications could be made.[22] It was found that all the investigated samples of diesel were distinguishable using CCDA, therefore proving that CCDA is a highly efficient data analysis approach for the classification of diesel fuels.[22]

Similarly, Hupp *et al.* and Marshall *et al.* used GC–MS to analyse diesel fuels before conducting chemometric analysis on the collected data.[23,24] The study conducted by Hupp *et al.* consisted of analysing 25 samples using GC–MS before applying Pearson product-moment correlation (PPMC) and PCA.[24] Results showed the successful clustering of samples by the different brands, with only a few samples being grouped with the wrong cluster. Marshall *et al.* analysed only five diesel samples, also using PPMC and PCA, achieving similar results to Hupp *et al.*[23]

A more complex analytical technique was used for the monitoring of diesel fuel adulteration in a study by de Aguiar *et al.*[25] It is a common practice in Brazil to illegally adulterate diesel fuels, and this study aimed to use nuclear magnetic resonance (NMR) spectroscopy and chemometric techniques to monitor this.[25] In this work, common dimensions analysis (ComDim) was used in combination with LDA and multiple linear regressions (MLR) to create a classification and quantification model, which showed good performance according to the evaluated parameters.[25]

4.3.6 Fire Debris

During the proceedings of a fire, a large amount of potential evidence can be created. Fire debris has been found to contain ILRs and other pyrolysis products, which can be used for analysis throughout forensic investigations. Research into the analysis of fire debris for forensic purposes is of great importance, and many studies focus on the application of chemometric methods in this area for detecting and classifying fire debris samples and the ILRs that can be found within them. For example, Gonzalez-Rodriguez *et al.* completed a study on analysing fire debris using Raman spectroscopy and chemometric techniques.[26] In this study, five polymer-based samples (carpet, nylon stockings, foam packaging, CD cases, and DVD cases) were burned with and without the presence of four ILs (petrol, diesel, kerosene, and ethanol) individually. The samples were then analysed using Raman spectroscopy before PCA was performed on the collected data. The use of PCA on the Raman spectra demonstrated that the different materials used in this study were able to be clustered, even with the use of different fuels.[26] This study provided a great basis for the discrimination and classification process that is possible through the application of chemometric techniques on analytical data for the analysis of fire debris.

Aliano-Gonzalez *et al.* conducted research that involved the application of chemometric techniques in the analysis of ILs from fire debris, using headspace gas chromatography-ion mobility spectrometry (HS-GS-IMS).[27] Similarly to Gonzalez-Rodriguez *et al.*'s study, four substrates (pinewood, cork, paper, and cotton sheet) were used, with each one being burned alone and then with each of the four ILs (gasoline, diesel, paraffin, and ethanol) that were chosen.[27] This resulted in 80 fire debris samples being analysed by HS-GC-IMS. The data collected was then normalised using maximum intensities before HCA and LDA were performed.[27] The HCA resulted in good clustering of some of the substrates tested; however, some substrates significantly overlapped. These results, however, did not hinder the ability of ILRs to be identified from the chromatograms. An LDA was then carried out to see if the substrates that had been burned with an IL could be discriminated, during which 60% of the samples were chosen as a calibration set, and the remaining samples were used as a validation set.[27] It was found that full classification was achieved for both the calibration and validation sets for this LDA. A second LDA was carried out to determine if classification between the types of ILs used to burn the fire debris was possible. This resulted in 98.4% of the calibration set being successfully calibrated and 98.4% of the validation set also being classified successfully.[27] This is a practical example of the results that can be achieved when chemometrics is applied in this forensic setting.

In a similar study by Barnett *et al.*, different ILs, including kerosene, diesel, Japan drier, charcoal starter fluid, naphtha, prestain, lighter fluid, and unleaded gasoline, were analysed on an array of substrates, which included carpet, wood flooring, sand, paper, and a cotton T-shirt.[28] The ILs on each substrate were analysed using thermal desorption DART–MS in replicate of three before a PLS-DA model was constructed. The classification with the PLS-DA model was successful for the ILs on the substrates, with an overall classification rate of 98%.[28]

Another example of the use of chemometrics in fire debris research is work by Falatová *et al.*, who investigated the discrimination of ILRs in burned petroleum derived substrates using HS–MS eNose and chemometrics.[29] In this study, vinyl flooring, linoleum flooring, polyester carpet, polyamide carpet, cotton sheet, and natural cork were the substrates of interest, with the ILs being gasoline, diesel, ethanol, and charcoal starter with kerosene. The total ion spectra data collected was standardised by assigning one unit to the maximum intensity before a HCA was constructed. The authors then performed an LDA on the data, resulting in a full discrimination of all the samples with and without ILR. An additional LDA also resulted in complete discrimination between each IL studied.[29] This study further shows the possible applications and ability that chemometrics have in this field.

Sigman *et al.* and Allen *et al.* both conducted research into the application of likelihood ratio approaches in the analysis of fire debris.[30,31] The study conducted by Sigman *et al.* involved the implementation of support vector machines (SVMs), LDA, QDA and *k*-NN methods of binary classification of ILs in fire debris samples, whereas Allen *et al.* used PLS-DA.[30] Overall, the SVM method was proved to demonstrate excellent performance when applied to *in silico* validation data. When tested with real fire debris validation data, the SVM, QDA, and *k*-NN methods decreased in performance, with LDA being calibrated relatively well for both sets of validation data.

Research by Waddell *et al.* involved an investigation into the progress toward the determination of correct classification rates in fire debris analysis in two parts.[32,33] Part one of this research included the use of PCA, LDA, and QDA to develop a multistep classification procedure for ILs in fire debris.[32] PCA was first applied to reduce the dimensionality of the data before LDA and QDA were applied separately. QDA was found to be the more appropriate analysis method, and the classification model resulted in the correct classification rate of 92.4% for fire debris samples.[32] Part two of Waddell *et al.*'s research involved the use of soft independent modelling of class analogy (SIMCA) as a chemometric method along with a two-class multistep classification scheme for the identification and classification of ILs in fire debris samples.[33] Correct classification using these techniques was successful, showing an overall improvement in the classification when using QDA and LDA.

4.3.7 Weathering Analysis

Chemometric methods can be utilised to not only classify and discriminate chemical data, as has been shown in the gasoline studies, but can also be implemented to increase an understanding of natural processes that may occur to ILs that may affect the forensic analysis. Turner *et al.* compared the effects of weathering and microbial degradation on gasoline using chemometric methods.[34] PCA was performed on GC–MS data collected on a range of varyingly weathered and degraded gasoline samples. PCA was performed on the combined dataset of both the weathered and degraded samples to determine potential causes for the weathering and microbial degradation. It was found through this process that compounds with boiling points larger than 155 °C are the most susceptible to weathering, whereas long-chain alkanes and lower substituted aromatics are susceptible to microbial degradation.

Another study that investigated the weathering process of gasoline, conducted by Aliano-Gonzalez *et al.*, involved the use of electronic nose analysis.[35] In this study, two different volumes of gasoline (40 μL and 80 μL) were applied to four substrates (pinewood, paper, cork, and cotton sheet) before being left for various times (0 h, 6 h, 12 h, 24 h, 72 h, 1 week, 15 days, 21 days, and 1 month) to experience natural evaporation and weathering processes.[35] The samples were analysed after these periods using an electronic nose before HCA and LDA were performed on the collected data, in the form of total ion spectra. The HCA resulted in successful discrimination of samples analysed at 0 h and the rest of the samples.[35] For the samples that spent longer before analysis, samples analysed at 6 h and 12 h were distinguished from samples analysed at 24 h or longer.[35] From these results, it was determined that the weathering process starts rapidly, and sudden changes occur after 6–12 h.[35] The LDA resulted in a successful classification of 100% of the samples between samples weathered for 0–12 h and longer.[35] This study not only determined that the weathering process of gasoline is critical after 6 h but also proved that the electronic nose is a suitable technique for the study of the gasoline weathering process.

The "weathering" of ILRs is not confined to their length of exposure to the environment, but also involves factors such as the maximum temperature of any resulting fire. Baerncopf *et al.* investigated just this in a study looking at the association of ILRs to neat ILs in the presence of matrix interferences.[36] This study involved burning 12 varying ILs such as diesel, kerosene, adhesive remover, *etc.*, on nylon carpet in differing conditions. One sample set were heavily burned through use of a blowtorch for 10 s, whilst the second set was burned for 20 s.[36] Through using PPMC along with PCA, each ILR studied was able to be associated with the respective neat IL, with it being established that increased weathering of the ILR had a minimal effect on associating the ILR to the neat IL.[36]

Mat-Desa *et al.* conducted two similar studies which involved the use of chemometric techniques such as PCA, HCA and self-organising feature map (SOFM) ANNs to investigate the classification of lighter fuels and medium petroleum distillates (MPDs) at different levels of evaporation.[37,38] In the lighter fuels study, GC–FID and GC–MIS were used to analyse 15 different lighter fluid samples, from five different brands, at six different levels of evaporation.[37] PCA and HCA were performed on the collected data, as well as SOFM-ANN analysis. PCA and HCA were unable to correctly classify the samples by brand, however, when using the normalised fourth root data set, HCA was capable of separating all of the pure and evaporated samples by brand.[37] SOFM-ANN was also able to separate the pure and evaporated samples by brand; however, SOFM proved to have a more robust sample linkage capacity.[37] In the MPD study, eight different MPD samples, at six different levels of evaporation, were analysed using GC–MS.[38] Once again, PCA, HCA, and SOFM-ANN were performed on the collected data for chemometric analysis. PCA resulted in the discrimination of three of the samples, with each sample cluster containing all levels of evaporation.[38] HCA resulted in four of the eight samples discriminated, with similar clustering as the PCA.[38] In a similar fashion, SOFM performed the best with all MPD samples being accurately classified, both fresh and evaporated, according to brand. These studies by Mat-Desa *et al.* have simultaneously shown that lighter fuels and MPDs can be classified at different levels of weathering, as well as that SOFM-ANN is a superior method for this type of classification.

4.4 Explosives

Explosions occur *via* the sudden and violent conversion of potential energy into kinetic energy along with the production and release of gases under high pressures. There are often high temperatures, loud noises and violent shocks produced at the scene of an explosion.[39] There are three main types of explosions: nuclear, mechanical, and chemical.[39] Nuclear explosions occur when there is the fission or fusion of atoms occurring, whereas

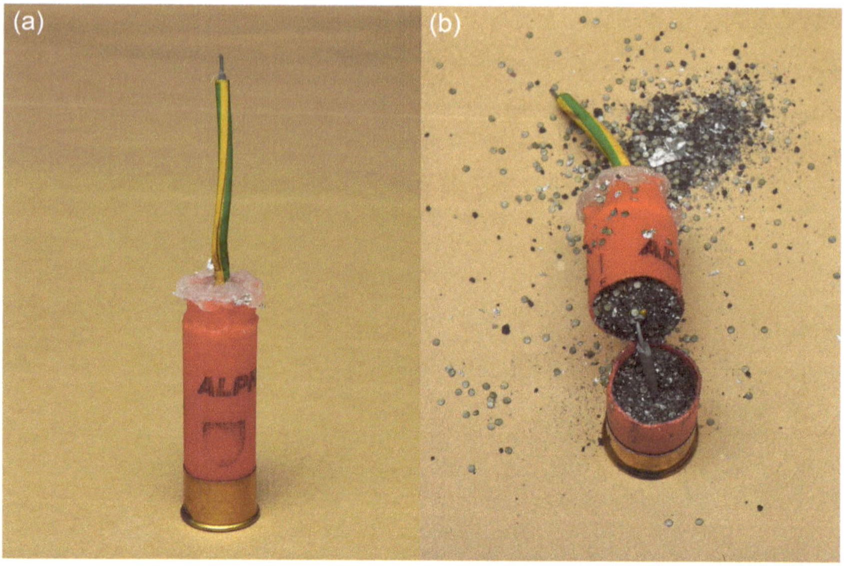

Figure 4.5 (a) Homemade IED using a shotgun shell and a sparkler fuse. (b) Cross section of the IED. Photos courtesy of ChemCentre.

mechanical explosions are a result of a gradual build-up of pressure in an enclosed container as the contents undergo a phase change, leading to the sudden failure of the vessel.[39] Chemical explosions result from the extremely rapid conversion of a solid, liquid or gas into reaction products with much greater volume than the originating substance.[39] The formulations for chemical explosives can be pure compounds or mixtures, inorganic or organic in composition, and either of military, commercial or improvised grade.[39] The chemical analysis of explosive materials can provide valuable information for forensic investigations. Improvised explosive devices, such as the example provided (see Figure 4.5), are commonly encountered by law enforcement agencies worldwide. ERs found at the scene of an explosion, when analysed using an array of analytical techniques, provide information that can then be used to trace back to the precursor materials used for the manufacture of the explosives. Recent analytical advancements have enabled the identification of specific brands of precursors used in the manufacture of explosives, which greatly assists an investigation.

4.4.1 Analytical Techniques

Many analytical techniques are available for the detection and identification of ERs collected throughout criminal investigations, each varying in the amount of forensic intelligence they provide. The techniques available fall under the main categories of separation and spectroscopic methods, along with a few exceptions, such as canine detection. Separation methods include techniques such as liquid chromatography (LC), gas chromatography (GC),

Table 4.1 TWGFEX categories of analytical techniques for the identification of explosive residues. Reproduced from ref. 40 courtesy of the National Institute of Standards and Technology, U.S. Department of Commerce. Not copyrightable in the United States.

Categories 1 & 2	Category 3	Category 4
Infrared spectroscopy	Gas chromatography	Burn test
Gas chromatography/mass spectrometry	Gas chromatography/thermal energy analyser	Flame test
Energy dispersive X-ray analyser	Liquid chromatography	Spot test
Raman spectroscopy	Ion chromatography	Melting point
X-ray diffraction	Capillary electrophoresis	
Liquid chromatography/mass spectrometry	Thin layer chromatography	
	Ion mobility spectrometry	
	Polarising light microscopy	

and ion chromatography. Spectroscopic methods include infrared and Raman spectroscopy, mass spectrometry (MS) and more advanced techniques. LC paired with MS (LC–MS) and GC paired with MS (GC–MS) are the most common analytical techniques used in detecting and identifying ERs.[40] The Technical Working Group for Fire and Explosions (TWGFEX) guidelines detail several analytical techniques suitable for the identification of ERs into four different categories, based on the level of specificity they provide.[40] Techniques under category 1 provide significant structural and/or elemental information, 2 provide limited structural or elemental information, 3 provide a high degree of selectivity, and 4 are those that are useful but do not fall into either of the other categories.[40] Many of the utilised techniques rely on matching the output data against a library database to confirm the identification of samples. A summary of the techniques under each category can be seen in the provided table (see Table 4.1).

4.4.2 Experimental Design

The detection and identification of ERs rely on successfully sampling and storing ERs from the scene of interest before analysis is conducted in the laboratory. Sampling and storage protocols widely vary depending on the scene's conditions and the materials being sampled. The methods used for sampling and storing ERs must be appropriate for optimal recovery for analysis. For example, swabbing is a common technique used for sampling ERs at crime scenes, and a great deal of research focuses on the efficiency of these techniques.[41,42] However, much of this research lacks the investigation into combinations of sampling conditions to explore optimal conditions. For example, a study may have investigated the effect of storage conditions on explosive recovery whilst only using a single swab type or extraction procedure. For proper optimisation of sampling and storage protocols, the combination of different factors involved in sampling and storing ERs for analysis must be included.

Chemometric methods can be utilised to aid in the experimental design for the optimisation of certain processes. In the context of explosive investigations, chemometric techniques such as central composite design (CCD) can be used to develop response surface models for target analytes as a function of the factors being investigated. For example, the sampling, storage and extraction factors can be investigated to identify the optimum parameters between all three for the maximum recovery of information.[43] The use of chemometric methods allows for multiple variables to be investigated alongside each other, which once may have been deemed unachievable due to the large amount of data that would need to be processed. Sauzier *et al.*, for example, used a face-centred CCD to investigate the optimal recovery protocols for double-base smokeless powder residues.[43] Varying sampling, storage, and extraction parameters were investigated, such as the type of swab, storage locations and duration and extraction times for the analysis of smokeless powders using TV-SPME GC–MS.[43] This resulted in 78 experiments throughout the investigation, which were then compared against each other using chemometric methods to find the optimal conditions for the recovery of the residues.[43] It was found that the storage temperature was the most important single factor in the recovery of the target analytes.[43] Using this chemometric approach, it was discovered that the optimal sampling method was to use iso-propanol pre-wetted swabs, with an extraction time of 15 min.[43]

Another example of the use of chemometric methods used for experimental design is research conducted by Fryš *et al.* that investigated the optimisation of focused ultrasonic extraction of propellant components determined by GC–MS.[44] Frys *et al.* investigated the performance of the extraction of propellant components, commonly found in smokeless powders or gunshot residues, using focussed ultrasonic extraction through changing three variables of the extraction method. These variables were the extraction time, sample weight and extraction power, represented by the amplitude of ultrasonic waves.[44] To find the conditions that would result in the maximum yield of the ultrasonic extraction, a three-factor CCD analysis was implemented. The collected data was fitted using multiple non-linear regression before second-order polynomial equations were used to calculate the response variables related to the coded factors for the target compounds. This data was then used to create response surface plots. This is a great example of how implementing chemometrics in the experimental design can optimise pre-analytical processes' performance.

Chang *et al.* also conducted research using chemometric approaches to aid experimental design by investigating the optimisation of headspace solid-phase microextraction of smokeless powder compounds in forensic applications.[45] The investigated variables throughout this research were the weight of the sample, incubation time, extraction time, and sample temperature. The multivariate optimisation design of the experiment involved two chemometric methods, including a two-level full-factorial first-order design and the Doehlert design. An analysis of variance (ANOVA) using a two-level full-factorial first-order design was used to evaluate the interaction between the studied variables and then determine the influential variables and significant potential interactions between them. It was found that the weight of the sample and the incubation time did not significantly affect the performance of the SPME extraction; therefore, these parameters were fixed at certain values for the remainder of the experiment. The Doehlert design was then used to optimise the method, looking at the sample temperature, studied in five levels, and the extraction time, studied in three levels. Chang *et al.* successfully evaluated the optimal value for sample temperature and extraction time; however, it should be noted that it was found that the optimal set of conditions was affected by the volatility of the target compounds. This research is a further example of how chemometrics can be utilised in the experimental design process.

Chemometrics is utilised not only in the optimisation of pre-analytical processes but also in the optimisation of the analytical techniques used to detect and identify ERs. Many different analytical techniques can be used for the detection and identification of ERs (see Table 4.1). Many factors can affect the data collected from certain analytical techniques, for example, the flow rate and type of column utilised in GC–MS. These factors can affect the outcome of the analysis of ERs, and therefore the use of chemometric methods for optimising this process can aid in this situation. Jimeneze-Pireze *et al.*, for example, used the chemometric methods PCA and non-iterative partial least squares-linear discriminant analysis (NIPALS-LDA) to aid in the design of a virtual sensor data array for the analysis of RDX, HMX, and DMNB using metal-doped screen-printed electrodes.[46] Jimeneze-Pireze *et al.* were able to use the mentioned chemometric techniques to determine which types of electrodes were best suited for the identification of the target analytes. The responses of the explosive investigated from the studied electrodes were first analysed with NIPALS as a method of reducing the data to principal components before inputting through to LDA. It was found, through PCA, that the single-metal doped screen-printed electrodes were not providing sufficient differentiation for accurate classification, so the collected voltammograms were combined to create a virtual array. This virtual array was then analysed using PCA, which successfully separated the different explosive substances, indicating that the virtual array is suitable to provide enough chemical information to obtain a sufficient separation.[46] The use of chemometrics in this study succeeded in designing a virtual sensor data array for the successful detection and classification of the investigated explosives.

Another study by Vanini *et al.* investigated the optimisation of inductively coupled plasma-optical emission spectrometry (ICP-OES) instrumental parameters for the measurement of lead, barium and antimony in gunshot residues.[47] The three variables studied were the radio frequency power, nebuliser gas flow and aspiration rate. These parameters were optimised by applying CCD throughout the research as each variable was studied at five levels in the CCD.[47] Overall, this study was successful in the use of chemometric techniques to optimise ICP-OES for the measurement of the metals lead, barium, and antimony, with the optimal conditions for the radio frequency power, nebuliser gas flow, and aspiration rate being 1300 W, 1.2 L min^{-1}, and 1.0 mL min^{-1} respectively.[47] This is a great example of the use of chemometrics in the optimisation of commonly used analytical techniques.

4.4.3 Pattern Recognition

Over recent years, with developments in chemometric methods, there has been an increasing amount of research into utilising chemometrics in combination with varying analytical techniques. When chemometric techniques are applied to these analytical techniques, it expands the potential for discrimination and classification of target analytes through pattern recognition. Chemometric pattern recognition can be achieved through the use of an unsupervised multivariate analysis technique, such as PCA, as previously described, along with the use of discriminant analysis processes. This area has been of increased interest in the forensic discipline due to the potential ability to create a pattern recognition model that can detect and classify target compounds of interest for criminal investigations. For example, in the event of an explosion, the ERs must be identified as part of the investigation. Creating a classification model for target compounds suitable for use in any analytical laboratory is complicated due to the differences in instrumentation, protocols, and interferences between different laboratories.[48] Therefore, there has been a volume of research into the possible applications of chemometric techniques for forensic purposes such as explosive analysis.

For example, D'Uva *et al.* investigated the source attributions of party sparklers using trace elemental analysis for an insight into the analysis of common homemade explosives found in Australia.[49] Throughout this study, eight different brands of sparklers were analysed *via* ICP–MS before PCA was implemented in the dataset to try and classify the data into different brands.[49] This was achieved through the generation of five principal components to visualise the distribution of the data until the maximum separation between classes was achieved.[49] Once the classification was refined, LDA was implemented to construct a discriminant model. The LDA model resulted in a 100% calibration accuracy and successfully matched 100% of the samples in the validation process.[49] These results are a great example of the ability that chemometric techniques have in the discrimination and classification of explosive materials, as PCA was successfully used to visualise the distribution of the data to classify the sparkler samples into their respective brands.

Similar research into the analysis of consumer fireworks post-blast residues rather than sparklers was conducted by Martin-Alberca *et al.*, who used ATR–FTIR to conduct the analysis.[50] Martin-Alberca *et al.* utilised PCA to successfully trace the post-blast residues of different consumer fireworks to their original pyrotechnic compositions. Along with this achievement, chemometrics was also used to show that different types of swabs resulted in varying results of the post-blast residues; however, further investigation using perhaps a CCD would be needed to optimise the pre-analytical swabbing process.

4.4.4 Smokeless Powders

Smokeless powders are commonly employed in the creation of improvised explosive devices, as they are readily available in many countries and, when confined, produce a powerful explosion.[43] Smokeless powders typically consist of nitrocellulose (single-base) or nitrocellulose and nitroglycerin (double-base) along with several additives, which are constituents of the propellants used in ammunition cartridges.[51] There has been research into the characterisation and classification of smokeless powders using varying analytical techniques with the aid of chemometric methods. Reese *et al.*, for example, investigated the characterisation of smokeless powders using liquid chromatography–atmospheric pressure chemical ionisation–time of flight mass spectrometry (LC–APCA–TOFMS) along with PCA and HCA.[51] In this study, unburned powders and fired powders were analysed to assess if an association between the two was possible. Using PCA and HCA, the statistical association and discrimination of the unburned powders were demonstrated, with only a degree of association of the fired residues to the unburned powders being achieved using HCA.[51] This work shows how the chemometric methods of PCA and HCA can be used to form a pattern recognition model for the classification of smokeless powders, with preliminary evidence into the possibility of associating post-blast residues of explosive materials with pre-blast samples.

Similarly, Lennert *et al.* investigated the classification of smokeless powders using GC–MS and DART–MS.[52] HCA was used to identify groupings within the data based on the chemical compositions before PCA was used to observe potential relationships between variables and maximise inter-cluster variation.[52] The use of PCA in this research allowed for the identification of major contributors to variance through analysis of the principal component loadings. PCA was used after HCA to provide a visual validation of the groupings HCA identified, for which all matched.[52] LDA was then employed to assess the accuracy of the PCA models classification, which resulted in a correct classification rate of higher than 95%.[52]

Other examples of investigating the classification and evidentiary value of smokeless powders include work by Perez *et al.*, who used PCA and LDA to classify smokeless powders analysed by laser electrospray mass spectrometry, and Dennis *et al.*, who used likelihood ratios and HCA to assess the evidentiary and investigative value of smokeless powders.[53,54]

4.4.5 Laser-induced Breakdown Spectroscopy

The use of laser-induced breakdown spectroscopy (LIBS) for the detection of ERs is an area of interest, specifically for the United States Army and Homeland Security, as it can be used for standoff residue analysis.[55,56]

Research by Gottfried *et al.* has implemented PLS-DA in the detection and discrimination of ERs on organic and inorganic substrates using LIBS.[55] Gottfried *et al.* created individual PLS-DA models for each of the substrates studied throughout the investigation. When testing these PLS-DA models, there was an overall rate of 95.7% true positives achieved, with a relatively low rate of false positives at 6.7%.[55] After this success, a combined PLS-DA model was created, with all the substrates included. The results for the combined model were slightly less accurate, with an overall rate of 88.6% true positive being achieved, with quite a high rate of false positives at 12.7%.[55] The combined model was then tested with a set of sample types not included in the model, which resulted in a rate of true positives of 72.8% and false positives of 2.2%. Overall, this research has shown how chemometric methods, such as PLS-DA, can be used to aid in the discrimination of explosive materials.

Similar research by de Lucia *et al.* has also shown the use of PLS-DA to help detect and discriminate ERs using LIBS.[56] Two PLS-DA models were created in this study: one using the whole LIBS spectra and the other using the atomic intensities and ratios of constituent elements as the variable inputs.[56] Both models were successful in the discrimination of ERs on car panels, with the second PLS-DA model presenting the advantage that the variables responsible for classification as an explosive were all in respect to the constituent elements of explosives.[56]

4.4.6 Raman Spectroscopy

Another analytical technique of particular interest due to the possible applications of standoff detection of ERs is Raman spectroscopy.[57] Hwang *et al.* studied fast and sensitive recognition of various explosive materials using Raman spectroscopy and PCA.[57] Eleven explosive materials, such as military-grade TNT, RDX, PETN, *etc.*, were first analysed using Raman spectroscopy before being pre-processed by smoothing methods to remove background noise.[57] PCA was then performed to reduce the dimensionality of the data before the maximum *a posteriori* probability (MAP) was used for discriminant analysis. Ten out of the 14 analytes assessed were perfectly classified, with the other four misclassified in 2–4% of cases.[57] The average classification rate for studied explosive compounds was roughly 99.3%.[57] These results show the success in pairing multivariate analysis techniques with Raman spectroscopy for the classification of explosive compounds, which proves the capability of implementing chemometric techniques in areas of forensic science.

Raman spectroscopy is also a technique that is useful in the detection of gunshot residues. Karahacane *et al.,* for example, investigated Raman spectroscopy and chemometric analysis of organic gunshot residues from two types of ammunition.[58] Organic gunshot residues, consisting of either nitrocellulose, nitrocellulose, and nitroglycerin, or a mixture of nitrocellulose, nitroglycerin, and nitroguanidine, may also be encountered during a criminal investigation.[58] In this study, Raman spectroscopy was performed with the subsequent application of PCA. The supervised statistical method of support vector machines (SVMs), which works through classifying data points closest to the separating hyperplane, was chosen in this study as the discrimination method.[58] This resulted in successfully discriminating the organic gunshot residues into the two types of ammunition that were investigated, with a 100% classification rate.

A similar study by Bueno *et al.* investigated the Raman analysis of gunshot residue in reference to calibre differentiation.[59] Bueno *et al.* analysed the gunshot residue of two different calibre samples using Raman spectroscopy before analysing the data using PCA, *k*-nearest neighbour (*k*-NN), PLS-DA, and SVMs.[59] PCA was used to analyse the collected data, with a generalised least-squares weighting algorithm used to reduce scattering, before classification analysis was performed.[59] *k*-NN, PLS-DA, and SVM methods were used to perform the classification analysis. The classification using *k*-NN resulted in one spectrum being misclassified; therefore, PLS-DA and SVM were also used. PLS-DA and SVM resulted in no false positive or false negative results, proving the increased accuracy of these classification methods in this context.[59] However, it was determined that a SVM provided the most promising results in the discriminatory analysis of the PCA model, as it achieved the most accurate calibre-weapon pair identification overall.

Further research by Bueno *et al.* investigated the analysis and discrimination of gunshot residues using combined Raman and FT-IR data.[60] The data from a total of 92 pre-processed Raman and FT-IR spectra were combined into 46 spectra before being analysed using PCA. PLS-DA was then used as the classification method, and it was found that there was an increased sensitivity and specificity for the statistical discrimination when combining Raman and FTIR spectra before chemometric analysis.[60]

4.4.7 Partial Least Squares-Discriminant Analysis

Partial least squares paired with discriminant analysis (PLS-DA) is a supervised classification chemometric technique used to aid in discriminating and classifying explosive compounds, much like LDA. For example, a study by Fraga *et al.* used PLS-DA to aid in investigating elemental source attribution signatures for calcium ammonium nitrate fertilisers, which are used in improvised explosives.[61] In this study, ICP–MS was used to collect data for a series of commercial calcium ammonium nitrate fertilisers, which were then processed using HCA and PCA to cluster the sample data.[61] After the data was clustered, PLS-DA was implemented using two data sets for training before a third set of data was processed for validation. This resulted in a 73% correct classification of calcium ammonium nitrate fertiliser samples from 11 stocks.[61] These results provided concrete evidence that the use of PLS-DA as a chemometric technique for classification is possible with explosive materials.

Research by Fernandez *et al.* on using near-infrared hyperspectral imaging to detect ERs is another example of research that uses PLS-DA to classify sample data.[62] Ammonium nitrate, black powder, smokeless powders and dynamite were analysed using near-infrared hyperspectral imaging, pre-processed through applying smoothing transformations and standard normal variation, and then processed using PLS-DA.[62] In this initial study, it was found that the PLS-DA model was successful in classifying the pixels from the collected images that related to ERs. Overall, this technique successfully detected only some of the explosives, including the smokeless powders and dynamite, and was proven to be a fast, non-destructive technique for detecting these ERs.

Fernandez *et al.* then conducted further research using this technique, investigating the analysis of dynamite on human handprints. As this experiment was limited due to the focus being on one supply of dynamite, the data processing was simple, as the model only had two options for classification: dynamite and not dynamite. The PLS-DA model was extremely successful, with an accuracy of 100% for the calibration and validation.[63]

4.4.8 Regression

Regression analysis methods are less common in studies relating to the detection and classification of ERs. Gallidabino *et al.* used multiple regression analysis techniques to investigate the estimation of the time since the discharge of 9 mm cartridges by headspace analysis.[64] This study involved the use of multiple multivariate regression analysis methods, including: partial least squares (PLS), multivariate adaptive regression splines (MARSs), artificial neural networks (ANNs), random forests (RFs), *k*-NN, and SVM.[64] Ammunition cartridges were aged for varying times before being sampled and analysed using thermal desorption (TD)-GC–MS. After analysis, the data was pre-treated using total sum normalisation (TSN), probabilistic quotient normalisation (PQN), and pairwise log-ratio normalisation (PLR) and then the six aforementioned multivariate regression methods were then tested on the pre-treated data in reverse regression mode.[64] It was found that PLR in combination with RFs and PLS regression presented satisfactory performance compared to the other models.[64] The accuracy of the obtained outcomes for this model demonstrated potential for estimating the time since discharge of the cartridges aged up to 48 h.[64]

4.5 Conclusion

The use of chemometric methods in the analysis process of fire and explosive evidence allows for enhanced detection and classification of the target analytes to be achieved. There is a vast amount of research into the application of chemometrics in the analysis of ignitable liquid residues throughout a forensic arson investigation, showing the successful classification and detection of ignitable liquid residues using a variety of different multivariate statistical analysis models. This also applies to the detection of explosive residues. Not only has it been shown that chemometrics can be used in the classification of these target substances, but it can also be used in optimising both pre-analytical and analytical processes so that the maximum recovery of evidence can be achieved throughout forensic investigations. Chemometrics has proven to provide a pathway to the efficient detection and classification of forensic evidence. With more research on this topic, it may even be used to create classification models that are suitable for use in different laboratories. There is a need for the use of chemometrics as a part of future method development and identification and classification of ILRs and ERs in criminal casework. Research to date has only scratched the surface on the applications of chemometrics in a forensic context, and further research must be undertaken in the hope of normalising the use of chemometrics in this setting.

Acknowledgements

We would like to thank Joshua D'Uva for providing chemometric data and ChemCentre for providing all other images included throughout this chapter.

References

1. V. K. Yadav, K. Nigam and A. Srivastava, *Med., Sci. Law*, 2020, **60**, 206–215.
2. G. Sauzier, PhD thesis, Curtin University, 2016.
3. R. A. Hites, *Handbook of instrumental techniques for analytical chemistry*, 1997, pp. 609–626.
4. N. Daéid and R. Waddell, *Talanta*, 2005, **67**, 280–285.
5. I. Talavera Bustamante, F. Silva Mata, N. Hernández González, R. González Gazapo, J. Palau and M. M. Ferreira, *Anal. Chim. Acta*, 2007, **595**, 43–50.
6. S. Ávila, P. S. Hornung, G. L. Teixeira, M. R. Beux, M. Lazzarotto and R. H. Ribani, *J. Near Infrared Spectrosc.*, 2018, **26**, 379–388.
7. M. Ferreiro-González, G. F. Barbero, M. Palma, J. Ayuso, J. A. Álvarez and C. G. Barroso, *Sensors*, 2016, **16**, 695.
8. M. J. Aliaño-González, M. Ferreiro-González, G. F. Barbero and M. Palma, *Talanta*, 2019, **199**, 189–194.
9. M. M. Grutters, J. Dogger and J. N. Hendrikse, *J. Forensic Sci.*, 2012, **57**, 1290–1298.
10. M. R. Williams and M. Sigman, *J. Forensic Sci.*, 2007, **52**, 579–585.
11. J. R. Almirall and K. G. Furton, *Analysis and interpretation of fire scene evidence*, CRC Press, 2004.
12. ASTM, ASTM International, West Conshohocken, PA, 2000.
13. R. T. Newman, W. Dietz and K. Lothridge, *J. Forensic Sci.*, 1996, **41**, 361–370.

14. P. Doble, M. Sandercock, E. Du Pasquier, P. Petocz, C. Roux and M. Dawson, *Forensic Sci. Int.*, 2003, **132**, 26–39.
15. I. Barnett and M. Zhang, *Forensic Chem.*, 2018, **10**, 58–66.
16. M. Monfreda and A. Gregori, *J. Forensic Sci.*, 2011, **56**, 372–380.
17. J. A. Ardila, F. L. F. Soares, M. A. dos Santos Farias and R. L. Carneiro, *Anal. Lett.*, 2017, **50**, 1126–1138.
18. N. A. Sinkov, P. M. L. Sandercock and J. J. Harynuk, *Forensic Sci. Int.*, 2013, **235**, 24–31.
19. V. A. Rudnev, A. P. Boichenko and P. V. Karnozhytskiy, *Talanta*, 2011, **84**, 963–970.
20. N. D. K. Petraco, M. Gil, P. A. Pizzola and T. A. Kubic, *J. Forensic Sci.*, 2008, **53**, 1092–1101.
21. M. de Figueiredo, C. B. Y. Cordella, D. Jouan-Rimbaud Bouveresse, X. Archer, J.-M. Bégué and D. N. Rutledge, *Forensic Sci. Int.*, 2019, **295**, 8–18.
22. M. Novák, D. Palya, Z. Bodai, Z. Nyiri, N. Magyar, J. Kovács and Z. Eke, *Forensic Sci. Int.*, 2016, **270**, 61–69.
23. L. J. Marshall, J. W. McIlroy, V. L. McGuffin and R. Waddell Smith, *Anal. Bioanal. Chem.*, 2009, **394**, 2049–2059.
24. A. M. Hupp, L. J. Marshall, D. I. Campbell, R. W. Smith and V. L. McGuffin, *Anal. Chim. Acta*, 2008, **606**, 159–171.
25. L. M. de Aguiar, D. Galvan, E. Bona, L. A. Colnago and M. H. M. Killner, *Talanta*, 2022, **236**, 122838.
26. J. González-Rodríguez, N. Sissons and S. Robinson, *J. Anal. Appl. Pyrolysis*, 2011, **91**, 210–218.
27. M. J. Aliaño-González, M. Ferreiro-González, G. F. Barbero, M. Palma and C. G. Barroso, *Separations*, 2018, **5**, 41.
28. I. Barnett, F. C. Bailey and M. Zhang, *J. Forensic Sci.*, 2019, **64**, 1486–1494.
29. B. Falatová, M. Ferreiro-González, J. L. P. Calle, J. Á. Álvarez and M. Palma, *Sensors*, 2021, **21**, 801.
30. M. E. Sigman and M. R. Williams, *Forensic Sci. Int.*, 2016, **264**, 113–121.
31. A. Allen, M. R. Williams and M. E. Sigman, *Forensic Chem.*, 2019, **16**, 100188.
32. E. E. Waddell, E. T. Song, C. N. Rinke, M. R. Williams and M. E. Sigman, *J. Forensic Sci.*, 2013, **58**, 887–896.
33. E. E. Waddell, M. R. Williams and M. E. Sigman, *J. Forensic Sci.*, 2014, **59**, 927–935.
34. D. A. Turner and J. V. Goodpaster, *J. Forensic Sci.*, 2012, **57**, 64–69.
35. M. J. Aliaño-González, M. Ferreiro-González, G. F. Barbero, J. Ayuso, M. Palma and C. G. Barroso, *Sensors*, 2018, **18**, 139.
36. J. M. Baerncopf, V. L. McGuffin and R. W. Smith, *J. Forensic Sci.*, 2011, **56**, 70–81.
37. W. N. S. Mat-Desa, N. N. Daéid, D. Ismail and K. Savage, *Anal. Chem.*, 2010, **82**, 6395–6400.
38. W. N. S. Mat-Desa, D. Ismail and N. NicDaeid, *Anal. Chem.*, 2011, **83**, 7745–7754.

39. J. Goodpaster, in *Forensic Chemistry*, 2015, pp. 175–227.
40. National-Institute-of-Standards-and-Technology, *Recommended Guidelines for Forensic Identification of Post-Blast Explosive Residues*, USA, 2007.
41. N. Song-im, S. Benson and C. Lennard, *Forensic Sci. Int.*, 2012, **222**, 102–110.
42. D. Fisher, R. Zach, Y. Matana, P. Elia, S. Shustack, Y. Sharon and Y. Zeiri, *Talanta*, 2017, **174**, 92–99.
43. G. Sauzier, D. Bors, J. Ash, J. V. Goodpaster and S. W. Lewis, *Talanta*, 2016, **158**, 368–374.
44. O. Fryš, P. Česla, P. Bajerová, M. Adam and K. Ventura, *Talanta*, 2012, **99**, 316–322.
45. K. H. Chang, C. H. Yew and A. F. L. Abdullah, *J. Forensic Sci.*, 2014, **59**, 1100–1108.
46. R. Jiménez Pérez, M. Baron, L. Elie and J.-G. Rodriguez, *Int. J. Electrochem. Sci.*, 2013, **8**, 3279–3289.
47. G. Vanini, M. O. Souza, M. T. W. D. Carneiro, P. R. Filgueiras, R. E. Bruns and W. Romão, *Microchem. J.*, 2015, **120**, 58–63.
48. M. E. Sigman and M. R. Williams, *WIREs Forensic Sci.*, 2020, **2**, e1368.
49. J. A. D'Uva, D. DeTata, C. D. May and S. W. Lewis, *Anal. Methods*, 2020, **12**, 4939–4948.
50. C. Martín-Alberca, F. Zapata, H. Carrascosa, F. E. Ortega-Ojeda and C. García-Ruiz, *Talanta*, 2016, **149**, 257–265.
51. K. L. Reese, A. D. Jones and R. W. Smith, *Forensic Sci. Int.*, 2016, **272**, 16–27.
52. E. Lennert and C. Bridge, *Forensic Sci. Int.*, 2018, **292**, 11–22.
53. D.-M. K. Dennis, M. R. Williams and M. E. Sigman, *Forensic Sci. Int.*, 2016, **259**, 179–187.
54. J. J. Perez, P. M. Flanigan, J. J. Brady and R. J. Levis, *Anal. Chem.*, 2013, **85**, 296–302.
55. J. L. Gottfried, F. C. De Lucia Jr and A. W. Miziolek, *J. Anal. At. Spectrom.*, 2009, **24**, 288–296.
56. F. C. De Lucia, Jr and J. L. Gottfried, *Appl. Opt.*, 2012, **51**, B83–B92.
57. J. Hwang, N. Choi, A. Park, J.-Q. Park, J. H. Chung, S. Baek, S. G. Cho, S.-J. Baek and J. Choo, *J. Mol. Struct.*, 2013, **1039**, 130–136.
58. D. S. Karahacane, A. Dahmani and K. Khimeche, *Forensic Sci. Int.*, 2019, **301**, 129–136.
59. J. Bueno, V. Sikirzhytski and I. K. Lednev, *Anal. Chem.*, 2012, **84**, 4334–4339.
60. J. Bueno and I. K. Lednev, *Anal. Methods*, 2013, **5**, 6292–6296.
61. C. G. Fraga, A. V. Mitroshkov, N. S. Mirjankar, B. P. Dockendorff and A. M. Melville, *Talanta*, 2017, **174**, 131–138.
62. M. Á. Fernández de la Ossa, J. M. Amigo and C. García-Ruiz, *Forensic Sci. Int.*, 2014, **242**, 228–235.
63. M. Á. Fernández de la Ossa, C. García-Ruiz and J. M. Amigo, *Talanta*, 2014, **130**, 315–321.
64. M. Gallidabino, F. S. Romolo and C. Weyermann, *Forensic Sci. Int.*, 2017, **272**, 171–183.

CHAPTER 5

Food Forensics

N. CAVALLINI,[a] A. BIANCOLILLO,[b] L. STRANI,[c] C. DURANTE[c] AND M. COCCHI*[c]

[a] Department of Applied Science and Technology, Polytechnic of Turin, Corso Duca degli Abruzzi 24, 10129 Torino, Italy; [b] Department of Physical and Chemical Sciences, University of L'Aquila, Via Vetoio 67100, Coppito, Italy; [c] Department of Chemical and Geological Sciences, University of Modena and Reggio Emilia, Via Campi 103, 41125 Modena, Italy
*Email: marina.cocchi@unimore.it

5.1 Introduction

It is well recognized that globalization impacts food production; not only do many of the products we consume come from foreign countries but most often they are processed in different and remote parts of the world with respect to the one in which the food products are sold and consumed. This globalization process raises the problem of guaranteeing quality and safety in an extremely complex context.

The European Union, since 1979, has set up a rapid alert system created for the rapid exchange of information on dangers associated with consumer products, *i.e.*, the Food and Feed Safety Alerts, and it issues a yearly report summarizing data and trends on safety alerts. In the latest published report,[1] it emerged that the number of alerts and non-compliance notifications for food products follows a significant increasing trend, especially for non-compliance. The top hazard categories reported were related to pathogenic micro-organisms, *e.g.*, salmonella in food commodities such as poultry meat and derivates, herbs/spices, chocolate, pesticides in fruits and vegetables, especially the ethylene oxide incident originating from sesame seeds coming from India, mycotoxins, food additives and flavouring. In addition, most issues relating to the migration of chemicals from food

Chemometric Methods in Forensic Science
Edited by Vishal Sharma, Georgina Sauzier and Simon W. Lewis

Published by the Royal Society of Chemistry, www.rsc.org

contact materials are about melamine and nylon. Concerning fraud, just to cite a recent few,[2] fake saffron modified with gardenia extract imported from China, seafood samples with incorrect species labelling, wine adulterated with water/sugar and fake geographical indication, meat with falsification of the expiry date, misuse of geographical indication logos, lack of traceability information, and clandestine production of fake olive oil and extra-virgin olive oil containing a mix of other vegetable oils and lacking traceability.

It is clear from this context that contrasting adulteration, counterfeiting and identifying contamination is very complex since there are so many types of unpermitted or harmful substances possibly added to food (always varying) and illegal food practices and processing. Consequently, there is an increasing need for reliable analytical techniques for food quality control and fraud detection as also testified by the attention paid by the European Community to this matter through funded projects and networks.[3,4] Moreover, verifying authenticity with respect to denomination of origin, organic food, animal and plant species, *etc.* poses more challenges since these attributes and their assessment are necessarily linked to the foodstuff as a whole and require a holistic approach to food characterization, *i.e.*, untargeted methodologies.

In this framework, food forensic research has been developing.

The term food forensics refers to the use of forensic techniques in the field of authenticity determination, detection of adulteration, food frauds and any other legal issues inherent in the food product.[5,6] In this regard, it largely overlaps with food quality, authenticity and safety research fields, while focusing on the pool of techniques and methodologies most suitable to identify counterfeiting and fraudulent practices in a court perspective. Indeed, most of the analytical methodologies currently employed in other forensic areas, such as evaluation of crime scenes, detecting doping practice in sport, *etc.*, are the same or are of interest to be applied to food authentication.[5–7] These techniques can be classified according to different criteria; the most relevant in the food forensic context are being destructive or not, rapid or not, and directly or indirectly linked to the origin, *i.e.*, capable of tracing the provenance or the animal/plant species, as illustrated in Figure 5.1.

They are thoroughly discussed in terms of pros and cons together with their main characteristics in recent reviews[5,6] and the Food Integrity Handbook[7] (including references therein). Here, we just review the main developments in the last 10 years concerning the spread of DNA-based techniques in food authenticity assessment, leading also to some routine implementation and adoption as reference methods and the breakthrough in untargeted methods, either chromatographic or spectroscopic based, the latter including near-infrared spectroscopy (NIR) and proton nuclear magnetic resonance (^{1}H-NMR) as the most prominent. This chapter will focus attention on the prominent role of chemometrics to elaborate information coming from these techniques and building reliable authenticity models[8–15] and to the methodological path from sampling to validation allowing the achievement of this goal.

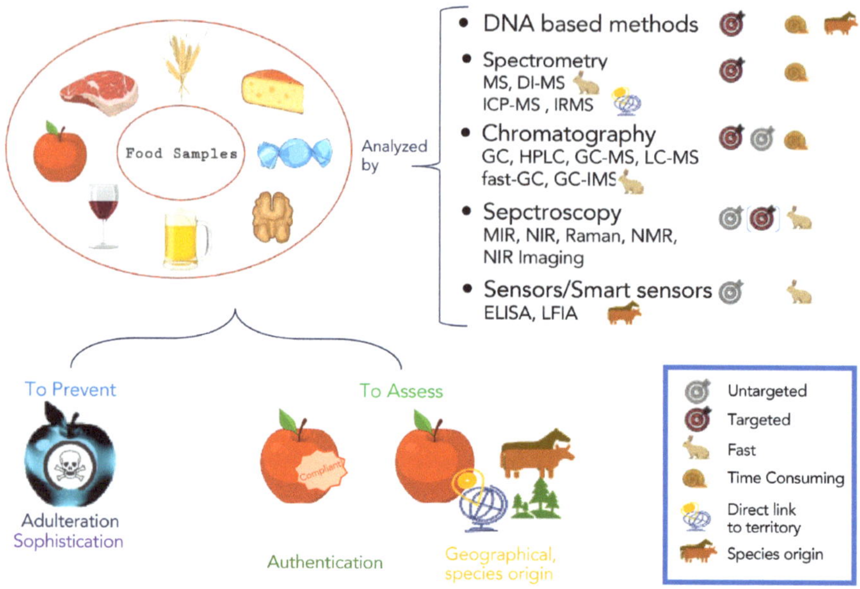

Figure 5.1 Analytical methods and prevention/assessment scope in the food forensics context. The analytical methods are: MS = mass spectrometry; DI–MS = direct infusion–mass spectrometry; ICP–MS = inductively coupled plasma–mass spectrometry; IRMS = isotope-ratio mass spectrometry; GC = gas chromatography; HPLC = high-performance liquid chromatography; GC–MS = gas chromatography–mass spectrometry; LC–MS = liquid chromatography–mass spectrometry; GC–IMS = gas chromatography–ion mobility spectrometry; MIR = medium infra-red; NIR = near infra-red; NMR = nuclear magnetic resonance; ELISA = enzyme-linked immunosorbent assay; LFIA = lateral flow immunoassay. The top-left sketch has been created with BioRender.com.

5.2 Food Authenticity and Fraud Issues

There are several aspects, which translate in different terms, associated with food authenticity and food fraud.[5,7] These have been revised and focused on by different bodies, aiming at having clear definitions to better combat the deceitful practices.

In particular: (i) a food fraud think tank set up with the support of the Global Food Safety Initiative (GFSI) published a document[16,17] providing some definitions of the different types of fraud, and (ii) a standardization initiative within the framework of a "CEN/CENELEC Workshop Agreement" (CWA) was launched to set up a consensus-based terminology of authenticity and food fraud leading to the CWA 17369:2019 standard norm "Authenticity and fraud in the feed and food chain. Concepts, terms, and definitions".[18]

The first document highlighted the difference between food fraud, food defence, food safety and food quality, while central to the CWA standard is the definition of authenticity exploiting the relationship between food product characteristics and food product claims. Figure 5.2 illustrates and summarizes these concepts and terms. A food fraud (Figure 5.2, left) implies a deliberate action motivated by economic gain; in the case of a criminal motivation to harm people a food defence issue arises. On the other hand, when food alteration is unintentional, *e.g.*, contamination, we face a food quality issue, when no health harm is provoked, and a food safety issue when health risk is present. The possible types of fraud are also illustrated in Figure 5.2, right and Table 5.1. Adulteration is one of the main types and encompasses several subtypes, with counterfeiting and mislabelling being the most common.

The absence of adulteration is a prerequisite for both food authenticity and the integrity of food products. Authenticity, according to the CWA, norm requires: "a food product where there is a match between the actual food product characteristics and the corresponding food product claims; when the food product actually is what the claim says that it is", thus implicitly considering all the attributes which are possibly specified in the label but remaining anchored to it. The Codex Alimentarius definition[19] distinguishes "food authenticity" and "food integrity", where the former is the state of being "not altered or modified with respect to expected characteristics including, safety, quality, and nutrition", while the latter is the state of being "genuine and undisputed in its nature, origin, identity, and claims, and to meet expected properties". Thus, being explicit on the aspects which are inherent to the identity of the product is not necessarily linked to the label/claim.

A clear definition of authenticity and fraud issues is also necessary for the selection of analytical and chemometrics methodologies.

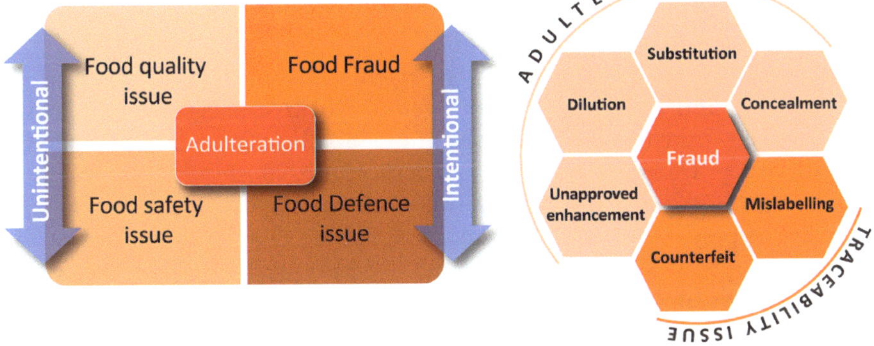

Figure 5.2 Food adulteration can be unintentional or intentional (left), depending on whether its motivation is a deliberate action (intentional) or not (unintentional). When the motivation leading to food adulteration is intentional, different types of food fraud can occur (right).

Table 5.1 Definitions of terminology used in food authentication.

Authenticity	A food product where there is a match between the actual food product characteristics and the corresponding food product claims, when the food product actually is what the claim says that it is.
Traceability	Traceability is the process for verifying the relation between a foodstuff and its source of the material. Traceability procedures study the chemical parameters to figure out the paths of the various phases within a production chain
Food fraud	
Adulteration	A type of food fraud which includes the intentional addition of a foreign or inferior substance or element, especially to prepare for sale by replacing more valuable with less valuable or inert ingredients
Substitution	Process of replacing a nutrient, an ingredient or part of a food (often one with high value), with another nutrient, ingredient or part of food (often one with lower value)
Dilution	The process of mixing a liquid ingredient (solute) with high value with a liquid of lower value
Unapproved enhancement	Process of adding unknown and undeclared compounds to food products in order to enhance their quality attributes
Concealment	Process of hiding the low quality of food ingredients or products
Counterfeit	Intellectual property right (IPR) infringement is in effect, *e.g.* copying the brand name, packaging concept or processing method, imitation with fake labels
Mislabelling	Concerns the process of putting false claims on packaging for economic gain, *i.e.*, distortion of the information provided on the label

Adulteration overall manifests by the presence (or absence/lower amount), of one or more substances that should not be (or should be) in the food product. In this case, depending on if we know or not which substance/s we resort to targeted or untargeted analysis; nonetheless in the case of untargeted analysis we are also interested in identification (at least putative identification) of possible adulterants.

Counterfeiting and mislabelling are intrinsically holistic features of the food product and thus require untargeted, fingerprinting methodologies. Traceability is linked to specific cases of mislabelling/counterfeiting such as geographical/territory origin and species origin; in those cases methodologies capable of establishing the link with the territory and then tracing back the supply chain are the most suitable. However, when for fast screening requirements, practical and/or economic reasons only the end food product is analysed, without determining specific origin markers, the sampling design should be robust and representative to ensure sound class models of the investigated food product/variety.

In the following section, data analysis strategies will be discussed keeping the focus on these relevant aspects.

5.3 Chemometrics Tools and Strategies

5.3.1 Sampling

Food forensics, as pointed out in previous sections, include several facets, but from the data analysis point of view whether the task is assessing authenticity, *i.e.*, that the food sample belongs to the category claimed on the label or verifying the absence of adulterants/contaminants, we can see the sampling step issues in a common perspective. In fact, the main assumption is that we can model the reference category, hence truly authentic samples belonging to that category (or unadulterated, genuine) must be available (collectable) and a statistically representative set should be sampled. Sampling is the most crucial step,[9,20,21] and depending on the task and the approach, sampling may concern food commodities, or it can encompass all the steps of the production chain, *e.g.* from soils to food or from raw material to finished product (traceability). Figure 5.3 illustrates the main issues linked to sampling. The first is ensuring representativeness when passing from the lot (in a broad sense) to be sampled to the laboratory sample (then a similar, albeit less severe, issue arises from passing from a laboratory to an analytical sample), which can be accomplished only if samples' variability sources are studied and heterogeneity with respect to them assessed.[22]

In the specific case of geographical origin traceability, there are two different sampling steps: the selection of the fields, where to take the samples, and the locations in the fields to collect them. Salient sources of variability affecting field selection are climate, geographical and geological features, the incidence in food commodities production from the field, agronomic practices, and varieties/species cultivated. Hence representativeness should

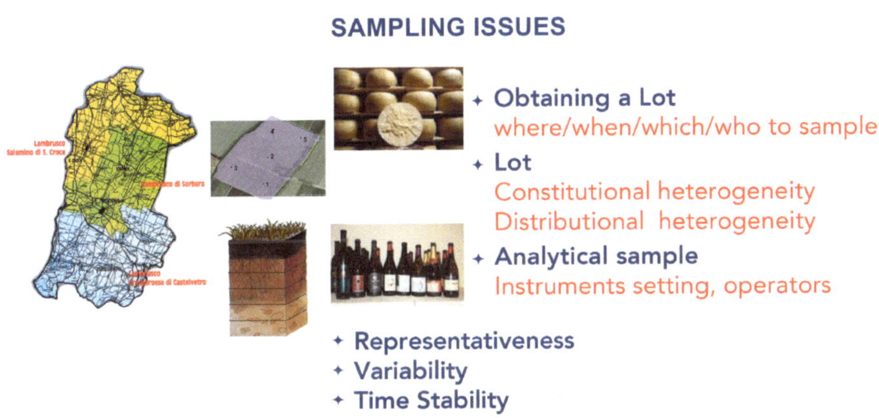

Figure 5.3 The main issues linked to sampling in terms of sampling procedure (obtaining a lot, lot characteristics, extracting an analytical sample from a lot), representativeness, variability, and time stability.

be ensured with respect to each of them.[23] In planning which location in the field to sample heterogeneity with respect to extension and depth has to be considered.

Similar reasoning and planning strategies can be extended to foodstuffs, *e.g.*, in all cases where authentication with respect to denomination, such as protected denomination of origin (PDO), mountain production denomination, organic food, wild species food, *etc.*, has to be verified, as well as when adulteration of a commodity has to be controlled, *e.g.*, eggs, chocolate, cow meat, *etc.* Two similar steps may be foreseen: (i) which farms/producers to sample, ensuring representativeness with respect to producers (also considering the incidence of their production on marketing), year/season, variety/species, including all elements, which can vary according to production protocols (*e.g.*, raw materials, production practices, *etc.*); and (ii) which and how many batches/lots have to be sampled to ensure representativeness. Sampling along the production chain similarly requires representativeness with respect to raw material employed, process settings in time, and with respect to how and how many (how often) intermediate samples are taken.

To ensure representativeness when passing from the lot to the laboratory sample, *i.e.,* the step of producing a first composite sample by pooling primary increments, the theory of sampling fundaments has to be adopted.[21,22,24]

Not the least important issue is ensuring the stability of analytical samples during the testing period and establishing variability sources inherent in the measurement process; with this aim, experimental design[25] is an invaluable tool that should always be used. In general, chemometric methods such as ANOVA simultaneous component analysis (ASCA)[26–28] are extremely powerful in assessing which sources of variability introduce significant effects in the analytical responses, including spectral profiles (which are the main responses in untargeted analysis).

Besides these general considerations, a further aspect to consider is the link between the task and the sampling strategy; in particular, in authenticity studies, a meta question is whether models should be based on sampling the food commodity (*e.g.* of course directly at the production site or involving producers' consortia to ensure the samples are authentic) with its inherent variability or to use reference samples *ad hoc* obtained by control authorities, *e.g.* the case of micro-vinification with grapes of authentic origin/varietal. The first approach is usually more suited to untargeted analysis or fingerprinting techniques, while the second may be sound for targeted analysis of specific adulterants when measuring specific markers, *e.g.*, light stable isotopes.

Finally, it has to be remembered that authenticity models are learned from samples. Thus, no matter how careful and accurate the sampling in the model-building phase, additional authentic samples (with the same sampling strategy) have to be collected at a future time to check that entities or sources of variability have not changed. In other words, model maintenance, or database curation, should be in place.

5.3.2 Modelling

In Figure 5.1 analytical techniques are distinguished as targeted and untargeted depending on their capability to allow the identification or putative identification of food constituents or adulterant/contaminant as well as of elemental composition, varietal, and animal species. However, this is not clear cut since, *e.g.*, a spectrum or chromatogram can be used as such or after a feature extraction step; this is illustrated schematically in Figure 5.4a, referring to the traceability task. Depending on the modelling approach and objectives a switch from targeted to untargeted analysis is possible.

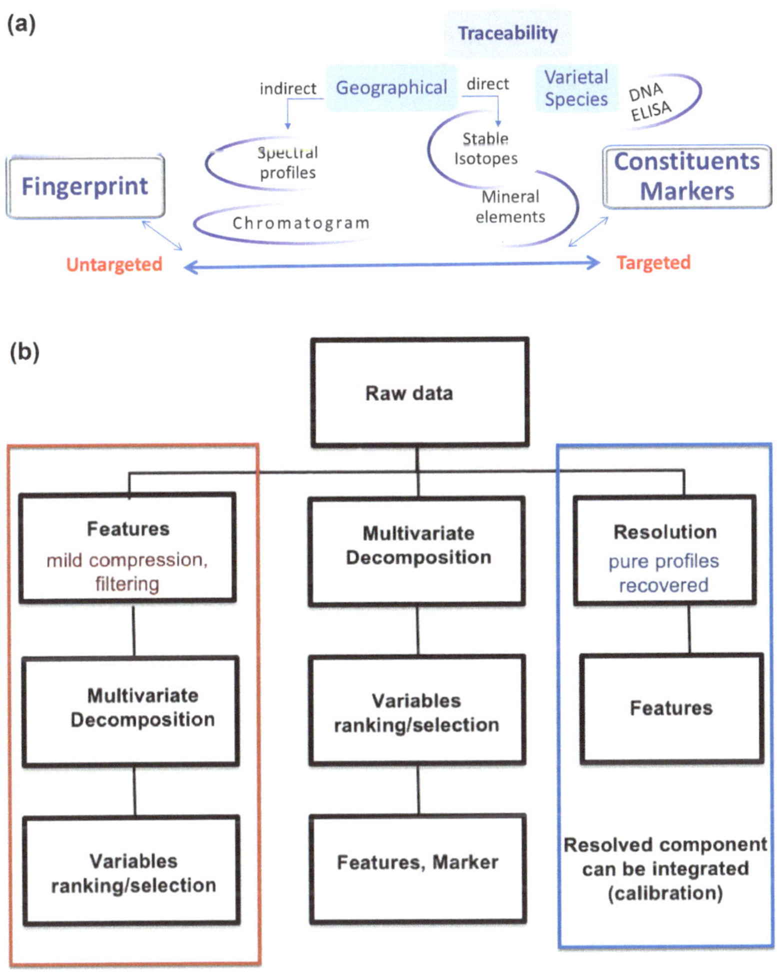

Figure 5.4 (a) The traceability task and its relationship with untargeted and targeted modelling approaches. (b) The three main modelling paths: (1) filtering and compression methods, (2) multivariate decomposition techniques (PCA, PLS, *etc.*), and (3) resolution techniques (MCR, PARAFAC).

Figure 5.4b shows the three main modelling paths which can be adopted: (1) the acquired signals are mildly compressed to obtain a set of features; an example is binning (widely adopted in NMR or in untargeted mass spectrometry, MS), automatic peak picking, and integration (as routinely implemented in some chromatographic software). Afterwards, classical decomposition techniques, such as principal components analysis (PCA[29,30]) and partial least squares (PLS[31]) are applied to the reduced feature data set, and finally variables ranking (or selection[32]) can be used to highlight the most relevant information, such as specific identity markers or identifying possible un-admitted/counterfeiting compounds or class of compounds. (2) Decomposition techniques (PCA, PLS, *etc.*) are applied, then the model interpretation and eventually features selection/ranking features could be used to isolate specific peaks, which afterwards can be integrated to provide authenticity markers, *etc.* (3) The use of resolution techniques, such as multivariate curve resolution (MCR[33]) or parallel factor analysis (PARAFAC[34]), as a first step, allows recovering the chemical profile of single constituents, and then features, such as the relative concentration of each recovered constituent, are extracted and can be used for linear calibration or fed into other multivariate methods (*e.g.* for classification).

A variant of the first approach above consists of using transform or filtering methods, such as Fourier (FT) or wavelet (WT) transforms,[35] to extract features (*e.g.* wavelet coefficients at different scales). In this case, a high degree of compression is achieved, while multivariate modelling is done in a different domain. Thus, the results for interpretation, have to be referred back to the original domain by inverse transforms.

None of these modelling approaches require *a priori* knowledge of which specific compounds are searched for (or are relevant) but different assumptions/conditions hold for each specific method.

While the reader is referred to other chapters of this book and the literature for a detailed method description, the main and most used chemometrics modelling tools in food forensics are summarised in Table 5.2.

As shown in Figure 5.5, different modelling approaches can be used depending on the task to be performed and the specific context/aims.

In adulteration detection, notwithstanding the type of fraud, the tasks reduce to identifying and/or quantifying the presence (or absence) of one or a few substances. This is mostly achieved by calibration (generally multivariate, unless a specific sensor responds to a single species or standard calibration following a separation technique is feasible because selective peaks are clearly identifiable). Nonetheless, classification models may be suitable either if only identification is required or when it is sufficient to determine at which level the adulterant is present. In the latter case, the objective is to assess the category below/over a threshold or whether it is absent, moderate, high, *etc.* A third approach is to apply chemometrics resolution methods, such as multivariate curve resolution, independent component analysis (less used), or parallel factor analysis to resolve spectra or chromatograms in distinct "chemical" meaningful contributions which

Table 5.2 Most used chemometrics modeling tools.

Name[a]	Basic principles	Development/extension
Class modeling[8–10,40]		
SIMCA[67]	Disjoint PCA models for each class Classification rule based on the distance to the class model	Multiway extension NSIMCA[10] Data driven SIMCA[36] Optimizing decision threshold[37] Using kernel transform prior to PCA[38]
UNEQ[8]	Assuming Gaussian distribution for each class. Classification rule based on Mahalonobis distance	Neural network based class modeling[39]
OCPLS[40]	Uses PLS with a *y*-variable which is a unitary vector.	Using kernel transformation of the data RBF-OCPLS[6]
Discriminant classification[8,9]		
PLS-DA[65,66]	PLS with dummy coded *y*-value for each category.	Multiway extension NPLS-DA[10] Multiblock PLS (SO-PLS-DA)[96] Multiway-multiblock SO-NPLS-DA
LDA,[64] QDA	Parametric class distribution. Bayes rule for classification	ECVA[8]
PFM[8]	Non-parametric class distribution	
KNN[8]	Decision rule based on neighbours' category	
CART[41]	Decision rule: varying single variables threshold to minimize error)	RF, boost-RF[42]
SVM[68] classification	Use "dense" points to define class boundary (maximize margins)	Extension to multi class, non-linear Kernel-trick SVM
ANN[43] classification	Artificial neural networks learning categories membership	
Resolution		
MCR[33]	Bilinear decomposition in pure profiles (constraint needed)	
PARAFAC[34]	Low rank tensor decomposition	PARAFAC2[44]
ICA[45]	Statistical independence of the components	
Multivariate calibration/regression		
PLS[31]	Maximizing *XY* covariance	Multiway extension NPLS[46] Multiblock extension SO-PLS, ROSA[47]

Table 5.2 (*Continued*)

Name[a]	Basic principles	Development/extension
MCR	A regression constraint is introduced in the alternating least squares	
ANN	Learning continuous response	

[a] Full text of acronyms not already present in the manuscript: UNEQ (unequal class modeling); OCPLS (one class PLS); RBF (radial basis functions); QDA (quadratic discriminant analysis); ECVA (extended canonical variate analysis); PFM (potential function methods); KNN (*k*-nearest neighbours); CART (classification and regression tree); ANN (artificial neural networks); ICA (independent component analysis); ROSA (recursive orthogonal sequential alternation).

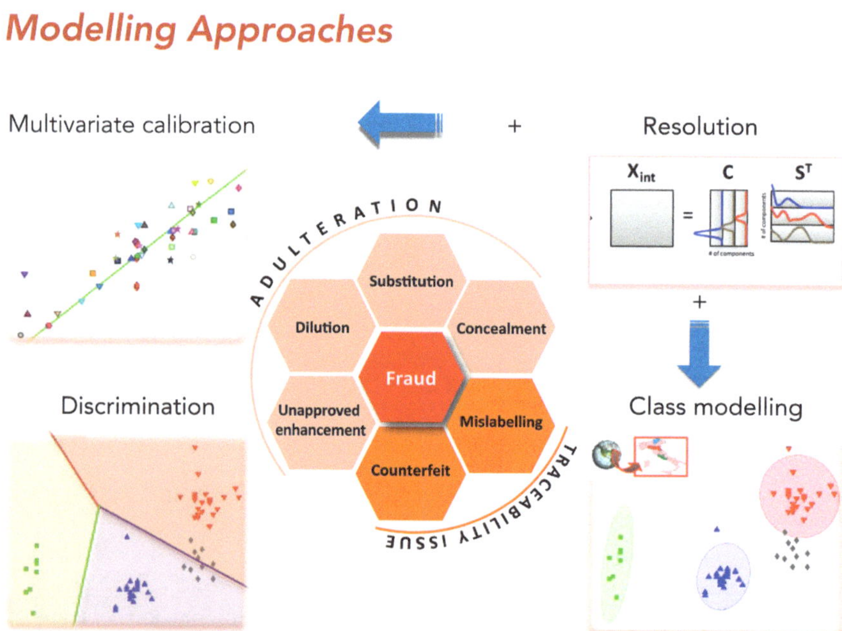

Figure 5.5 Modelling approaches recap: depending on the task to be performed and specific context/aims a calibration (*i.e.*, prediction of a specific quantitative property) or a classification (*i.e.*, identification and assignment to a specific class) modelling approach could be required.

may be putatively identified through library search and then confirmed by analysis of standards. In particular, two recent toolboxes are worth mentioning MCR-ROI[48,49] to resolve complex high-resolution liquid or gas chromatography–mass spectrometry data (HR/LC–MS and HR/GC–MS), and PARADISe[50,51] for LC–MS and GC–MS. Afterwards resolution peak areas or concentrations (if quantified) can be fed to calibration or classification models.

Authentication tasks mainly deal with assessing a category to which the food of interest belongs; hence modelling consists of providing classification models.[8]

From the modelling point of view, a class is by definition an ensemble of samples that are similar with respect to the pool of features used to describe them. As we have discussed before, the representativeness of the sampling is mandatory for obtaining a robust classification model.

In particular, class modeling[8,15,52] has to be applied anytime the focus is on the uniqueness of the food product, *i.e.* on assessing its identity and that it possesses specific traits or attributes, rather than contrasting it with other products, which may define different categories. Typical tasks are assessing the quality label specifications, such as the denomination of origin (PDO), protected geographical indication (PGI), mountain indication, and so on. Class-modelling methods are designed to capture the similarities among samples from the same category, *i.e.*, they focus on the uniqueness of the class rather than on the differences among the classes. The presumption of samples similarity and representativeness holds only for the ones belonging to the class being modelled.

More than one class can be modelled but each one is modelled individually, independently, and irrespective of the presence of others (as shown in Figure 5.5 left bottom, each category has its own confidence ellipse). Typically, a components model (generally a PCA model) is built for each category, thus from the modelling point of view, a reference hyper-volume is sought in the features space which defines the space of the modelled category. Hence a sample is recognized as belonging to the class if its position, once projected on the PCA model of the class, is within the class space. When there are more categories, as many models as categories are built and a test sample is projected in turn on each of the single-category models. It may be assigned to: (i) only one class, meaning that it is authentic for that category; (ii) more than one; in this case, it could be a mixture or a borderline case, *i.e.*, when categories are defined based on continuously varying features, for example, the scores attained by sensory evaluation; or (iii) none of them, in this case, maybe a non-authentic sample differently counterfeited/adulterated or a sample belonging to a different class from the modelled ones.

Instead, class discrimination is focused on finding the differences among samples belonging to different classes; the classes are modelled altogether, and a hyper-surface is sought in the variables (features) space to unequivocally assign to each class a region in that space. In practice, discriminant models define directions of maximal separation (or minimum overlap) among the classes (in Figure 5.5 bottom right, the category boundaries are represented by the lines, and the partition of space according to each class is coloured accordingly) by minimizing the misplacement of the samples in a region different from that of the class they belong to. In other words, a sample is assigned to a class if its coordinates are such that it falls in the region labelled as that class and as a consequence each sample is univocally assigned to one of the modelled classes. Thus, all classes must be known, and hence a set of samples sharing similar characteristics and being statistically representative should be available for all the modelled classes.

This may often not be the case in authentication tasks, where only one category may be defined and/or it is possible to collect representative authentic samples for that class only. Often, these situations are also dealt with discriminant modelling, *i.e.*, reformulated as a two-class discrimination issue, where class 1 is, *e.g.*, the PDO category and class 2 the rest of the world. Actually, this is an ill-posed question[53] since class 2, in this case, lacks the "sharing similar characteristics" assumption. The samples that are not authentic for that PDO category cannot be expected to have common features; these will be typically heterogeneous and thus not share a common region in features space. At the same time, representative sampling of the rest of the world is somehow an unattainable target.

Difficulties also arise when discriminant analysis is applied to contrast several known (hence representative sampling is possible) categories, respecting the similarity assumption. This is because there will always be in the samples to be tested at a future time some non-authentic samples, *i.e.*, differently counterfeited or adulterated, that do not belong to any of the modelled classes but nonetheless will be assigned to one, resulting in an incorrect decision. This situation is represented by the grey diamond symbols in Figure 5.5, which are correctly not assigned in class modelling (plot bottom left), but are assigned partly to the red and partly to the blue categories in discriminant classification (plot bottom right).

Thus, discrimination is only suited when it is clear which is/are the category/categories to contrast with, *e.g.*, recognizing wild from bred salmon.

5.3.3 Validation

Validation is a core concept in chemometrics and, needless to say, without validation any model is useless. Nevertheless, there is still limited awareness of both the conceptual and factual implications of validation. A recent tutorial[54] reviewed validation methodologies and problems, while a specific focus on the case of food authentication can be found in the Riedl *et al.* review.[55]

It is mostly underestimated that all data analysis steps starting from raw data acquisition to the routine use of a model have to be validated. Figure 5.6 provides a schematic view of the main aspects of concern, subdivided along the data acquisition and analysis pipeline.

Phase 1 goes from sampling to data acquisition: this encompasses sampling design, sample preparation, measurement (one or more platforms), data acquisition and assembly. This first phase mainly pertains to the optimization of the analytical procedures and methods, but it is not disconnected from the further analysis steps. In fact, the raw data that will be acquired for new/future samples, *i.e.*, modelling validation samples, are assumed to be affected by the same sources of variability with respect to the calibration data, as well as the analytical measurements are assumed to remain of the same quality. Thus, at this very first stage one has to be aware and take care of sampling issues (Section 5.3.1), sample stability with respect to chemical/biological variability

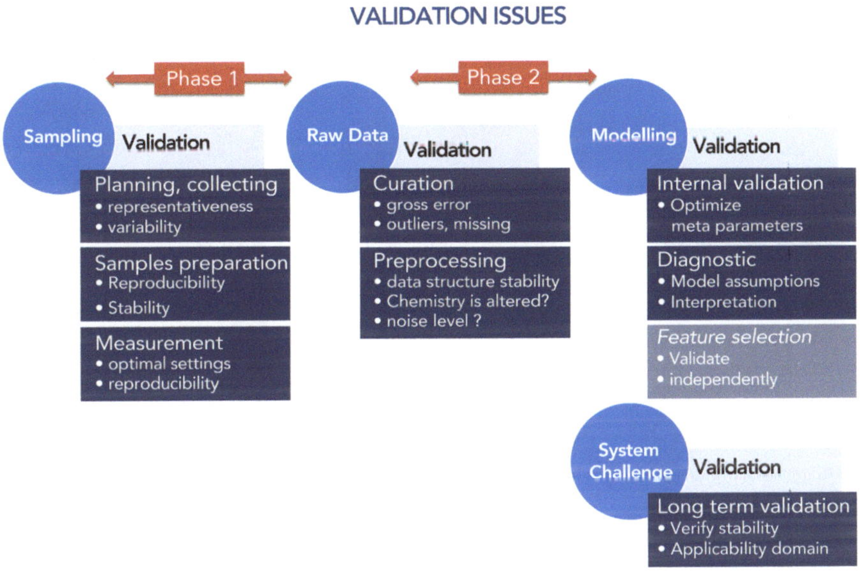

Figure 5.6 Validation issues and strategy.

over time, matrix effects, instrument stability, the influence of different operators/labs, *etc.* In other words, when planning sampling and optimizing analytical procedures it is of utmost importance to foresee and mimic the conditions in which future samples will be acquired.

Phase 2 goes from raw data to model building including data pre-processing, model diagnostics and validation in the calibration step, model interpretation, model refinement, and finally assessing model stability (long-term validation). Revising the authenticity models proposed in the literature, a first remark is that most often validation is applied to the sole model-building step and not always properly, while the other steps undertaken are poorly described.[55]

The following are lists of guidelines stressing the relevance of an all-steps validation strategy:

- Pre-processing validation: consider both spectral/signal pre-processing, centring, and scaling pre-treatments. Evaluate which effects pre-processing removes from the raw data and how it transforms it. Is it compatible with the underlying chemical/biological behaviour of the samples,[29,56] are, by any chance, artefacts introduced?
- Internal validation during model building: take into deep consideration that proper validation schemes have to be adopted as a function of sample stratification (replicates, batches, production years, locations, *etc.*). This internal validation allows assessing the model parameters, *e.g.*, the model dimensionality in the latent variables model, not the predictive capability of the final selected model. Moreover, different

approaches may be needed to validate the stability/significance of variable regression coefficients, loadings, and weights.[54]

- Features selection validation: in the case that variable selection is one of the data analysis steps, be aware that just repeating internal cross-validation schemes (or even worse not using a specific validation for this step) may be misleading. In this case model validations approaches are recommended.[57]
- Once all meta parameters have been assessed and a "final" model is selected, its validation should cover three aspects: (i) residuals diagnostic, *i.e.*, checking that methods assumptions hold; (ii) the use of an independent validation set (test set or external test set) in order to evaluate the model predictive capability; (iii) validation of the model by the soundness of results interpretation.
- Long-term model validation: a model to be safely used should be robust against the so-called system challenge,[55] *i.e.*, the long-term use of the model in routine situations. The difference here with respect to the step of assessing the predictive model capability is that we expect that at a certain point in the future food samples to be tested may not happen to be in the domain we have calibrated and model updating may be needed. A related issue is calibration transfer among instruments laboratories, and nowadays many methodological developments are in place in the chemometric community.[58,59] Validation at this step allows realizing when an update is needed and/or samples are not in the applicability domain of the model.

5.4 Illustrative Cases of Study

5.4.1 Authenticity and Adulteration Detection

As previously clarified, there are several occasions in which the authenticity of a food product can be undermined. For instance, this can happen when the product is adulterated or when there is intentional mislabelling of a commodity. Clearly, this type of fraudulent deception is applied by selling a product of lower commercial value in place of one that has a higher one (for example, a non-PDO product on behalf of a PDO one). To prevent this type of scam, the development of tools that allow the authentication of a specific product of interest is of unequivocal importance. In this regard, chemometric methods, in particular classification approaches, serve as an invaluable aid to contrast these frauds.

It is therefore not surprising that chemometric methods are widely used to assess the authenticity of food products, and that a plethora of applications in this regard can be found in the literature. A brief search on the Scopus database (www.scopus.com), combining "chemometrics", "authentication" and "food" as keywords (conducted in May 2022) reveals that, since 1996, 399 documents simultaneously associable with these three topics have been indexed, and that, of these, the greatest part (359) have been published in the last ten years (time range: 2012 to May 2022).

A brief investigation of the most recent ones shows that many untargeted analytical methods are used: mainly, among others, liquid chromatography–mass spectrometry (LC–MS), ultra-high-performance liquid chromatography coupled with hybrid triple quadrupole time-of-flight mass spectrometry (UHPLC-Q-TOF/MS), inductively coupled plasma–optical emission spectrometry (ICP–OES), and headspace solid-phase microextraction (HS–SPME) coupled with gas chromatography and mass spectrometry (GC–MS[60–66]). Nevertheless, these approaches, despite being extremely accurate, present relevant drawbacks, mainly associated with cost, complex sample preparation procedures, and the use of environmentally unfriendly chemicals. For this reason, in recent years, rapid and non-destructive/mini-invasive untargeted methods have played a key role in the authentication of food products. As expected, the main players in this area are mid and near infrared (MIR/NIR), Raman and ultraviolet-visible (UV-Vis) spectroscopy, are applied for the authentication of the most diverse types of food matrices.[67–75] Regardless of the nature of the exploited analytical methodology, common classifiers in this context are the discriminant ones, such as linear discriminant analysis (LDA[76]), partial least squares-discriminant analysis (PLS-DA[77,78]) or principal component analysis coupled with LDA (PCA-LDA), but also soft independent modelling by class analogy (SIMCA[79]) and support vector machines (SVMs[80]) are often encountered.

Chemometrics is also extensively used to detect food adulteration. An analogous search on Scopus (performed in May 2022) showed how 155 papers associated with "chemometrics", "adulteration" and "food" keywords have been indexed since 1996, of which the majority, *i.e.*, 129, have been published in the last ten years.

Aside from the many chromatographic applications,[81–89] one of the most common analytical techniques for the identification of adulterated foods is ^{1}H-NMR, due to its ability to assess quantitatively the composition of complex mixtures, namely the fingerprint of the analyte.[90–97] However, again, the highly expensive instrumentation of NMR, coupled with the use of chemicals (deuterated solvent at least) and the need for accurate sample preparation (for food commodities not viable to be analysed as such) moved the scientists to look for cheaper and more versatile techniques. Indeed, the majority of most recent papers in the area of food adulteration have been conducted through fast, cheap, non-destructive and non-invasive spectroscopic techniques, such as MIR,[74,98] NIR[99,100] and Raman.[101,102] Moreover, some documents investigated the use of e-tongue and e-nose sensors,[103,104] capable of being used, as well as NIR portable spectrometers, in in-field applications.

The most used chemometric tools to highlight differences between adulterated and non-adulterated samples are both discriminant classification methods, such as PLS-DA, multilinear partial least squares discriminant analysis (NPLS-DA), unfolded partial least-squares discriminant analysis (UPLS-DA[105]), and class modelling methods, mainly SIMCA.[106] The partial least squares (PLS) regression method is also widely used to quantify traces of adulterants in samples.[64]

In this framework, a strong contribution, which made further improvement in the accuracy of the predictive models possible, was certainly provided by multi-block (or data fusion) analysis. In fact, data fusion (DF) techniques allow a synergic extraction of the information from diverse sources of variability and support a refining of predictive models. As for classic classification methods, multi-block approaches allow product authentication by means of DF-classifiers. Despite the relatively recent development of these strategies, DF approaches are spreading in the literature, thanks to the numerous benefits they provide from the predictive and interpretative standpoint. The advantages associated with the two DF methods will be better highlighted in the illustrative example presented in Section 5.4.1.1 on saffron authentication.

Another aspect where chemometrics can greatly improve the development of authenticity models suitable for large-scale screening and routine analysis is assessing the potentiality and soundness of the model based on miniaturized spectrometers. The case study presented in Section 5.4.1.2 focuses on this and presents a challenging discrimination between fresh and frozen-thawed fish products, a type of fraud that is very common.

Finally, Section 5.4.1.3 presents a case of quantitative adulteration detection, namely the presence of crumb in grated PDO Parmigiano cheese.

5.4.1.1 *Multi-block Authentication of Italian Saffron – A Case Study*

Saffron is a well-known spice cultivated in many countries, mainly in Europe and Asia, but consumed all over the world.

Several precautions can be used during the cultivation and harvesting of saffron, generally linked to the tradition of the production areas and the know-how of the farmers. The different cultivation/harvesting processes ensure that the final product has different characteristics and, consequently, this means that different quality grades of saffron can be found on the market with a price ranging from about €7 (per gram) up to about €35 per gram. Due to its market value, this spice is often the subject of scams, hence the need to have tools to ensure its quality. One of the possible frauds that can be found in this product is intentional mislabelling, *i.e.*, a lower quality saffron is sold in place of a higher quality one. This can happen, for example, when a saffron that has a high added value associated with the cultivation area (due to particular soil and climatic conditions or the know-how of the producer) is replaced with one grown in a different locality. From this derives the need to authenticate the product by guaranteeing the area of origin and the discipline to which it has been subjected.

In the light of these considerations, Biancolillo and collaborators[107] have developed a multi-platform tool for the authentication of saffron grown in diverse Italian locations. In particular, they have analysed 114 different samples of this spice, harvested in four different areas (in accordance with

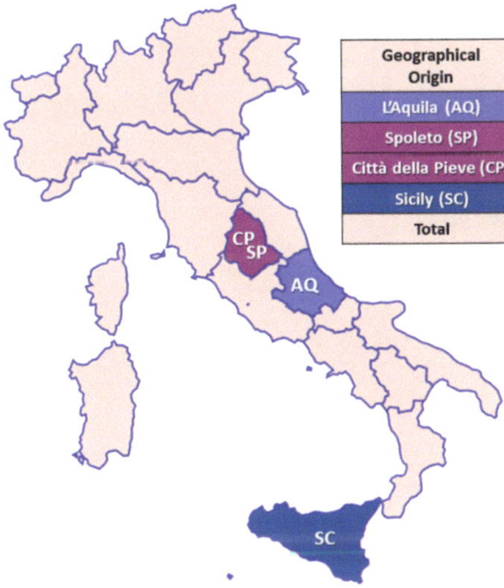

Geographical Origin	Training Samples	Test Samples	Total Samples
L'Aquila (AQ)	11	5	16
Spoleto (SP)	26	8	34
Città della Pieve (CP)	35	10	45
Sicily (SC)	11	8	19
Total	83	31	114

Figure 5.7 Map of Italy highlighting the four different geographical origins and table depicting the numerosity of samples. Reproduced from ref. 95, https://doi.org/10.3390/molecules25102332, under the terms of the CC BY 4.0 license, https://creativecommons.org/licenses/by/4.0/.

the numbers and origins described in Figure 5.7) by means of infrared (IR) and ultraviolet (UV) spectroscopy.

Prior to the analysis, stigmas were freshly ground and then analysed by a PerkinElmer Spectrum Two™ (PerkinElmer, Waltham, MA, USA) FTIR spectrometer equipped with a universal attenuated total reflectance (uATR), and by a Cary 50 Probe (Agilent Technologies, Santa Clara, CA, USA) UV-Vis spectrophotometer (more details on the instrumental analyses can be found in the original publication[107]).

After spectra collection, two sequential multi-block classifiers, sequential and orthogonalized-PLS-linear discriminant analysis (SO-PLS-LDA)[108] and sequential and orthogonalized covariance selection-linear discriminant analysis (SO-CovSel-LDA)[109] were exploited in order to distinguish samples according to the diverse origins. These approaches (whose algorithm can be found in the relevant literature) present a number of benefits from both a predictive and an interpretative point of view.[110] Furthermore, they are scale-invariant, and therefore allow the extraction of information from blocks of different nature without requiring block-scaling. The algorithms of the two approaches are similar; in both cases, the information is sequentially extracted from the data blocks after possible redundancies (among data matrices) are removed by an orthogonalization step.

As in all predictive models, the authentication of the saffron samples requires validation; consequently, samples were divided into two representative sets, one for training the classification models, and one (the test set)

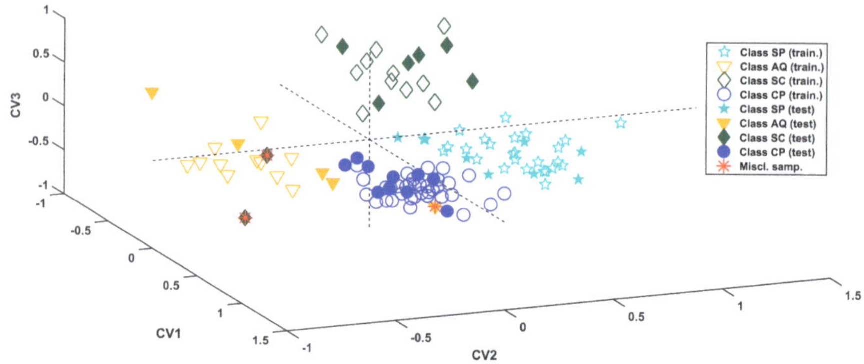

Figure 5.8 SO-PLS-LDA analysis. Samples projected onto the space spanned by canonical variates. Reproduced from ref. 95, https://doi.org/10.3390/molecules25102332, under the terms of the CC BY 4.0 license, https://creativecommons.org/licenses/by/4.0/.

for validating them (*i.e.*, to estimate the predictive capability of the models). The number of samples per class in the respective sets is shown in Figure 5.7. Both the aforementioned approaches were applied and provided notable results. From the predictive standpoint, the tested tools were demonstrated to be a suitable solution; in fact, SO-PLS-LDA and SO-CovSel-LDA provided 90% and 87% correct classification rates on the test set, respectively, corresponding to the misclassification of only three and four test samples. In Figure 5.8, samples are projected onto the space spanned by the canonical variates (CV) defined by the SO-PLS-LDA model. This graphical representation shows that the objects present a clear grouping tendency according to the origin. In fact, along CV1, samples belonging to class AQ (downward orange triangles), falling at negative/slightly positive scores, can be discriminated from those appertaining to class CP (blue dots) and SP (aquamarine stars) at positive values of CV1. On the other hand, the discrimination between CP and SP takes place along CV2. Eventually, samples belonging to class SC clearly discriminate from all the others along CV3, where they fall at positive values, presenting an opposite trend with respect to all the other individuals. The misclassified samples, highlighted in the plot by red asterisks, are two SC samples, clearly falling closer to the AQ group, and an AQ object which has been associated with the CP class.

Despite the similarity between the samples, the developed models allow an excellent distinction among saffron grown in the different zones (the predictive outcome of SO-CovSel-LDA is not discussed here, but it resembles the results provided by SO-PLS-LDA). The most accurate model, *i.e.*, SO-PLS-LDA, describes an expected situation. The fact that two Sicilian samples (class SC) are misclassified is reliable, as they come from a very large area, and therefore have greater internal variability, unlike the other objects, which are collected in small, delimited locations. The confusion of an AQ sample (predicted as belonging to class CP) is probably caused by the fact the two areas are not far from one another and present similar pedoclimatic conditions (the two locations have similar altitudes).

From the interpretative standpoint, SO-PLS-LDA and SO-CovSel-LDA grant a deep interpretation of the system under study. In fact, the former approach, being a PLS-based method, allows the calculation of variable importance in projection (VIP) indices,[111] which represents a suitable strategy in this context.[112] On the other hand, the latter classifier exploits a variable selection approach (covariance selection) for dimensionality reduction; consequently, it naturally provides indications about the features contributing the most to the solution of the classification problem. The inspection of the spectral variables ranked as the most meaningful by VIP and of those selected by covariance selection reveals both approaches selected features in the same spectral ranges. In Figure 5.9, is possible to observe which variables have been selected by covariance selection (highlighted by green squares and bars in the plots). The figure clearly shows one of the main interpretation advantages of SO-CovSel-LDA, *i.e.*, it provides a very parsimonious selection, reducing a lot the number of relevant variables.

In the left plot of Figure 5.9, the mean FT-IR spectrum is shown. Only two variables have been selected on this data block, 1057 cm^{-1} and 966 cm^{-1}. The former one belongs to the intense band ascribable to the stretching of C–O in C–O–C in glycosidic linkages of sugar rings, whereas the latter one can be associated with the bending (out of plane) of C–H groups. This could represent an indication that a different composition in terms of sugars can be observed among saffron from different origins. This is a valid hypothesis, considering the close link between the sugars' metabolism in plants and the climate/composition of the soil of the different areas. In the right subplot in Figure 5.9, the average UV-Vis spectrum is shown, and green squares and bars highlight the variables most meaningful for the solution of the classification problem. In this case, more features have been selected (12), indicating a richer contribution from this block. The investigation of the CovSel outcome reveals that crocins, picocrocins and flavonoids play a key role in the authentication of saffron. In fact, these compounds mainly absorb in the area between 400 nm and 500 nm. A further inspection of the spectra unveiled that the main differences among classes are not linked to the total concentration of these analytes, but to their relative concentrations.

In general, despite a small difference in predictive terms, both SO-PLS-LDA and SO-CovSel-LDA have proven to be able to extract relevant information from the spectral blocks resulting in efficient tools for the authentication of the spices under study, representing a green tool against intentional mislabelling.

5.4.1.2 *Authentication of Frozen-thawed Cephalopods – A Case Study*

Fish products are among the most defrauded food commodities[113] since they can be easily counterfeited and mislabelled,[114–117] especially regarding

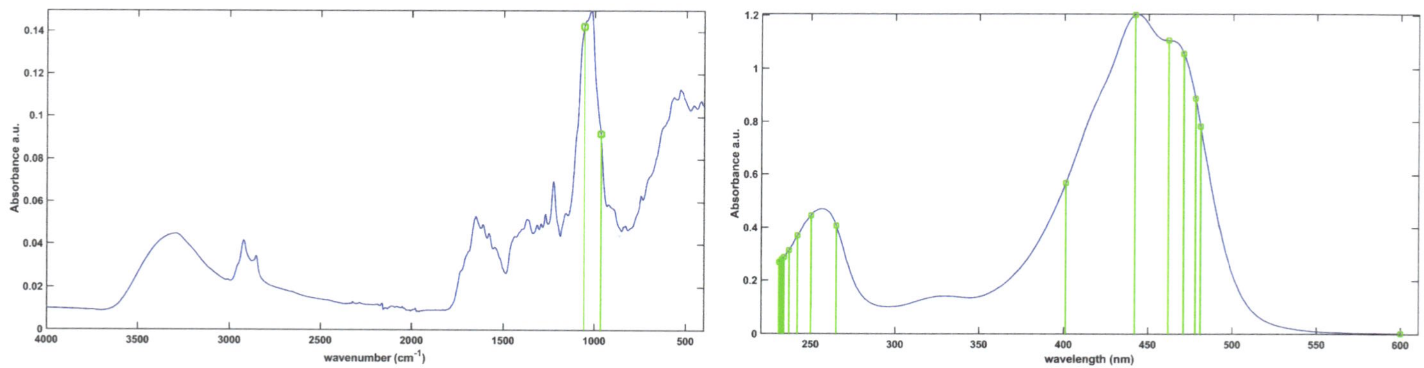

Figure 5.9 Variables selected by SO-CovSel-LDA. Left plot: FT-IR mean spectrum (blue line). Right plot: UV-Vis mean spectrum (blue line). Green squares and bars represent selected variables. Reproduced from ref. 95, https://doi.org/10.3390/molecules25102332, under the terms of the CC BY 4.0 license, https://creativecommons.org/licenses/by/4.0/.

the declaration of freshness, which in many cases can be faked by selling frozen-thawed products as fresh. From a legal point of view, this type of conduct infringes European law (European Parliament Legislative Resolution No. 1169/2011), which regulates how information about the defrosted status of products is presented to the final consumer, requiring clear labelling.

From the point of view of health safety, it has been proven that freezing-thawing cycles impair food wholesomeness by promoting the proliferation of bacteria and other microorganisms, causing significant risks for its consumption. In this context, different methods have been developed and investigated over the past years,[118] for instance histology,[119] protein extraction and proteomics,[120] evaluation of erythrocyte and haematocrit, measurement of dielectric properties, and enzymatic essays. All these methods are widely used and have proven their applicability and efficacy, but are generally rather time-consuming and labour-intensive, making them difficult to apply on an industrial scale due to the large amount of resources required for sample treatment, management, and obtaining the actual analytical results.

As an attempt to overcome these issues, Pennisi *et al.*[121] carried out a study regarding the identification of frozen-thawed cephalopods using a combination of near-infrared (NIR) spectroscopy and chemometrics. Even though cephalopods have extensive and diversified consumption all over the world and are therefore very exposed to fraud, only a few studies about the rapid discrimination of fresh/thawed cephalopods by using NIR spectroscopy[122] can be found in the literature.

The aim of this case study by Pennisi *et al.* was to set up a quick and easy-to-use method for distinguishing fresh from frozen-thawed cuttlefish (*Sepia officinalis*) and musky octopus (*Eledone* spp.). Three NIR instruments were used: a benchtop instrument (Multi Purpose Analyzer – MPA, by Bruker Optics, Ettlingen, Germany), a portable medium/high-cost device (MicroNIR Pro Es, by VIAVI Solutions) and a handheld, low-cost, and ultra-fast device (SCiO, by Consumer Physics). The inclusion of two portable instruments allowed quick analysis of large numbers of samples directly on site, while the benchtop instrument was used as a reference for the experimental acquisitions.

Fifty fresh cephalopod samples for each species (cuttlefish and musky octopus) were analysed at refrigeration temperature (2 ± 2 °C), then frozen at -20 °C for at least 10 days, to be finally thawed to room temperature (20 °C) and analysed again with the same three instruments. All acquired NIR datasets were processed and analysed with chemometric methods, from exploratory analysis with PCA to classification modelling using partial least squares-discriminant analysis (PLS-DA[78,123]). Regarding the classification step, first one model for each cephalopod species was built, and then a global model considering both species together was also computed.

PLS-DA was chosen for the rather simple and straightforward outputs that it provides, in the perspective of keeping the classification modelling approach as simple as possible, and also aimed at communicating and

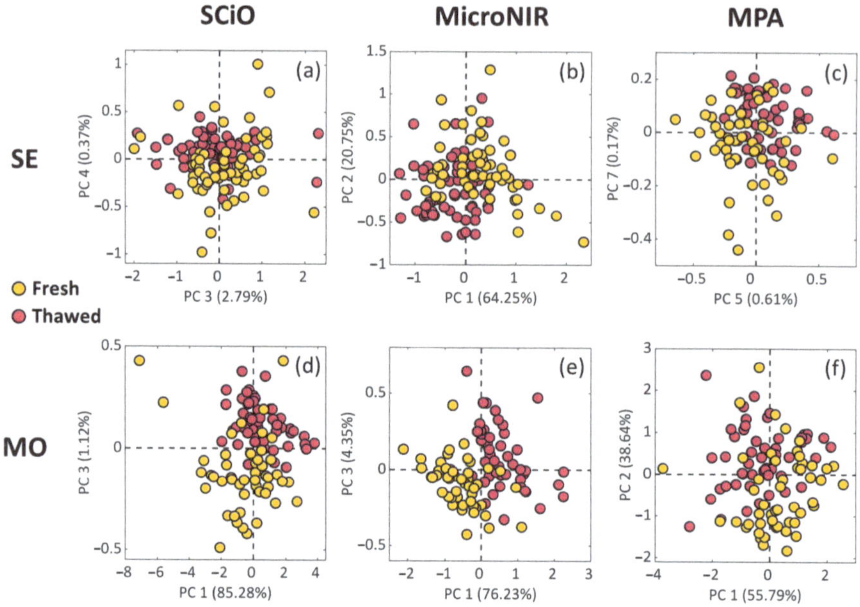

Figure 5.10 Most relevant PCA scores plots of cuttlefish (SE; a–c) samples and musky octopus (MO; d–f) samples. Colour coding as follows: fresh samples in yellow, frozen-thawed samples in red. Reproduced from ref. 109, https://doi.org/10.3390/foods10030528, under the terms of the CC BY 4.0 license, https://creativecommons.org/licenses/by/4.0/.

applying the results of the study in real-world environments (thus, together with the regulatory authorities). To validate the models, a calibration-test 70–30% split of each dataset was performed using the Duplex algorithm.[124] To ensure a correct and balanced sampling of the frozen and thawed groups of samples, the splitting procedure was applied to each class separately. From the interpretation point of view, variable importance in projection (VIP[111]) scores were used for identifying the most influential wavelengths for each experimental dataset (Figure 5.10), allowing for the interpretation based on classes of chemical compounds. All spectral datasets were pre-processed with the standard normal variate (SNV[125]) method and mean centring, once again in the perspective of "keeping it simple".

The cornerstone of any data analysis workflow is the exploratory step. In the study, PCA was used to gain knowledge about the information content of all spectral datasets, also with the aim of estimating the feasibility of subsequent classification models.

The most relevant PCA scores are reported in Figure 5.10, from which it is possible to see that all datasets contain some grouping information to different extents: interesting separation tendencies between the fresh (yellow) and frozen-thawed (orange) groups can be seen, especially in the case of musky octopus (Figure 5.10d–f). In the case of cuttlefish, even if the PCA

Table 5.3 Classification results fresh *vs.* frozen-thawed of the individual (cuttlefish – SE, musky octopus – MO) and of the global models. All values are percentages and those in bold are above 90%. Reproduced from ref. 109, https://doi.org/10.3390/foods10030528, under the terms of the CC BY 4.0 license, https://creativecommons.org/licenses/by/4.0/.

		SCiO					MicroNIR					MPA				
		LVs	Spec	Sens	NER	Acc	LVs	Spec	Sens	NER	Acc	LVs	Spec	Sens	NER	Acc
SE	Cal		**97.0**	87.9	**92.4**	**92.4**		**97.0**	**100**	**98.5**	**98.5**		**100**	**97.0**	**98.5**	**98.5**
	CV	9	84.8	69.7	77.3	77.3	6	**97.0**	**97.0**	**97.0**	**97.0**	8	87.9	81.8	84.8	84.8
	Test		70.6	**94.1**	82.3	82.3		**94.1**	**94.1**	**94.1**	**94.1**		82.3	88.2	85.3	85.3
MO	Cal		**97.0**	**97.0**	**97.0**	**97.0**		**100**	**100**	**100**	**100**		**97.0**	**93.9**	**95.4**	**95.4**
	CV	9	**97.0**	93.9	**95.4**	**95.4**	7	**100**	**100**	**100**	**100**	6	**93.9**	**90.9**	**92.4**	**92.4**
	Test		94.1	94.1	94.1	94.1		94.1	**100**	**97.1**	**97.1**		88.4	**94.1**	**91.2**	**91.2**
Global	Cal		**93.9**	75.8	84.8	84.8		**98.5**	**95.4**	**97.0**	**97.0**		**90.9**	84.8	87.9	87.9
	CV	6	**95.4**	71.2	83.3	83.3	5	**97.0**	**93.9**	**95.4**	**95.4**	5	87.9	81.8	84.8	84.8
	Test		85.3	88.2	86.8	86.8		**97.1**	**94.1**	**95.6**	**95.6**		**91.2**	**97.1**	**94.1**	**94.1**

representation of the separation tendency (Figure 5.10a–c) appears less clear, the classification results still turned out satisfactorily.

Regarding the classification step, a total of nine PLS-DA models were fitted, all providing good results, suggesting that the tested instrument and general approach combining chemometrics and spectroscopy constitute a suitable solution for the practical authentication problem at hand. In fact, the obtained accuracy values in prediction were: cuttlefish (SE) 82.3–94.1%, musky octopus (MO) 91.2–97.1%, and global model 86.8–95.6%. The classification performances figures[123] are reported in Table 5.3 from the point of view of the correct classification of thawed samples (the "positive" outcomes, in more general classification terminology):

- **Specificity** (Spec): the ability to avoid fresh samples being wrongly classified as thawed (false positives).
- **Sensitivity** (Sens): the ability to avoid thawed samples being wrongly classified as fresh (false negatives).
- **Non-error rate** (NER): the mean of the sensitivities (one for each class), the model's capability to correctly classify any sample.
- **Accuracy** (Acc): an estimation of the model's error computed as the sum of the true positives (TP, correctly classified thawed samples) and true negatives (TN, correctly classified fresh samples) divided by the total number of samples.

The best classification performances for the cuttlefish (SE) data were obtained with the MicroNIR spectrometer, with which 94.7% prediction accuracy was achieved. The other two instruments provided significantly lower but still good classification performances. Regarding the musky octopus (MO) data, also in this case the best classification performances were obtained with the MicroNIR spectrometer, with 97.1% prediction accuracy. The performances obtained with the other two instruments are in this case not much lower, and the handheld SCiO sensor provided the second-best result, with 94.1% accuracy in prediction.

The global models were built to test whether the discriminative information shared between the cuttlefish and musky octopus datasets was robust enough to allow for good fresh-thawed classification, even in the case of modelling different cephalopod species together. In the cases of the two portable instruments, the accuracy values in prediction of the global model resulted in an accuracy between the two individual classes' accuracies, suggesting that the better-performing species positively influenced the classification effort. The benchtop MPA instrument resulted in an exception since its prediction accuracy was higher than the two individual models. As for the individual cases, the MicroNIR instrument provided the best classification performance, with an accuracy of 95.6%. The MPA outcomes resulted were comparably good with 94.1% accuracy, while the SCiO achieved 86.8% accuracy.

Regarding the interpretation of the results in terms of original variables, the most influential wavelengths/wavenumbers are represented as VIP

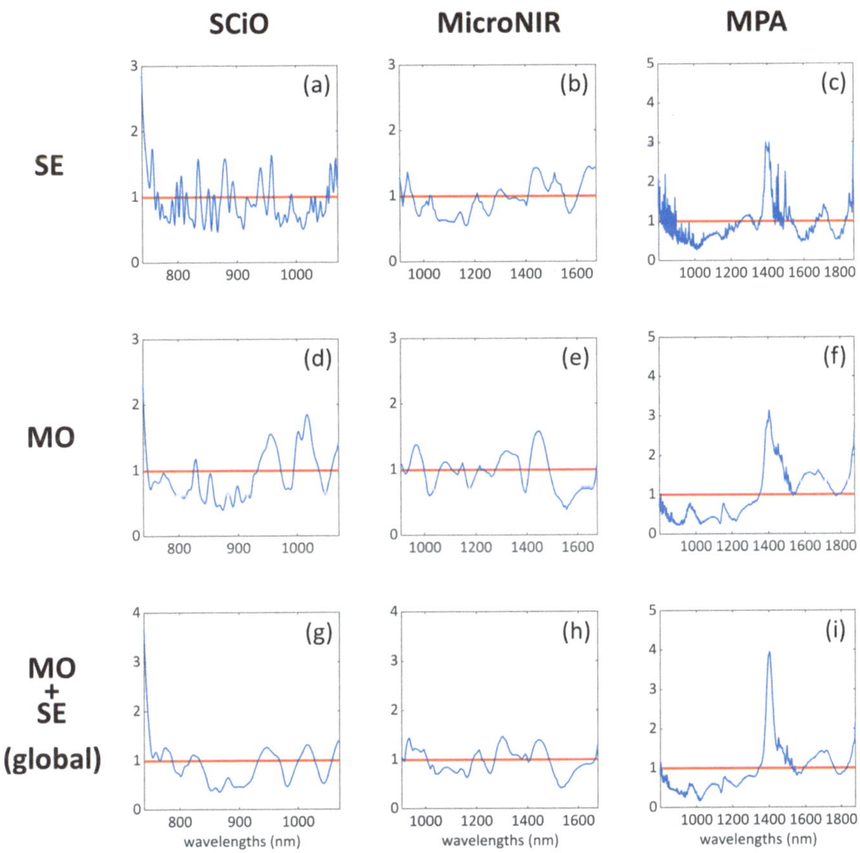

Figure 5.11 PLS-DA VIP scores of cuttlefish (SE, a–c), musky octopus (MO, d–f) and global (g–i) models. Reproduced from ref. 109, https://doi.org/10.3390/foods10030528, under the terms of the CC BY 4.0 license, https://creativecommons.org/licenses/by/4.0/.

scores in Figure 5.11. The VIP scores suggest that the most influential spectral variables for the global models are rather consistent with those selected for the individual SE and MO models.

The results were interpreted according to the phenomena occurring in the specimens' cells during the freezing-thawing process. For instance, protein denaturation affects the water interaction in terms of the ratio between free and bound water and this has been found to be higher in thawed samples than in fresh ones.[126] As a consequence, the 1400–1500 nm wavelength range is related to the combination of O–H stretching and bending bands[127] and can be essentially linked to an increase in free water species during thawing.[122,126,128] Regarding the protein content and degradation, proteolysis activity and relaxation of the lipid structure during thawing give rise to signals in the spectral region between 900 and 1400 nm, and these can be related to the vibrational modes of O–H and C–H bonds.[129,130]

Further chemical alteration of components such as lipids, proteins, amino and fatty acids can be seen by looking at the wavelength absorptions at 1600 nm, 1800 nm and 2200 nm, which correspond, in turn, to N–H first and second overtone, first overtone of the C–H aliphatic group stretching, and the combination of N–H and C=O bonds.[129]

In conclusion, in this case study, the two portable instruments (SCiO and MicroNIR) provided good results, thus providing a solid starting point for further implementations and developments of on-site applications for fraud detection. More specifically, the MicroNIR instrument was found to be the best solution among those investigated, suggesting that the chemical information gathered in the first part of the NIR range, and also in the range between 1000 and 1600 nm, is directly related to the phenomena happening in cephalopods during thawing. Despite the spectral width limitations of the SCiO instrument, its rather good results demonstrate how this simple device can be effectively used for the discrimination between fresh and thawed cephalopod samples, possibly taking advantage of useful chemical information in the last part of the visible range (Figure 5.11a, d and g, on the left part of the plots).

5.4.1.3 Adulteration Detection of Rind Content in PDO Parmigiano Reggiano Cheese – A Case Study

Parmigiano Reggiano (PR) is one of the most well-known, consumed and appreciated hard cheeses worldwide and gained PDO denomination in 2012 (EC 1151/2012). PDO-PR must meet specific technological, compositional and organoleptic requirements, as defined by the specifications of Parmigiano Reggiano Cheese protocol (https://www.parmigianoreggiano.com/consortium/rules_regulation_2/default.aspx last accessed on 20 May 2022). Given the commercial success of Parmigiano Reggiano, this product is particularly subject to counterfeiting, especially when it is sold in the form of grated cheese. Grated Parmigiano Reggiano must be obtained directly from an original Parmigiano Reggiano cheese wheel, and must be put in packaging immediately after grating, while still in the production area of origin. Moreover, additional parameters have to be met, like a minimum of 12 months of ripening, fat content no lower than 32% in proportion to dry matter, moisture between 25% and 35% and rind content not higher than 18% (w/w).

In this case study, Li Vigni *et al.*[131] evaluated the potential use of a handheld Raman device (which allows acquiring spectra directly through packaging) coupled with multivariate data analysis methods, with the aim of modelling the authenticity of Parmigiano cheese (great results were obtained, with sensitivity and specificity of 100% for the test set). Moreover, the authors built a PLS calibration model to determine the percentage of rind in grated cheese, obtaining a prediction error of around 5%, with 18% being the maximum content allowed by the official production protocol to be compliant with the protected denomination of origin.

The calibration model was developed by simulating the situation closest to a confirmation analysis protocol, *i.e.*, by acquiring the Raman spectra on intact packaged grated cheese samples.

To this aim a total of 80 Parmigiano Reggiano samples were analysed, and all collected Raman spectra were split into a calibration set (46 samples) and a test set (34 samples) using the Duplex algorithm. The resulting fitted PLS model was very parsimonious with only three latent variables, also showing consistent values of error in fit and prediction. An additional test set of initially unknown samples was also used for validation and when their rind content became available the authors could compute an error equal to 5.1%, consistent with the other test set samples.

In Figure 5.12a, the measured *vs.* predicted rind content is shown for all predicted samples. Even if a 5% of error in prediction is clearly not satisfactory (and can be explained by considering variability due to ripening, grinding, *etc.*), the model can quite well distinguish compliant from non-compliant samples. As a matter of fact, only six samples were wrongly predicted, three as compliant, and three as non-compliant.

Figure 5.12b shows the VIP scores of the PLS model for predicting the rind content, and three main spectral regions can be identified: the highlighted areas are coherent with the chemical composition of cheese and dairy products,[132–134] with signals ascribable to compounds like aliphatic amino acids, phosphate, lipids, amides, α-helix structures, fatty acids, carbohydrates, and aliphatic amino acids side chains. Regarding the possible contributions to the overall signal by the plastic packaging, the authors decided that, since all samples had the same bag and packaging material, a very reduced effect could be assumed. Further developments would of course be needed to consider different packaging materials.

In conclusion, even if the resulting prediction error of 5% is higher than models based on NIR (3.5%[135]) and NIR imaging (2–2.5%[136]), the practical advantage of measuring directly through packaging (*i.e.*, without opening it) and with no interference due to sample humidity (a strong interferent with NIR-based instruments) makes the application of Raman spectroscopy with a simple calibration method like PLS a very interesting and efficient tool for adulteration detection.

5.4.2 Geographical Traceability

European legislation (EC 178/2002) introduced food traceability,[137] defining it as the ability to trace and follow the history of food by means of recorded objective identification procedures, starting from every level of the agri-food chain.

Verifying geographical traceability, or denomination of origin in general, is a special case of authentication. The task is the same, *i.e.*, releasing a compliance or non-compliance notification; however, beyond that we may wish to evaluate whether the link with the territory of origin can be established in term of some characteristic parameters which are possible to

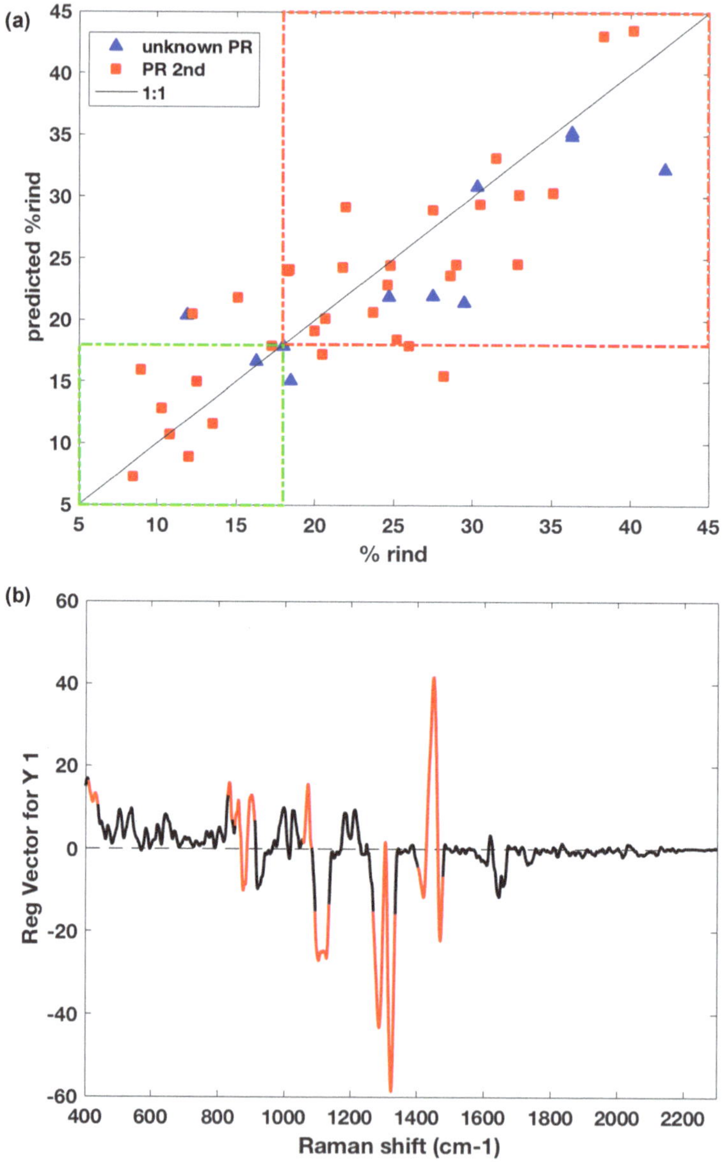

Figure 5.12 (a) PLS calibration model for percentage rind content based on the updated calibration set (PR first sampling + 6 PR second sampling). Only predicted samples are shown, measured *vs.* predicted rind values. PR second samples (test): red squares; unknown-PR: blue triangles. Samples falling in the green framed area are correctly seen as compliant and samples in the red frame are correctly seen as non-compliant. (b) PLS regression coefficients with region with variable importance in projection (VIP) scores higher than one highlighted in red. Reproduced from ref. 119, https://doi.org/10.3390/foods9111563, under the terms of the CC BY 4.0 license, https://creativecommons.org/licenses/by/4.0/.

measure in all the transformation steps from soil to foodstuff (mediated by feed and metabolism in the case of food of animal origin). In this case, these parameters, if they can be determined, are called primary or direct geographical indicators,[14,138] *i.e.*, elemental composition, radiogenic isotope ratio, and stable isotopc ratio can be directly linked to the same determinations in soil samples or referred to a particular geographical area. Of course, variables related to food composition/making procedure and any type of spectroscopic techniques can as well furnish a fingerprint that could be used to establish an identity profile for the studied foodstuff. These variables, called secondary indicators in this context, may allow building classification models and distinguishing authentic foodstuff from food products of different geographic origins, but cannot be directly linked to the production territory.

Another aspect that is worth noting is that the robustness of traceability models is strictly related to the extension of the region where the denomination insists. In fact, typical products, with protection indication labels referring to extensive geographical areas, require models developed on a "large scale", which should consider analytical parameters (indicators) and systematic knowledge of the investigated foodstuffs as well as of their territory of origin (mapping). Food, for which the disciplinary regulations involve "more restricted" areas and/or a specific brand, will have a more restricted sampling domain and bring a lower variability of the investigated indicators.

The case study presented here regards the characterization of a specific area, the Modena district in Italy, on which three different PDO Lambrusco wine denominations are produced. On the one hand, from a marketing point of view, there is the need to protect the Lambrusco from these three denominations from Lambrusco wine produced in regions different from the Modena district, as well as from counterfeiting and sounds-like products. On the other hand, from a valorisation perspective, producers are interested in differentiating the three PDO categories among them. Different objectives were thus pursued and will serve to highlight that different chemometrics tools are used to achieve each of them: (i) a geographical traceability model for the Lambrusco of Modena district based on direct geographical indicators, (ii) an indirect NMR spectroscopic fingerprint, and finally, (iii) classification models to distinguish the three PDOs.

5.4.2.1 *Sampling Strategy and Geographical Traceability Modelling by Direct Indicators*

The first step of the study concerned the sampling issue, which was solved by the application of a DoE methodology.[14,139] Figure 5.13 illustrates the strategy adopted to achieve representative soil sampling, to establish the link between territory and foodstuff with protected denomination of origin (PDO) when using direct geographical indicators, namely metal content and strontium isotopic ratio. Territory variability was accounted for by the

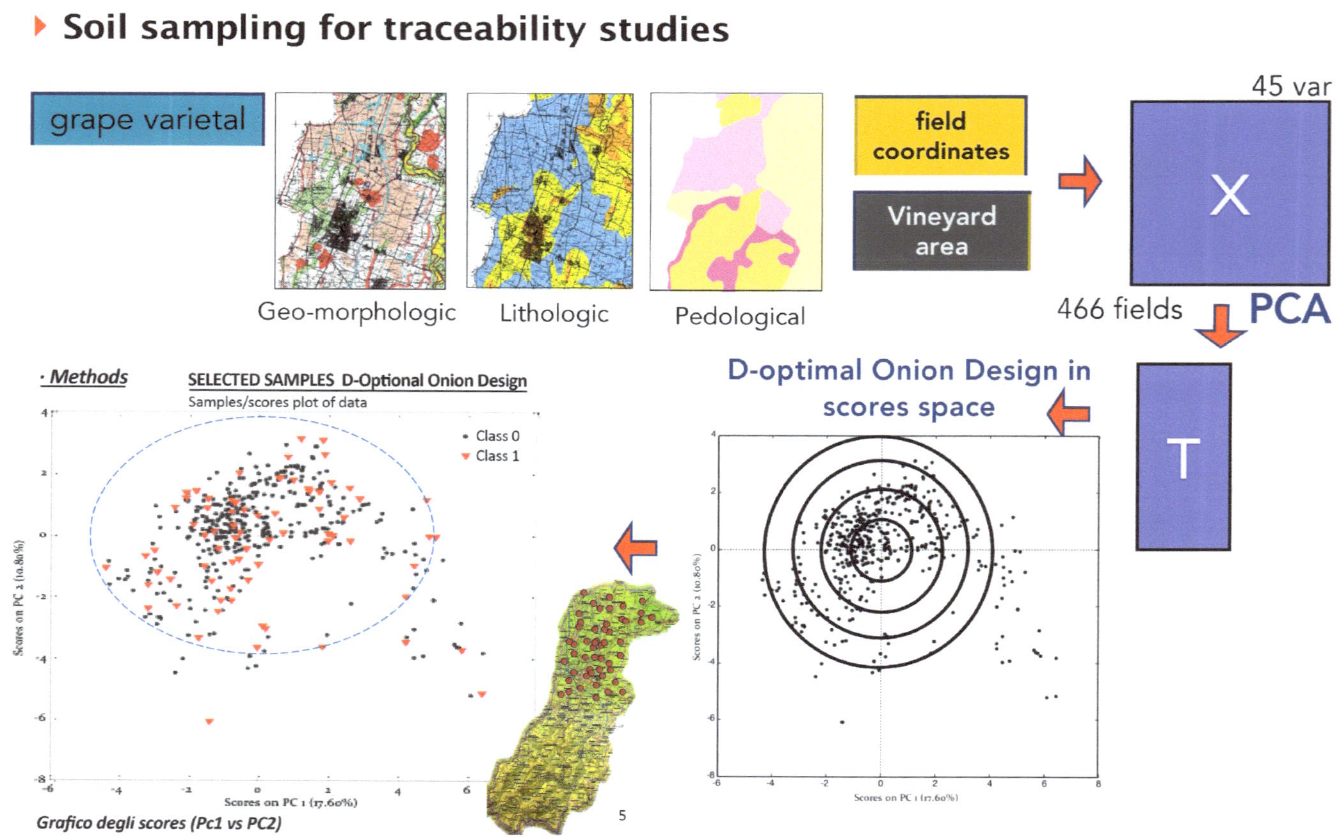

Figure 5.13 Field sampling planning in the geographical traceability study for Lambrusco wines of the Modena district. From top-right to bottom-left the selection data flow is shown: collected territorial and production data, numerical codification, PCA compression, d-optimal onion selection in scores space (selected fields red triangles).

geo-morphological, pedological, lithological maps, vineyard location, as well as with respect to productivity (estimated by the vineyard area cultivated with the specific grape variety compliant to the PDO). Since map information is qualitative (colour coding descriptive properties) a binary codification was adopted to translate this information into categorical variables, *e.g.*, the presence/absence of a given type of characteristics. In this way, a data table was obtained that can be explored by PCA to depict the variability of the territory of interest. Most importantly, compression in PC space aids the use of experimental design methodology, since PCs are orthogonal factors. In this case, an onion D-optimal design[140] was employed to ensure maximum coverage of the experimental domain (*i.e.*, the studied territory).

This strategy led to an affordable number of vineyard sites, where to sample soil, and then depth and field heterogeneity was studied in order to decide how many holes, and at which depths, are needed and how to proceed to assemble an operative sample. Overall, 100 soil samples were acquired and, by using the global position coordinates (GPS), vine branches in correspondence with the holes were also acquired. Regarding the Lambrusco wine samples, since the grapes are conferred to wineries that keep the PDO varieties distinct but not the vineyards, the sampling was done per variety respecting the numerosity of the production lots and ensuring the representativeness of the whole production area.

The elemental composition, the light stable isotopes and the strontium isotopic ratio (as reported in the labels in Figure 5.14) were determined on each soil, branch and wine sample.[141,142]

In Figure 5.14 (top) from left to right are shown the strontium isotopic ratio distribution maps for soils and branches, and the frequency histograms of the wine samples for each PDO, on the same colour scale. A good correspondence among the three matrices can be observed, even if the variability in the wines is reduced. As expected, the denomination *Grasparossa* extending from in-plane to hill areas shows the greatest range of values according to the soils and branches maps. The link with the territory is evident, thus the strontium isotopic ratio could be considered a good marker for geographical traceability; however, this parameter alone does not allow for a clear authenticity assessment. To this aim, a one-class class-modelling model was built with the SIMCA method whose results are shown in Figure 5.14 (bottom). The calibration set comprise all three PDOs samples (blue circles) and was validated with samples of the same three PDOs denominations (grey triangles), PDO Modena samples (grey circles), Lambrusco samples produced in Mantova and Reggio Emilia districts (green circles), which are close in the same Italian region, Spanish Lambrusco samples (red squares), and Lambrusco wines produced in a different Italian region, *i.e.*, Puglia (brown squares). The red curve represents the class boundary (Figure 5.14 bottom right) and the model could be considered satisfactory, both in terms of sensitivity and specificity, the most important indicators are assessed by looking at the discriminant power (Figure 5.14 bottom left).[79]

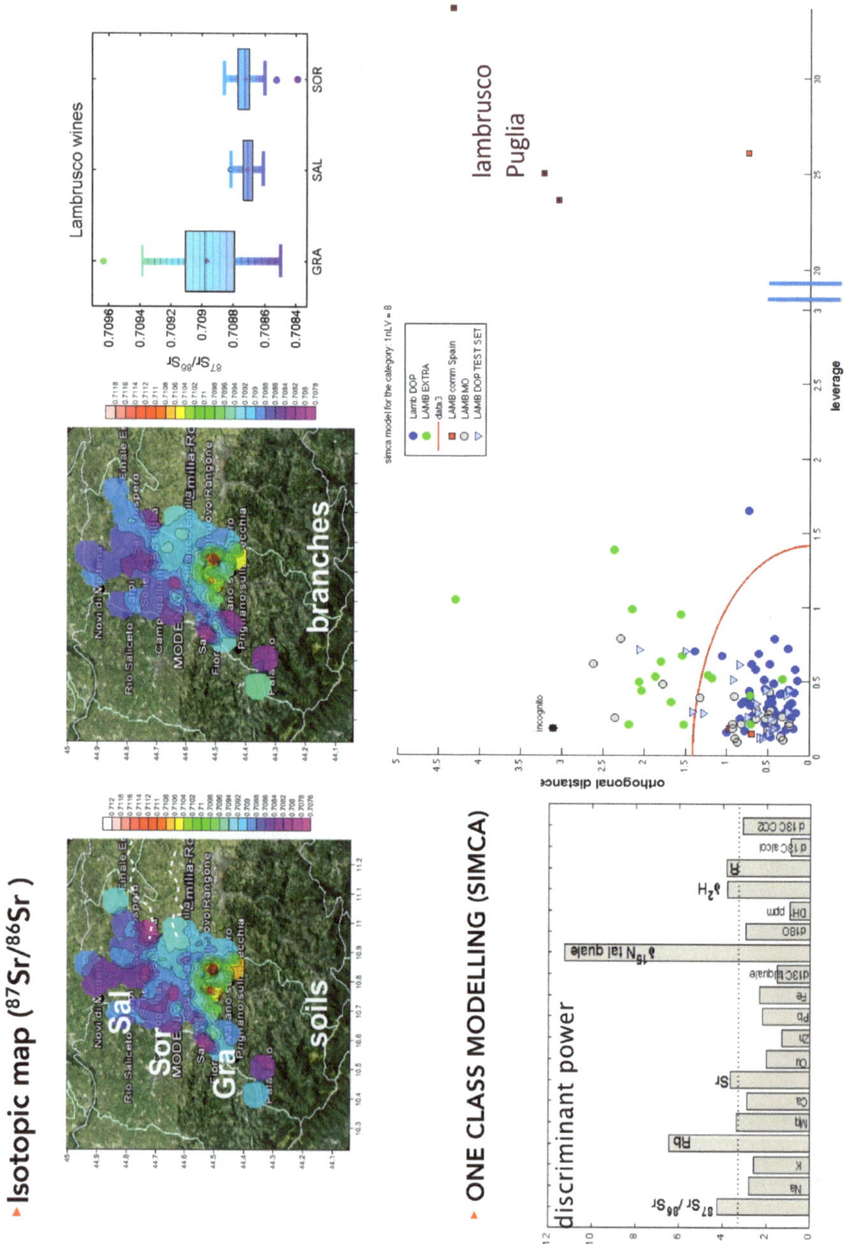
Isotopic map ($^{87}Sr/^{86}Sr$)
Sal
Sor
Gra
soils
branches
Lambrusco wines
GRA
SAL
SOR
ONE CLASS MODELLING (SIMCA)
discriminant power
orthogonal distance
leverage
lambrusco
Puglia
Lamb DOP
LAMB EXTRA
LAMB comm Spain
LAMB MO
LAMB DOP TEST SET

5.4.2.2 *Authentication Models by NMR (Indirect Origin Indicators)*

The Lambrusco wines with protected denomination of origin (PDO) were characterized by ^{1}H-NMR as the fingerprinting technique.[143] According to their regulation, there are three main PDO typologies, namely: *Lambrusco Grasparossa di Castelvetro* PDO, *Lambrusco di Sorbara* PDO, *Lambrusco Salamino di Santa Croce* PDO, and in addition a *Lambrusco di Modena* PDO (referred to as MO) denomination that does not correspond to a specific grape but is a mixture of the other three denominations. Notwithstanding the three different denominations, they are not forced to be made from 100% of the grape corresponding to the name but could contain up to 15% of other specified grapes.

As previously pointed out, it is of interest both to distinguish the three main denominations *Grasparossa*, *Salamino* and *Sorbara*, which are located in different areas of the Modena district and are differently appreciated and valued, and to have a general model for geographical traceability of *Lambrusco* wine of the Modena district (to fight counterfeiting frauds).

The first objective is better pursued by discrimination (PLS-DA in this case); all three categories are known and can be representatively sampled. The second is a typical case of distinguishing one category (*Lambrusco* di Modena) from the rest of the world, hence class modelling has to be used (in this application SIMCA was used).

The results are described in detail in ref. 9 and 143 and summarized in Figure 5.15.

The data were split into a calibration and a validation set, albeit modest, and centred prior to PLS-DA modelling, while spectra were aligned and block-scaled. The model dimensionality, four latent variables, was established according to minimum classification error in cross-validation, as shown in the top right plot in Figure 5.15, while the bottom right plot reports the PLS-DA predicted y-values *vs.* the number of samples. Each predicted y (a category) is shown as a line plot in different colours and symbols; vertical solid lines separate the samples belonging to the different classes. Validation samples are shown in paler colours. Samples are assigned to the class for which the predicted y-value is the highest. The number of misclassified

Figure 5.14 Geographical traceability model based on direct indicators for Lambrusco wines of the Modena district. Top: $^{87}Sr/^{86}Sr$ strontium isotopic ratio map in colour coded scale for soils (left) and branches (middle) and $^{87}Sr/^{86}Sr$ boxplots for the Grasparossa, Salamino and Sorbara PDOs wine samples, respectively (right). Bottom: SIMCA results using $^{87}Sr/^{86}Sr$, content of Na, K, Rb, Ca, Sr, Zn, Pb, Fe, and light element stable isotopes (δ^1H, δ^2H, $\delta^{18}O$, $\delta^{15}N$, $\delta^{13}C$, $\delta^{13}C(CO_2)$ as variables. On the right, in the SIMCA reduced distance plot, the red curve is the class boundary, symbols according to legend, on the left the SIMCA discriminant power variables bar plot, the dashed horizontal line corresponds to the 95% percentile value.

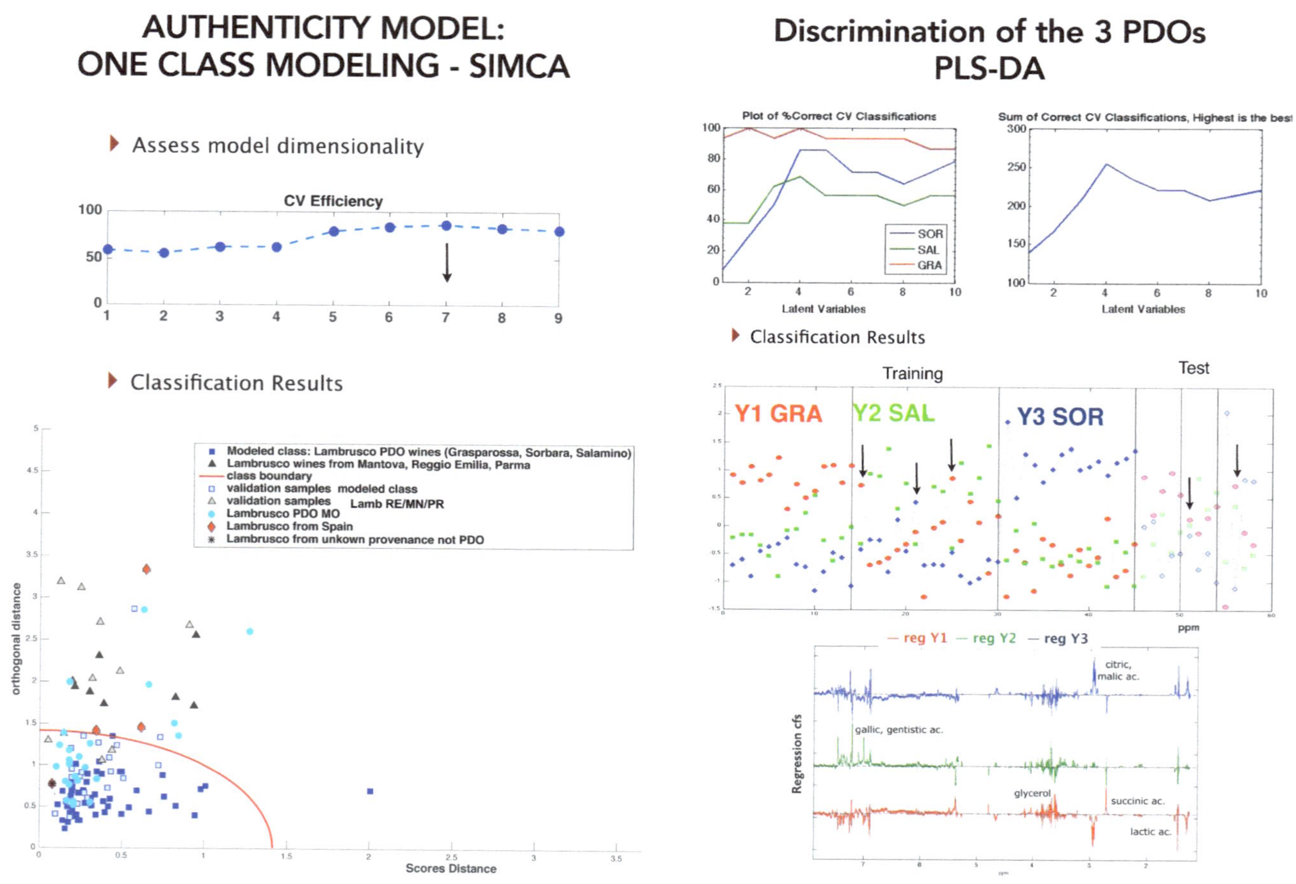
AUTHENTICITY MODEL:
ONE CLASS MODELING - SIMCA
Assess model dimensionality
CV Efficiency
Classification Results
Modeled class: Lambrusco PDO wines (Grasparossa, Sorbara, Salamino)
Lambrusco wines from Mantova, Reggio Emilia, Parma
class boundary
validation samples modeled class
validation samples Lamb RE/MN/PR
Lambrusco PDO MO
Lambrusco from Spain
Lambrusco from unkown provenance not PDO
orthogonal distance
Scores Distance
Discrimination of the 3 PDOs
PLS-DA
Plot of %Correct CV Classifications
SOR
SAL
GRA
Latent Variables
Sum of Correct CV Classifications, Highest is the best
Latent Variables
Classification Results
Training
Test
Y1 GRA
Y2 SAL
Y3 SOR
ppm
— reg Y1 — reg Y2 — reg Y3
Regression cfs
citric, malic ac.
gallic, gentistic ac.
glycerol
succinic ac.
lactic ac.

samples in fit, cross-validation, and validation is similar; the higher errors are for *Salamino* samples. The regression coefficients plot (bottom right plot in Figure 5.15) shows which ones are the main compounds responsible for the discrimination.

The authenticity model for *Lambrusco* PDO wines of the Modena district was developed considering calibration set samples of the three main denominations, *Grasparossa*, *Salamino*, and *Sorbara*, representative of two years of production, setting apart some samples as a validation set. The same pre-processing as for PLS-DA was applied.

The number of components was chosen according to the best efficiency in cross-validation (Figure 5.15, top left plot), and to this aim nine samples of Lambrusco wines of different districts, but close to Modena and sharing similarity with respect to soil geomorphology and climate, were used. This means that these samples were not modelled as a class but were projected onto the PCA space of the modelled class and then class specificity was assessed. The classification results are shown in the bottom left plot in Figure 5.15, the red semicircle is the class acceptance boundary. To validate the model in addition to the validation set belonging to the modelled class (light blue square symbol), the same samples as in the case of the PLS-DA model above were used (the symbols are described in the figure legend). Class sensitivity resulted in good, *i.e.*, 91% for the validation set (only two samples refused), and overall specificity resulted in acceptable, considering that of the five samples that were wrongly accepted, four originated from close districts. All Spanish samples were correctly rejected.

5.5 Conclusion

Food forensics encompasses verification of authenticity, adulteration detection, and traceability issues. Assessing food authenticity is a rather complex task that can be achieved only by efficient integration of analytical and chemometric methodologies. These are established and mature methodologies both from the analytical and the data analysis points of view. The literature is rich in successful case studies, nonetheless, much work is still needed to transfer these results to routine analysis and regulatory practice.

Figure 5.15 Left: Geographical traceability model based on ^{1}H-NMR data, indirect indicators, for Lambrusco wines of the Modena district. At the top is the efficiency in cross validation (CV) *vs.* number of components plot. At the bottom, in the reduced distance SIMCA plot, the red curve indicates class boundary. Right: Authenticity model based on ^{1}H-NMR data for the three Lambrusco wine PDOs of the Modena district. At the top, classification error *vs.* number of PLS-DA components for the single categories (left) and the average (right). At the bottom are the predicted *y*-values for each category (colour according to the label in the plot) *vs.* the number of samples. The arrows indicate the misclassified samples. The PLS-DA regression coefficients plot is also reported.

In our opinion, there is still an educational issue, *i.e.*, the knowledge and practice of chemometrics are not consolidated and widespread enough. In addition, more efforts are required at least in two main directions, such as establishing reference guidelines for each subfield and objective, and the diffusion of best data analysis practices. The main bottlenecks are appropriate sampling strategies and correct model validation, including long-term model assessment (system challenge). Overcoming these issues may also lead to an increase in the level of method standardization and confidence.

In this chapter, we gave an overview of the most critical issues faced in food forensics and of the suitability and correct use of the main chemometric methods involved in such analyses. The methods themselves were not presented in detail, for which the reader is referred to the literature and the other chapters in this book. A few representative and challenging benchmark applications (Section 5.4) were discussed, aiming at pointing out the most critical issues, such as the proper setting of experimental plans/data acquisition, the importance of the data preprocessing step, the underestimation of the need for and importance of exploratory data analysis as a necessary first/foundational step, and, again, the validation of all the steps in a data analysis workflow.

References

1. RASFF – The Rapid Alert System for Food and Feed – Annual Report 2020, http://ec.europa.eu/food/safety/rasff/index_en.htm (accessed 17 May 2022).
2. Food_fraud_newsletter_04-2022, https://knowledge4policy.ec.europa.eu/publication/food-fraud-summary-april-2022_en (accessed 17 May 2022).
3. Food Integrity – Assuring quality and authenticity in the food chain. Funded by the European Union's Seventh Framework Programme for research, technological development and demonstration under grant agreement no. 613688, 2014–2018, https://foodintegrity.fera.co.uk/ (accessed 17 May 2022).
4. Authent-Net project, H2020 coordination and support action. Grant agreement no. 696371, http://www.authent-net.eu/ (accessed 17 May 2022).
5. S. Saadat, H. Pandya, A. Dey and D. Rawtani, Food forensics: Techniques for authenticity determination of food products, *Forensic Sci. Int.*, 2022, **333**, 111243.
6. M. Aceto, in *Comprehensive Analytical Chemistry*, ed. Y. Picó, Elsevier, 2015, vol. 68, pp. 441–514.
7. *Food Integrity Handbook*, ed. M. Lees and J. M. Morin, Eurofins Analytics France, Nantes Cedex 3, 2018.
8. M. Bevilacqua, R. Bucci, A. D. Magrì, A. L. Magrì and R. Nescatelli, Chemometrics in Food Chemistry, *Data Handl. Sci. Technol.*, 2013, vol. 28, pp. 171–233.

9. M. Cocchi, in *Encyclopedia of Analytical Chemistry*, ed. R. A. Meyers, John Wiley & Sons, Ltd, 2017.
10. E. Salvatore, M. Bevilacqua, R. Bro, F. Marini and M. Cocchi, in *Comprehensive Analytical Chemistry*, ed. A. D. L. G. Miguel and A. Gonzálvez, Elsevier B.V., 2013, vol. 60, pp. 339–382.
11. M. Bevilacqua, R. Bro, F. Marini, Å. Rinnan, M. Arendt and T. Skov, Trends in Analytical Chemistry Recent chemometrics advances for foodomics, *Trends Anal. Chem.*, 2017, **96**, 42–51.
12. S. Sentellas and J. Saurina, in *Chromatographic And Related Separation Techniques In Food Integrity And Authenticity*, ed. O. Núñez and G. Campmajó, World Scientific Publishing Co., 2021, pp. 167–200.
13. D. Caballero, R. Rios-Reina and J. M. Amigo, in *Comprehensive Foodomics*, ed. A. Cifuentes, Elsevier, 2020, pp. 387–406.
14. L. Bertacchini, M. Cocchi, M. Li Vigni, A. Marchetti, E. Salvatore, S. Sighinolfi, M. Silvestri and C. Durante, *The Impact of Chemometrics on Food Traceability*, Copyright © 2013 Elsevier B.V. All rights reserved, 1st edn, 2013, vol. 28.
15. P. Oliveri and R. Simonetti, in *Advances in Food Authenticity Testing*, ed. G. Downey, Woodhead Publishing, Woodhead P., 2016, pp. 701–728.
16. GFSI, *Tackling Food Fraud Through Food Safety*, 2018.
17. GFSI, *GFSI Position on Mitigating the Public Health Risk of Food Fraud*, 2014.
18. Authenticity and fraud in the feed and food chain – Concepts, terms, and definitions, European Committee for Standardization, 2019, pp. 1–10.
19. Codex Alimentarius (2018), – Discussion paper on food integrity and food authenticity – Joint FAO/WHO food standards programme, Codex committee on food import and export inspection and certification systems, Twenty – Fourth Session, Brisbane, Australia, 22, 2018.
20. J. Donarski, F. Camin, C. Fauhl-Hassek, R. Posey and M. Sudnik, Sampling guidelines for building and curating food authenticity databases, *Trends Food Sci. Technol.*, 2019, **90**, 187–193.
21. C. Paoletti and K. H. Esbensen, Distributional assumptions in food and feed commodities – Development of fit-for-purpose sampling protocols, *J. AOAC Int.*, 2015, **98**, 295–300.
22. K. H. Esbensen and C. A. Ramsey, QC of sampling processes – A first overview: From field to test portion, *J. AOAC Int.*, 2015, **98**, 282–287.
23. S. Totaro, P. Coratza, C. Durante, G. Foca, M. Li Vigni, A. Marchetti, M. Marchetti and M. Cocchi, Soil sampling planning in traceability studies by means of Experimental Design approaches, *Chemom. Intell. Lab. Syst.*, 2013, **124**, 14–20.
24. K. H. Esbensen and L. P. Julius, in *Comprehensive Chemometrics*, ed. S. D. Brown, R. Tauler and B. Walczak, Elsevier B.V., 1st edn, 2009, pp. 1–20.
25. R. Leardi, Experimental design in chemistry: A tutorial, *Anal. Chim. Acta*, 2009, **652**, 161–172.
26. A. Rust, F. Marini, M. Allsopp, P. J. Williams and M. Manley, Application of ANOVA-simultaneous component analysis to quantify

and characterise effects of age, temperature, syrup adulteration and irradiation on near-infrared (NIR) spectral data of honey, *Spectrochim. Acta, Part A*, 2021, **253**, 119546.
27. F. Marini, D. de Beer, E. Joubert and B. Walczak, Analysis of variance of designed chromatographic data sets: The analysis of variance-target projection approach, *J. Chromatogr. A*, 2015, **1405**, 94–102.
28. A. D'Alessandro, D. Ballestrieri, L. Strani, M. Cocchi and C. Durante, Characterization of Basil Volatile Fraction and Study of Its Agronomic Variation by ASCA, *Molecules*, 2021, **26**(13), 3842.
29. M. Li Vigni, C. Durante and M. Cocchi, *Exploratory Data Analysis*, 2013, vol. 28.
30. R. Bro and A. K. Smilde, Principal component analysis, *Anal. Methods*, 2014, **6**, 2812.
31. S. Wold, M. Sjöström and L. Eriksson, PLS-regression: A basic tool of chemometrics, *Chemom. Intell. Lab. Syst.*, 2001, **58**, 109–130.
32. C. M. Andersen and R. Bro, Variable selection in regression-a tutorial, *J. Chemom.*, 2010, **24**, 728–737.
33. A. de Juan and R. Tauler, Multivariate Curve Resolution (MCR) from 2000: Progress in Concepts and Applications, *Crit. Rev. Anal. Chem.*, 2006, **36**, 163–176.
34. R. Bro, PARAFAC. Tutorial and applications, *Chemom. Intell. Lab. Syst.*, 1997, **38**, 149–171.
35. M. Li Vigni and M. Cocchi, in *Resolving Spectral Mixtures – With Applications from Ultrafast Time-Resolved Spectroscopy to Super-Resolution Imaging*, ed. C. Ruckebusch, Elsevier, 2016, pp. 409–451.
36. Y. V. Zontov, O. Y. Rodionova, S. V. Kucheryavskiy and A. L. Pomerantsev, DD-SIMCA – A MATLAB GUI tool for data driven SIMCA approach, *Chemom. Intell. Lab. Syst.*, 2017, **167**, 23–28.
37. R. Vitale, F. Marini and C. Ruckebusch, SIMCA Modeling for Overlapping Classes: Fixed or Optimized Decision Threshold?, *Anal. Chem.*, 2018, **90**, 10738–10747.
38. Z. Małyjurek and B. Walczak, The scope of applicability of the selected class-modelling methods, *Chemom. Intell. Lab. Syst.*, 2021, **218**, 104427.
39. F. Marini, A. L. Magrì and R. Bucci, Multilayer feed-forward artificial neural networks for class modeling, *Chemom. Intell. Lab. Syst.*, 2007, **88**, 118–124.
40. L. Xu, S.-M. Yan, C.-B. Cai and X.-P. Yu, One-class partial least squares (OCPLS) classifier, *Chemom. Intell. Lab. Syst.*, 2013, **126**, 1–5.
41. L. Breiman, J. H. Friedman, R. A. Olshen and C. J. Stone, *Classification and Regression Trees*, Wadsworth, Pacific Grove, California, 1984.
42. J. B. O. Mitchell, Machine learning methods in chemoinformatics, *Wiley Interdiscip. Rev.: Comput. Mol. Sci.*, 2014, **4**, 468–480.
43. F. Marini, Artificial neural networks in foodstuff analyses: Trends and perspectives A review, *Anal. Chim. Acta*, 2009, **635**, 121–131.
44. L. G. Johnsen, M. Amigo, T. Skov and R. Bro, Automated resolution of overlapping peaks in chromatographic data, *J. Chemom.*, 2014, 71–82.

45. F. Westad and M. Kermit, in *Comprehensive Chemometrics*, Elsevier, 2009, pp. 227–248.
46. A. G. E. K. Smilde, Comments on multilinear PLS, *J. Chemom.*, 1997, **11**, 367–377.
47. K. H. Liland, T. Næs and U. G. Indahl, ROSA—a fast extension of partial least squares regression for multiblock data analysis, *J. Chemom.*, 2016, **30**, 651–662.
48. E. Gorrochategui, J. Jaumot, S. Lacorte and R. Tauler, Data analysis strategies for targeted and untargeted LC–MS metabolomic studies: Overview and workflow, *TrAC, Trends Anal. Chem.*, 2016, **82**, 425–442.
49. R. Tauler, E. Gorrochategui, J. Jaumot and R. Tauler, A protocol for LC–MS metabolomic data processing using chemometric tools, *Protoc. Exch.*, 2015, 1–41.
50. G. Baccolo, B. Quintanilla-Casas, S. Vichi, D. Augustijn and R. Bro, From untargeted chemical profiling to peak tables – A fully automated AI driven approach to untargeted GC–MS, *TrAC, Trends Anal. Chem.*, 2021, **145**, 116451.
51. L. G. Johnsen, P. B. Skou, B. Khakimov and R. Bro, Gas chromatography–mass spectrometry data processing made easy, *J. Chromatogr. A*, 2017, **1503**, 57–64.
52. M. Forina, P. Oliveri, S. Lanteri and M. Casale, Class-modeling techniques, classic and new, for old and new problems, *Chemom. Intell. Lab. Syst.*, 2008, **93**, 132–148.
53. O. Ye, A. V Titova and A. L. Pomerantsev, Trends in Analytical Chemistry Discriminant analysis is an inappropriate method of authentication, *Trends Anal. Chem.*, 2016, **78**, 17–22.
54. F. Westad and F. Marini, *Anal. Chim. Acta*, 2015, **893**, 14–24.
55. J. Riedl, S. Esslinger and C. Fauhl-Hassek, Review of validation and reporting of non-targeted fingerprinting approaches for food authentication, *Anal. Chim. Acta*, 2015, **885**, 17–32.
56. P. Filzmoser and B. Walczak, What can go wrong at the data normalization step for identification of biomarkers?, *J. Chromatogr. A*, 2014, **1362**, 194–205.
57. J. S. Urban Hjort, *Computer Intensive Statistical Methods: Validation, Model Selection and Bootstrap*, CRC Press, London, UK, 1994.
58. P. Mishra, R. Nikzad-Langerodi, F. Marini, J. M. Roger, A. Biancolillo, D. N. Rutledge and S. Lohumi, Are standard sample measurements still needed to transfer multivariate calibration models between near-infrared spectrometers? The answer is not always, *TrAC, Trends Anal. Chem.*, 2021, **143**, 116331.
59. J. Workman Jr. and H. Mark, Calibration transfer chemometrics, Part II: A review of the subject, *Spectroscopy*, 2018, **33**, 22–26.
60. X. Wang, Y. Li, L. Chen and J. Zhou, Analytical Strategies for LC–MS-Based Untargeted and Targeted Metabolomics Approaches Reveal the Entomological Origins of Honey, *J. Agric. Food Chem.*, 2022, **70**, 1358–1366.

61. A. S. Kritikou, R. Aalizadeh, D. E. Damalas, I. V. Barla, C. Baessmann and N. S. Thomaidis, MALDI-TOF–MS integrated workflow for food authenticity investigations: An untargeted protein-based approach for rapid detection of PDO feta cheese adulteration, *Food Chem.*, 2022, **370**, 131057.
62. M. Tabbassum, F. Zeeshan and K. H. Low, Discrimination and Recognition of Bentong Ginger Based on Multi-elemental Fingerprints and Chemometrics, *Food Anal. Methods*, 2022, **15**, 637–646.
63. G. Xue, S. Su, P. Yan, J. Shang, J. Wang, C. Yan, J. Li, Q. Wang, Y. Du, L. Cao, L. Cao and H. Xu, Quality control of Zingiberis Rhizoma and its processed products by UHPLC-Q-TOF/MS-based non-targeted metabonomics combining with SIBDV method, *Food Res. Int.*, 2022, **154**, 111021.
64. M. Kharbach, H. Yu, R. Kamal, I. Marmouzi, K. Alaoui, J. Vercammen, A. Bouklouze and Y. Vander Heyden, Authentication of extra virgin Argan oil by selected-ion flow-tube mass-spectrometry fingerprinting and chemometrics, *Food Chem.*, 2022, **383**, 132565.
65. M. Zhu, J. Sun, H. Zhao, F. Wu, X. Xue, L. Wu and W. Cao, Volatile compounds of five types of unifloral honey in Northwest China: Correlation with aroma and floral origin based on HS–SPME/GC–MS combined with chemometrics, *Food Chem.*, 2022, **384**, 132461.
66. A. Biancolillo, R. Aloia, L. Rossi and A. A. D'Archivio, Organosulfur volatile profiles in Italian red garlic (Allium Sativum L.) varieties investigated by HS–SPME/GC–MS and chemometrics, *Food Control*, 2022, **131**, 108477.
67. G. D. A. Mendes, M. A. L. de Oliveira, M. P. Rodarte, V. de Carvalho dos Anjos and M. J. V. Bell, Origin geographical classification of green coffee beans (Coffea arabica L.) produced in different regions of the Minas Gerais state by FT-MIR and chemometric, *Curr. Res. Food Sci.*, 2022, **5**, 298–305.
68. P. Cazón, D. Cazón, M. Vázquez and E. Guerra-Rodriguez, Rapid authentication and composition determination of cellulose films by UV-VIS-NIR spectroscopy, *Food Packag. Shelf Life*, 2022, **31**, 100791.
69. I. Batubara, S. Husna, M. Rafi, T. Sumaryada, S. Uchiyama, B. Juliandi, S. P. Putri and E. Fukusaki, A Combination of UV-Vis Spectroscopy and Chemometrics for Detection of Sappanwood (Caesalpinia sappan) Adulteration from Three Dyes, *Sains Malays.*, 2022, **51**, 775–781.
70. M. L. da Silva Medeiros, J. P. Cruz-Tirado, A. F. Lima, J. M. de Souza Netto, A. P. B. Ribeiro, D. Bassegio, H. T. Godoy and D. F. Barbin, Assessment oil composition and species discrimination of Brassicas seeds based on hyperspectral imaging and portable near infrared (NIR) spectroscopy tools and chemometrics, *J. Food Compos. Anal.*, 2022, **107**, 104403.
71. J. Florián-Huamán, J. P. Cruz-Tirado, D. Fernandes Barbin and R. Siche, Detection of nutshells in cumin powder using NIR hyperspectral imaging and chemometrics tools, *J. Food Compos. Anal.*, 2022, **108**, 104407.

72. R. Gunarathne, N. Marikkar, C. Yalegama and E. Mendis, FTIR spectral analysis combined with chemometrics in evaluation of composite mixtures of coconut testa flour and wheat flour, *J. Food Meas. Charact.*, 2022, **16**, 1796–1806.
73. A. M. Jiménez-Carvelo, A. Arroyo-Cerezo, S. Bikrani, W. Jia, A. Koidis and L. Cuadros-Rodríguez, Rapid and non-destructive spatially offset Raman spectroscopic analysis of packaged margarines and fat-spread products, *Microchem. J.*, 2022, **178**, 107378.
74. Q. Ye and X. Meng, Highly efficient authentication of edible oils by FTIR spectroscopy coupled with chemometrics, *Food Chem.*, 2022, **385**, 132661.
75. A. Biancolillo, M. Foschi, M. Di Micco, F. Di Donato and A. A. D'Archivio, ATR-FTIR-based rapid solution for the discrimination of lentils from different origins, with a special focus on PGI and Slow Food typical varieties, *Microchem. J.*, 2022, **178**, 107327.
76. R. A. Fisher, The Use of Multiple Measurements in Taxonomic Problems, *Ann. Eugen.*, 1936, **7**(2), 179–188.
77. M. Sjöström, S. Wold and B. Söderström, in *Pattern Recognition in Practice*, 1986.
78. M. Barker and W. Rayens, Partial least squares for discrimination, *J. Chemom.*, 2003, **17**, 166–173.
79. M. Wold and S. Sjöström, in *Chemometrics, Theory and Application*, ed. B. R. Kowalski, American Chemical Society, Washington, DC, 1977, pp. 243–282.
80. R. G. Brereton and G. R. Lloyd, Support Vector Machines for classification and regression, *Analyst*, 2010, 230–267.
81. S. Yeganeh-Zare, K. Farhadi and S. Amiri, Rapid detection of apple juice concentrate adulteration with date concentrate, fructose and glucose syrup using HPLC-RID incorporated with chemometric tools, *Food Chem.*, 2022, **370**, 131015.
82. H. Sun, W. Lu and B. Gao, Non-targeted detection of butter adulteration using pointwise UHPLC-ELSD and UHPLC-UV fingerprints with chemometrics, *Food Chem.*, 2021, **356**, 129604.
83. T. Shi, G. Wu, Q. Jin and X. Wang, Detection of camellia oil adulteration using chemometrics based on fatty acids GC fingerprints and phytosterols GC–MS fingerprints, *Food Chem.*, 2021, **352**, 129422.
84. A. Amirvaresi, M. Rashidi, M. Kamyar, M. Amirahmadi, B. Daraei and H. Parastar, Combining multivariate image analysis with high-performance thin-layer chromatography for development of a reliable tool for saffron authentication and adulteration detection, *J. Chromatogr. A*, 2020, **1628**, 461461.
85. W. L. Cheah and M. Fang, HPLC-Based Chemometric Analysis for Coffee Adulteration, *Foods*, 2020, **9**, 880.
86. M. Esteki, Z. Shahsavari and J. Simal-Gandara, Gas Chromatographic Fingerprinting Coupled to Chemometrics for Food Authentication, *Food Rev. Int.*, 2019, **36**, 384–427.

87. P. Morozzi, A. Zappi, F. Gottardi, M. Locatelli and D. Melucci, A Quick and Efficient Non-Targeted Screening Test for Saffron Authentication: Application of Chemometrics to Gas-Chromatographic Data, *Molecules*, 2019, **24**, 2602.
88. W. Lu, J. Liu, B. Gao, X. Lv and L. (Lucy) Yu, Technical note: Non-targeted detection of adulterated plant proteins in raw milk by UPLC-quadrupole time-of-flight mass spectrometric proteomics combined with chemometrics, *J. Dairy Sci.*, 2017, **100**, 6980–6986.
89. H. Jabeur, M. Drira, A. Rebai and M. Bouaziz, Putative Markers of Adulteration of Higher-Grade Olive Oil with Less Expensive Pomace Olive Oil Identified by Gas Chromatography Combined with Chemometrics, *J. Agric. Food Chem.*, 2017, **65**, 5375–5383.
90. R. M. Alonso-Salces, L. Á. Berrueta, B. Quintanilla-Casas, S. Vichi, A. Tres, M. I. Collado, C. Asensio-Regalado, G. E. Viacava, A. A. Poliero, E. Valli, A. Bendini, T. Gallina Toschi, J. M. Martínez-Rivas, W. Moreda and B. Gallo, Stepwise strategy based on ^{1}H-NMR fingerprinting in combination with chemometrics to determine the content of vegetable oils in olive oil mixtures, *Food Chem.*, 2022, **366**, 130588.
91. C. H. Yong, S. A. Muhammad, F. A. Aziz, F. I. Nasir, M. Z. Mustafa, B. Ibrahim, S. D. Kelly, A. Cannavan and E. K. Seow, Detecting adulteration of stingless bee honey using untargeted ^{1}H NMR metabolomics with chemometrics, *Food Chem.*, 2022, **368**, 130808.
92. L. Marchetti, F. Pellati, S. Benvenuti and D. Bertelli, Use of ^{1}H NMR to detect the percentage of pure fruit juices in blends, *Molecules*, 2019, **24**(14), 2592.
93. T. Shi, M. T. Zhu, Y. Chen, X. L. Yan, Q. Chen, X. L. Wu, J. Lin and M. Xie, ^{1}H NMR combined with chemometrics for the rapid detection of adulteration in camellia oils, *Food Chem.*, 2018, **242**, 308–315.
94. R. Dowlatabadi, F. Farshidfar, Z. Zare, M. Pirali, M. Rabiei, M. R. Khoshayand and H. J. Vogel, Detection of adulteration in Iranian saffron samples by ^{1}H NMR spectroscopy and multivariate data analysis techniques, *Metabolomics*, 2017, **13**, 1–11.
95. W. Zhu, X. Wang and L. Chen, Rapid detection of peanut oil adulteration using low-field nuclear magnetic resonance and chemometrics, *Food Chem.*, 2017, **216**, 268–274.
96. E. A. Petrakis, L. R. Cagliani, M. G. Polissiou and R. Consonni, Evaluation of saffron (Crocus sativus L.) adulteration with plant adulterants by ^{1}H NMR metabolite fingerprinting, *Food Chem.*, 2015, **173**, 890–896.
97. Y. B. Monakhova, T. Kuballa and D. W. Lachenmeier, *J. Anal. Chem.*, 2013, **68**, 755–766.
98. C. Cavdaroglu and B. Ozen, Detection of vinegar adulteration with spirit vinegar and acetic acid using UV-visible and Fourier transform infrared spectroscopy, *Food Chem.*, 2022, **379**, 132150.
99. R. Khodabakhshian, M. R. Bayati and B. Emadi, Adulteration detection of Sudan Red and metanil yellow in turmeric powder by NIR

spectroscopy and chemometrics: The role of preprocessing methods in analysis, *Vib. Spectrosc.*, 2022, **120**, 103372.
100. A. Antonio D'archivio, A. Biancolillo, C. Kappacher, B. Trübenbacher, K. Losso, M. Rainer, G. K. Bonn and C. W. Huck, Portable *vs.* Benchtop NIR-Sensor Technology for Classification and Quality Evaluation of Black Truffle, *Molecules*, 2022, **27**, 589.
101. H. Wang, Y. Xin, H. Ma, P. Fang, C. Li, X. Wan, Z. He, J. Jia and Z. Ling, Rapid detection of Chinese-specific peony seed oil by using confocal Raman spectroscopy and chemometrics, *Food Chem.*, 2021, **362**, 130041.
102. M. Varnasseri, H. Muhamadali, Y. Xu, P. I. C. Richardson, N. Byrd, D. I. Ellis, P. Matousek and R. Goodacre, Portable through Bottle SORS for the Authentication of Extra Virgin Olive Oil, *Appl. Sci.*, 2021, **11**, 8347.
103. J. L. Z. Zaukuu, Z. Gillay and Z. Kovacs, Standardized Extraction Techniques for Meat Analysis with the Electronic Tongue: A Case Study of Poultry and Red Meat Adulteration, *Sensors*, 2021, **21**, 481.
104. X. Hong and J. Wang, Use of Electronic Nose and Tongue to Track Freshness of Cherry Tomatoes Squeezed for Juice Consumption: Comparison of Different Sensor Fusion Approaches, *Food Bioprocess Technol.*, 2015, **8**, 158–170.
105. D. C. Antônio, D. C. S. de Assis, B. G. Botelho and M. M. Sena, Detection of adulterations in a valuable Brazilian honey by using spectrofluorimetry and multiway classification, *Food Chem.*, 2022, **370**, 131064.
106. M. Özbay, F. N. Arslan and G. Görür, Qualitative and Quantitative Detection of Monofloral, Polyfloral, and Honeydew Honeys Adulteration by Employing Mid-Infrared Spectroscopy and Chemometrics, *Food Anal. Methods*, 2022, 1–16.
107. A. Biancolillo, M. Foschi and A. A. D'Archivio, Geographical Classification of Italian Saffron (*Crocus sativus* L.) by Multi-Block Treatments of UV-Vis and IR Spectroscopic Data, *Molecules*, 2020, **25**, 2332.
108. A. Biancolillo, I. MÅge and T. Næs, Combining SO-PLS and linear discriminant analysis for multi-block classification, *Chemom. Intell. Lab. Syst.*, 2015, **141**, 58–67.
109. A. Biancolillo, F. Marini and J.-M. Roger, *J. Chemom.*, 2020, **34**, e3120.
110. A. Biancolillo and T. Næs, in *Data Fusion Methodology and Applications*, ed. M. Cocchi, Elsevier, Amsterdam, 2019, vol. 31, pp. 157–177.
111. S. Wold, E. Johansson and M. Cocchi, in *3D QSAR Drug Design*, 1993, pp. 523–550.
112. A. Biancolillo, K. H. Liland, I. MÅge, T. Næs and R. Bro, Variable selection in multi-block regression, *Chemom. Intell. Lab. Syst.*, 2016, **156**, 89–101.
113. A. Reilly, *Overview of Food Fraud in the Fisheries Sector*, 2018.
114. M. C. Calosso, J. A. B. Claydon, S. Mariani and D. M. Cawthorn, Global footprint of mislabelled seafood on a small island nation, *Biol. Conserv.*, 2020, **245**, 108557.

115. C. J. Donlan and G. M. Luque, Exploring the causes of seafood fraud: A meta-analysis on mislabeling and price, *Mar. Policy*, 2019, **100**, 258–264.
116. D. C. Carvalho, D. Guedes, M. da Gloria Trindade, R. M. S. Coelho and P. H. de Lima Araujo, Nationwide Brazilian governmental forensic programme reveals seafood mislabelling trends and rates using DNA barcoding, *Fish. Res.*, 2017, **191**, 30–35.
117. A. Di Pinto, A. Mottola, P. Marchetti, M. Bottaro, V. Terio, G. Bozzo, E. Bonerba, E. Ceci and G. Tantillo, Packaged frozen fishery products: Species identification, mislabeling occurrence and legislative implications, *Food Chem.*, 2016, **194**, 279–283.
118. G. Duflos, B. Le Fur, V. Mulak, P. Becel and P. Malle, Comparison of methods of differentiating between fresh and frozen–thawed fish or fillets, *J. Sci. Food Agric.*, 2002, **82**, 1341–1345.
119. E. Bozzetta, M. Pezzolato, E. Cencetti, K. Varello, F. Abramo, F. Mutinelli, F. Ingravalle and E. Teneggi, Histology as a Valid and Reliable Tool To Differentiate Fresh from Frozen-Thawed Fish, *J. Food Prot.*, 2012, **75**, 1536–1541.
120. C. Guglielmetti, M. Manfredi, S. Brusadore, S. Sciuto, G. Esposito, P. G. Ubaldi, L. Magnani, S. Gili, E. Marengo, P. L. Acutis and M. Mazza, Two-dimensional gel and shotgun proteomics approaches to distinguish fresh and frozen-thawed curled octopus (Eledone cirrhosa), *J. Proteomics*, 2018, **186**, 1–7.
121. F. Pennisi, A. Giraudo, N. Cavallini, G. Esposito, G. Merlo, F. Geobaldo, P. L. Acutis, M. Pezzolato, F. Savorani and E. Bozzetta, Differentiation between fresh and thawed cephalopods using nir spectroscopy and multivariate data analysis, *Foods*, 2021, **10**, 1–14.
122. M. Sannia, L. Serva, S. Balzan, S. Segato, E. Novelli and L. Fasolato, Application of near-infrared spectroscopy for frozen-thawed characterization of cuttlefish (Sepia officinalis), *J. Food Sci. Technol.*, 2019, **56**, 4437–4447.
123. D. Ballabio and V. Consonni, Classification tools in chemistry. Part 1: linear models. PLS-DA, *Anal. Methods*, 2013, **5**, 3790–3798.
124. M. Daszykowski, B. Walczak and D. L. Massart, Representative subset selection, *Anal. Chim. Acta*, 2002, **468**, 91–103.
125. D.-W. Sun, Å. Rinnan, L. Nørgaard, F. van den Berg, J. Thygesen, R. Bro and S. B. Engelsen, in *Infrared Spectroscopy for Food Quality Analysis and Control*, ed. D.-W. Sun, Elsevier, 2009, pp. 29–50.
126. M. Ottavian, L. Fasolato, L. Serva, P. Facco and M. Barolo, Data Fusion for Food Authentication: Fresh/Frozen-Thawed Discrimination in West African Goatfish (Pseudupeneus prayensis) Fillets, *Food Bioprocess Technol.*, 2014, **7**, 1025–1036.
127. R. Tsenkova, Z. Kovacs and Y. Kubota, Aquaphotomics: Near infrared spectroscopy and water states in biological systems, *Subcell. Biochem.*, 2015, **71**, 189–211.

128. L. Fasolato, S. Balzan, R. Riovanto, P. Berzaghi, M. Mirisola, J. C. Ferlito, L. Serva, F. Benozzo, R. Passera, V. Tepedino and E. Novelli, Comparison of visible and near-infrared reflectance spectroscopy to authenticate fresh and frozen-thawed swordfish (Xiphias gladius L.), *J. Aquat. Food Prod. Technol.*, 2012, **21**, 493–507.
129. S. Grassi, E. Casiraghi and C. Alamprese, Handheld NIR device: A non-targeted approach to assess authenticity of fish fillets and patties, *Food Chem.*, 2018, **243**, 382–388.
130. M. M. Reis, E. Martínez, E. Saitua, R. Rodríguez, I. Pérez and I. Olabarrieta, Non-invasive differentiation between fresh and frozen/thawed tuna fillets using near infrared spectroscopy (Vis-NIRS), *LWT*, 2017, **78**, 129–137.
131. M. Li Vigni, C. Durante, S. Michelini, M. Nocetti and M. Cocchi, Preliminary Assessment of Parmigiano Reggiano Authenticity by Handheld Raman Spectroscopy, *Foods*, 2020, **9**(11), 1563.
132. V. Caponigro, F. Marini, R. M. Dorrepaal, A. Herrero-Langreo, A. G. M. Scannell and A. A. Gowen, Raman and Fourier transform infrared hyperspectral imaging to study dairy residues on different surface, *J. Spectrosc. Imaging*, 2019, **8**, a3.
133. P. Firmani, R. Vitale, C. Ruckebusch and F. Marini, ANOVA-Simultaneous Component analysis modelling of low-level-fused spectroscopic data: A food chemistry case-study, *Anal. Chim. Acta*, 2020, **1125**, 308–314.
134. K. De Sá Oliveira, L. De Souza Callegaro, R. Stephani, M. R. Almeida and L. F. C. De Oliveira, Analysis of spreadable cheese by Raman spectroscopy and chemometric tools, *Food Chem.*, 2016, **194**, 441–446.
135. C. Cevoli, A. Gori, M. Nocetti, L. Cuibus, M. F. Caboni and A. Fabbri, FT-NIR and FT-MIR spectroscopy to discriminate competitors, non compliance and compliance grated Parmigiano Reggiano cheese, *Food Res. Int.*, 2013, **52**, 214–220.
136. R. Calvini, S. Michelini, V. Pizzamiglio, G. Foca and A. Ulrici, Exploring the potential of NIR hyperspectral imaging for automated quantification of rind amount in grated Parmigiano Reggiano cheese, *Food Control*, 2020, **112**, 107111.
137. H. Przyrembel, in *Trends in Food Science and Technology*, 2004, vol. 15, pp. 360–365.
138. C. A. Georgiou and G. P. Danezis, *Advanced Mass Spectrometry for Food Safety and Quality*, 2015, vol. 68.
139. S. Totaro, P. Coratza, C. Durante, G. Foca, M. Li Vigni, A. Marchetti, M. Marchetti and M. Cocchi, Soil sampling planning in traceability studies by means of Experimental Design approaches, *Chemom. Intell. Lab. Syst.*, 2013, **124**, 14–20.
140. L. Eriksson, T. Arnhold, B. Beck, T. Fox, E. Johansson and J. M. Kriegl, Onion design and its application to a pharmaceutical QSAR problem, *J. Chemom.*, 2004, 188–202.

141. C. Durante, L. Bertacchini, M. Cocchi, D. Manzini, A. Marchetti, M. Cecilia, S. Sighinol and L. Tassi, Development of $^{87}Sr/^{86}Sr$ maps as targeted strategy to support wine quality, *Food Chem.*, 2018, **255**, 139–146.
142. C. Durante, C. Baschieri, L. Bertacchini, M. Cocchi, S. Sighinolfi, M. Silvestri and A. Marchetti, Geographical traceability based on $^{87}Sr/^{86}Sr$ indicator: A first approach for PDO Lambrusco wines from Modena, *Food Chem.*, 2013, **141**, 2779–2787.
143. M. Silvestri, A. Elia, D. Bertelli, E. Salvatore, C. Durante, M. Li Vigni, A. Marchetti and M. Cocchi, A mid level data fusion strategy for the Varietal Classification of Lambrusco PDO wines, *Chemom. Intell. Lab. Syst.*, 2014, **137**, 181–189.

CHAPTER 6

Cosmetics and Personal Hygiene Products

C. BURNIER*[a,b] AND K. BRUCE[c]

[a] Ecole des Sciences Criminelles, University of Lausanne, Switzerland; [b] Agilent Inc. Technologies, Basel, Switzerland; [c] School of Chemistry and Forensic Science, Division of Natural Sciences, University of Kent, UK
*Email: c.burnier@live.fr

6.1 Introduction

6.1.1 The Forensic Interest

Forensic trace evidence is generated from different sources and can be found in varied forms and locations. Some are universally accepted by the criminal justice system as tried-and-tested methods of proving a link between suspect and scene, or suspect and victim, such as DNA or fingerprint evidence.[1–3] Yet, even in cases when DNA is found, best forensic practice and courts increasingly require complementary evidence to proceed. This is best illustrated by the Farah Jama case in Victoria (Australia), in which the defendant was wrongfully convicted of a sexual assault based on DNA evidence alone. The inquiry into the circumstances that led to this miscarriage of justice (the Vincent Report) stipulates that no case should proceed on DNA evidence alone without specific sign-off by the Attorney General.[4] Amongst some complementary evidence to consider are cosmetic and personal hygiene products. These sources of evidence are the subject of significant research interest and peer review.

Cosmetics and personal hygiene products, namely, condoms and lubricants, are routinely used in current society. The use of cosmetics dates back

Chemometric Methods in Forensic Science
Edited by Vishal Sharma, Georgina Sauzier and Simon W. Lewis

Published by the Royal Society of Chemistry, www.rsc.org

to approximately 12 000 years ago, to the ancient Sumerian and Egyptian civilisations.[5] Then, cosmetics were primarily used to indicate membership in a tribe or to protect against the elements, rather than to beautify. From the Bronze Age right through to the Edwardian era, cosmetic use fluctuated in terms of popularity and social acceptance.[6] At the beginning of the 20th century, the creation of the cosmetic powerhouses Maxfactor, Elizabeth Arden, L'Oreal, and Nivea, amongst others, resulted in a vast array of new and exciting products.[6] The cosmetics market has seen sustained growth and had a global worth of €280.4 billion in 2019, with Europe being the largest cosmetics market in the world.[†]

Today, cosmetics are formally divided into seven categories: sun care, oral care, skin care, hair care, body care, perfumes and decorative cosmetics.[‡] They are defined as "any substance or mixture intended to be placed in contact with the various external parts of the human body (...) with a view exclusively or mainly to cleaning them, perfuming them, changing their appearance and/or correcting body odours and/or protecting them or keeping them in good condition".[§] The prevalence and use of cosmetics have increased and their traces may be left behind on many objects, such as cigarette butts, glasses or clothing.[7] Cosmetics have traditionally been particularly relevant for the forensic investigation of crimes against women; however, the expansion of cosmetic use amongst the overall population is challenging the stereotype. Cosmetic use is independent of gender,[¶] age, ethnicity, or socio-economic background, which makes it a globally interesting source of trace evidence, to add to the classic traces usually investigated in casework.

Condoms and personal hygiene products are more case-specific evidence, pertaining to sexual assault cases. Condoms are defined as a "medical device, not sterile, used by consumers, which covers and stands on the penis during sexual activity, for the purpose of contraception and prevention of sexually transmitted infections".[8] They are the most widely used and widely known contraceptive devices, regardless of an individual's socio-economic background, age or education.[9–11] However, the condom was originally designed and marketed to prevent the spread of sexually transmitted diseases.[12] Sexual behaviour as well as the perception of the risks of pregnancy or transmission of sexually transmitted diseases (STDs) will significantly affect the use of such a device.[9–11] Most recent communications about condom traces highlight an increase in condom use among victims of sexual assault. In 2013, Los Angeles Police Forces statistics indicated that traces of condoms were found in 2% of cases that were brought to the police.[13] In 2017, in Switzerland, the Centre Universitaire Romand de Médecine Légale (CURML) reported approximately 15 samples that could potentially

---

[†] https://www.ctpa.org.uk/eu-and-worldwide
[‡] https://cosmeticseurope.eu/cosmetic-products/
[§] https://www.ctpa.org.uk/definition-ofa-cosmetics
[¶] https://www.mensmake-up.co.uk

contain traces of condoms, out of 300 samples collected over a year between Lausanne and Geneva,[‖] representing 5% of the samples. In Western Australia, in 2020, the Sexual Assault Resource Center (SARC) reported carried out at least one analysis a week for condom traces.[**]

Literature research reveals that for both condoms and cosmetics as forensic evidence there is little guidance available as to how one might approach analysis of these traces, and comparatively few criminal cases involving such evidence. Even though these traces are found more and more frequently in casework, very little research is carried out regarding the detection and interpretation of these traces, and this is likely to cause issues in casework.

6.1.2 Casework Use and Issues

Both cosmetics and condom traces aim to be used as associative evidence in casework. Their purpose is to establish the *corpus delicti*[14,15] and, in the case of condom traces, to verify penetration.[16–18] At that stage, it is not the occurrence of an act but the way it happened that is questioned.[19] The first case of cosmetics being used as trace evidence is the strangulation and murder of Marie Latelle, in which "pink dust" was found underneath her partner's nails. These were chemically identified as being a cosmetic preparation coming from Marie's face powder and, thus, provided the important physical link between the suspect and victim.[20] Cosmetics in the form of glitter traces were used as associative evidence when found on the clothing of women who had been sexually assaulted.[15] They have also been reported as trace evidence in a case of child homicide, where glitter was found in the bedsheets,[15] and on car upholstery after a disappearance.[15] Makeup deposits on a deployed airbag have been used as complementary evidence with DNA evidence to prove the identity of the driver.[14] Similarly, there are examples where lipstick has been used as physical evidence, for example, lipstick used to scrawl a message onto a mirror at a murder scene[††] or lip and lipstick impressions being used as biometric identifiers.[21] Cosmetics have been used to conceal victims' injuries[‡‡] to disguise suspects during bank robberies,[22] and have been recovered from vehicles following hit-and-run road collisions with pedestrians.[22]

While all the aforementioned traces have contributed towards successful convictions for cosmetics, the lack of research in the area of condom evidence may equally lead to wrongful convictions, such as in the Malkinson case. In this case, a silicone trace was attributed to a condom when there was no existing model available to reliably assess this evidence.[23] Cases reported in the literature[16,17,24] show that experts tend to draw conclusions based on the presence or the absence of traces, giving a black-and-white answer: the presence of

[‖]Personal communication, Dr Frédéric Grosjean, CURML, CHUV, Lausanne.
[**]Personal communication, Dr Maire Kelly, SARC, Perth, WA.
[††]https://www.newspapers.com/newspage/88468768/.
[‡‡]https://www.asteetrace.org/cosmetics.

lubricants indicates the use of a condom while its absence suggests no use of a condom. However, as illustrated in some recent Australian case studies,[25] both the absence and presence of lubricants may be due to other factors and several hypotheses need to be considered when interpreting such results. The analysis and interpretation of traces of condoms is a real challenge worldwide.[25–30] The limited number of reliable and valid protocols as well as the absence of interpretative models underline the need to carry out research in this area. The establishment of suitable casework protocols requires objective approaches for sample profiling or source discrimination. The use of chemometrics can assist with this by identifying data patterns relevant to sample discrimination or classification. These methods can also be used to study the impact of sample treatment or analytical conditions on discrimination, allowing statistical validation of methods.

6.1.3 Research and Chemometrics Approaches Developed

Amongst the various research papers obtained while researching the present book chapter, several different but complementary approaches were identified in the literature regarding analysis and chemometric studies.

The main tool encountered is chemometric pattern recognition, which is used to assist in the distinction samples where certain components are present, while regression methods can be used for component quantification. Most of the research can be grouped into one of the four following areas:

- Studying the composition of the samples: this type of study usually aims to list the components of a sample of interest, to define which chemicals are present, or the patterns to look for during an investigation. The chemometric target tool here is regression method and pattern recognition.
- Discrimination of various samples: this research usually uses market studies to determine whether *n* samples bought on the market, which may seem to be undistinguishable, can be differentiated, based on the presence of additives, different resins or viscosities of the material. The chemometric target tool here is pattern recognition.
- Identification of chemical markers on different trace supports: this type of study usually simulates casework samples and investigates the effects of the support on the detection of the traces (to determine which compounds should be targeted and what could be recovered from casework samples). The chemometric target tool here is pattern recognition.
- Interpretation of the evidence: in order to analyse the alteration of the traces over time; parameters such as background interference, transfer and persistence must be considered. The chemometric target tool here is regression method and pattern recognition.

All of these approaches are essential to forensic investigation, the aim of which is to determine if a trace is present (detection), if it is possible to link this trace to a sample class (classification), and if it is possible to interpret this result with regards to the context of the case, in particular, if questions of persistence come into consideration (interpretation).[31] Significant challenges for forensic scientists are outlined from these new types of evidence, especially when it comes to the identification, discrimination and attribution of a source in a specific casework. Amongst the various chemometrics methods available, the one most used in this area of forensic science is principal component analysis (PCA), further coupled to discriminant analysis, usually linear (LDA). If these techniques are used for the discrimination of samples, they can also be used for other purposes, notably in the case of understanding key parameters such as background, transfer and persistence of residues. The various roles of chemometrics in the data analysis for each of the aforementioned categories in the areas of cosmetics and condom evidence are discussed with examples hereafter.

6.2 Condoms and Personal Hygiene Products

6.2.1 Composition and Components of Forensic Interest for Discrimination

Condoms are mass-produced products whose production is highly regulated by international norms, such as ISO4074.[8,32–34] The classical composition of a condom is a latex body, covered with solid particles (*e.g.* cornstarch, polyethylene powder), lubricants, spermicides (nonoxynol-9) and sometimes other additives such as benzocaine, aromas and/or colorants.[35] Three types of lubricant can be found on condoms, covering two classes of chemicals: silicone-based lubricants, based on polydimethylsiloxane (PDMS) and water-based lubricants based on glycerin and polyethylene/polypropylene glycol (PEG/PPG). Silicone-based lubricants are the most prevalent lubricants contained in over 90% of condoms.[17,35,36]

However, PDMS, glycerin, PEG and PPG are also major components in other products like sexual lubricants, which may be recovered from sexual assault cases,[35,37] as well as from personal hygiene products (such as creams or soaps) which are used daily on the skin.[18,38] The composition of sexual lubricants is very close to condom lubricant composition, but the prevalence of silicone is significantly lower: only 20% of the lubricants on the market contained silicone,[36,39–42] either alone or in a mixture with water-based lubricants. Sexual lubricants mostly contain PPG and glycerin, but other components can be detected, such as 2-phenoxyethanol or 1,2-octanediol. These mixtures of glycerin and phenoxyethanol or octanediol are also frequently observed in the composition of creams and hygiene products (*e.g.*, intimate soaps). Manufacturers, therefore, offer the forensic scientist a wide range of different product compositions. These are highly discriminative when investigating the total composition of the various products

Table 6.1 Composition of condoms, sexual lubricants, creams and hygiene products based on the composition obtained in the literature.[35–37,40–43] Columns are separated based on the major compounds commonly found in all the samples and minor compounds specific to different types of samples. Abr: PDMS = polydimethylsiloxane, PEG = polyethyleneglycol, PPG = polypropyleneglycol.

Sample type	Major compounds		Minor compounds
Condoms	PDMS	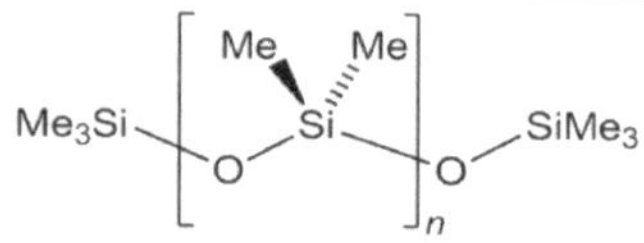	Nonoxynol-9, cornstarch, polyethylene powder, benzocaine, flavourings, aromas, triglycerides
Sexual lubricants	PEG	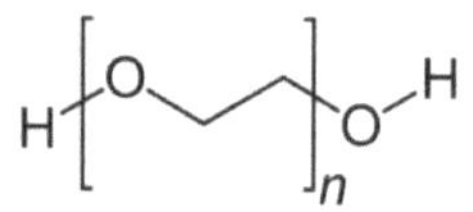	2-Phenoxyethanol, 1,2-octanediol, 2-oxybispropanol
Creams	PPG	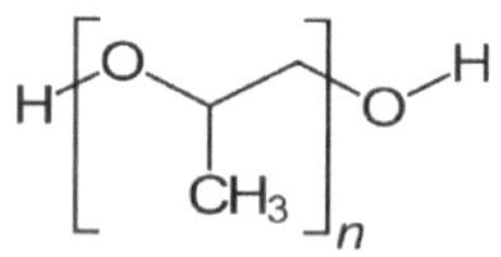	2-Phenoxyethanol, 1,2-hexanediol, 1,2-octanediol, octadecene, alkane patterns, (cyclo-) alkene patterns oxybisoctane
Hygiene products	Glycerin	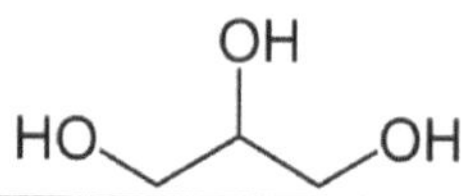	Benzoic acid, alkane patterns, 1,2-octanediol, triethylcitrate

that may be encountered in sexual assault casework (*i.e.*, condoms, sexual lubricants, intimate hygiene products, or daily creams/skin care products). Table 6.1 summarises the different compositions of these products based on their major and minor components.

The discrimination of the four main classes of samples presented in Table 6.1 is likely to firstly depend on the major compounds' composition, as they are the core active principle, and then within each grouping the creation of subgroups will likely be due to the minor compounds' composition. These discriminations are highly dependent on the analytical methods used (see Section 6.2.2). Discrimination based on major compounds may not require the use of chemometric algorithms, as different patterns would be obvious to see (see Section 6.2.2), which leads to simple visual discrimination and allows the creation of a generic cluster. However, as highlighted in both the National Academy of Sciences (NAS)[44,45] and more recently in the PCAST[46] report, sole visual interpretation of forensic evidence can suffer from operator bias or subjectivity. Chemometrics is therefore required for all types of discrimination and will allow the discrimination between different condoms from different brands which are likely to exhibit different chemical profiles (see Section 6.2.3).

6.2.2 Analysis and Discrimination of Condoms and Personal Hygiene Products

As previously mentioned, the efficiency of chemometrics for the discrimination of samples highly depends on which compounds are detected and, therefore, is dependent upon the analytical method used. In the case of condom lubricants and personal hygiene products, a wide range of different methods have been used, and these can be divided into three categories: spectroscopic techniques (*e.g.*, FTIR, Raman), separation techniques (*e.g.*, GC/MS, LC/MS), and mass spectrometric techniques (*e.g.*, MALDI–MS, DART–MS). The next sections will present the analytical and chemometric results obtained for each category.

6.2.2.1 *Spectroscopy Coupled to Chemometrics*

Easily accessible in most forensic laboratories, spectroscopic techniques, more specifically FTIR, were the first techniques used initially solely for the detection of condom residues in casework.[16,17,24] Mostly performed using FTIR,[17,36,39,47,48] and more rarely, Raman,[31,49] it is not before 2009 with the NAS report being published, and further in 2016 with the PCAST report, that any of the results obtained were processed using chemometrics. Within this chapter, only publication reported spectroscopy coupled to chemometrics will be presented (with no pretention of exhaustivity). Within the different possible FTIR modes (transmission, attenuated total reflectance (ATR–FTIR) and diffuse reflectance (DRIFTS–FTIR)), studies have highlighted that ATR and DRIFTS were the most relevant for condom evidence[29,39,48,50,51] but their use depends on the support of traces used, with DRIFTS being more suitable if extracted samples are to be analysed.[50] The focus of the following paragraphs will be on these two techniques as well as Raman coupled to chemometric tools.

ATR–FTIR coupled to chemometrics was performed on one of the biggest available published datasets, with 166 samples covering four major classes of products: condoms, sexual lubricants, creams and personal hygiene products.[39] The first discrimination patterns observed were based on the major compound content, with chemometrics providing two major groups: a silicone lubricant group and a very disparate water-based lubricant group, which are visually very different (Figure 6.1).[39,51] If subgroupings were not successful in the water-based group using principal component analysis (PCA), some interesting subgroupings were observed within the silicone cluster, with condoms and lubricants being clearly distinguished. After testing various classification models, the authors found that coupled to support vector machine (SVM) models, discrimination accuracies yielded over 90%. No discrimination of samples according to their brand, model or batch number was observed. Greater confidence in chemical discrimination of traces coming from these products was observed when developing a

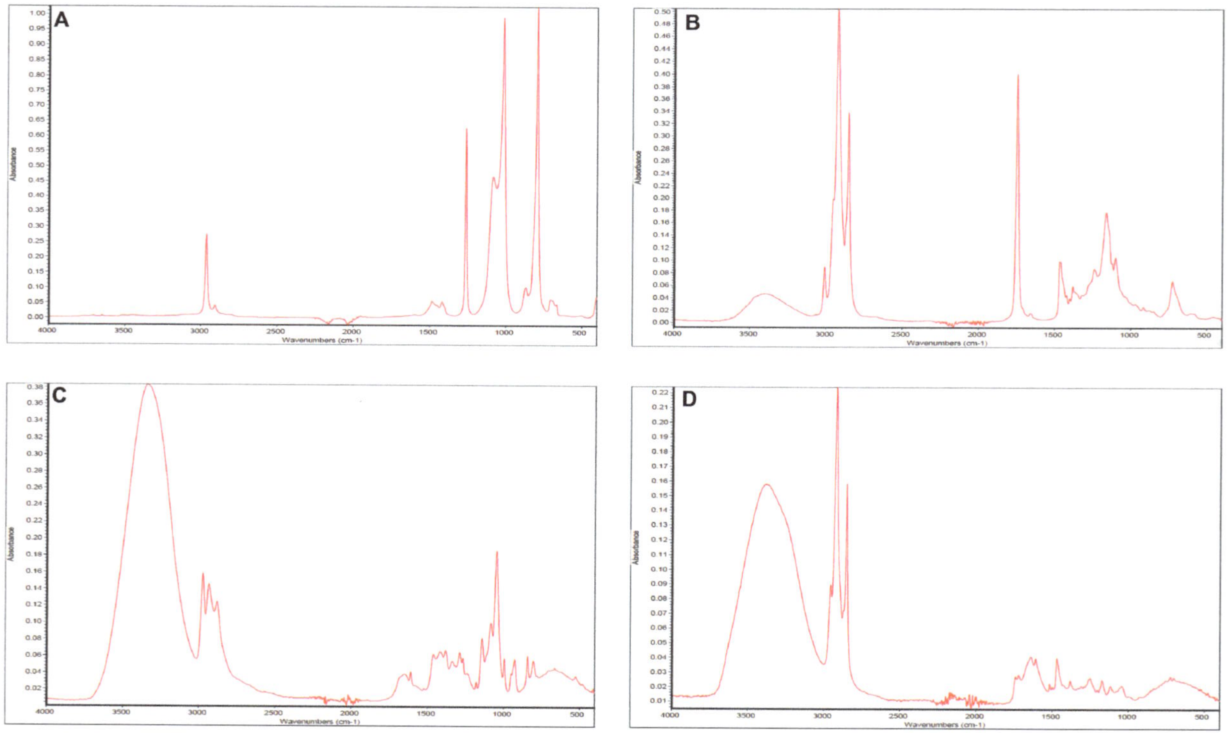

Figure 6.1 Illustration of the four major classes of spectra obtained when analysing condoms, sexual lubricants, creams and personal hygiene products using ATR–FTIR spectroscopy. Spectra are representing (A) silicone lubricants, (B) oil lubricants, (C) PEG lubricants and (D) glycerin lubricants.

two-step approach, which enabled the authors to reach a discrimination accuracy of 100%.[39] Similar results were obtained later with another set of 50 samples analysed using ATR–FTIR coupled to PCA for discrimination,[51] highlighting the relevance of ATR for the analysis and discrimination of sexual lubricants. Although one model was validated using blind validation,[39] one can regret the absence of model testing using real samples, collected from human volunteers.

To palliate the lack of casework relevance of the ATR studies, DRIFTS, which was found to be more adequate for extracted samples coming from casework,[50] was coupled to PCA and investigated as a discriminative tool for condom lubricants.[48] Despite being a smaller sample set, focusing on only 18 condoms, some simulated casework samples were added to the discrimination models.[48] The discrimination pattern observed in the ATR–FTIR publication[39] was found to be transposable to their data, which were samples extracted or diluted in hexane, the most adequate extraction solvent for silicone lubricants.[36,48,52] If only 86% of the variance was explained in the model, simulated casework samples were found to cluster within the silicone condom lubricant group. These results illustrated the ability of the technique to cluster trace evidence with initial raw samples. Chemometrics was also used for two additional reasons other than sample discrimination. At first, it was used as a tool for experimental design, to investigate what the best analytical conditions were for DRIFTS analysis of silicone lubricants, in terms of repeatability, reliability and sensitivity.[48] In a second attempt to provide the forensic community with quantification possibilities of DRIFTS to diluted silicones, chemometrics was used under the form of principal component regression (PCR) and partial least squares regression (PLSR) in addition to PCA, but the limits of the instrumental and statistical tools were reached, and no quantification or discrimination based on the concentration was obtained.[53]

Raman spectroscopy, on the other hand, was first reported for condom lubricants by Coyle and Anwar in 2009.[31] While detection was possible on pure raw material (liquid reference PDMS), the detection on simulated samples was found to be challenging. Criticised in 2020 for its lack of relevance and multiple issues for simulated casework samples,[50] Raman regained interest in 2021 with Thomas *et al.*'s work: 50 samples covering three major types of products (lubricants, condoms, personal hygiene products) were extracted in DCM/MeOH, analysed using Raman, and processed using Pearson's correlation coefficient, PCA, HCA and LDA. The reported results are similar to the ones obtained with ATR–FTIR,[39] namely, that the silicone and oil-lubricant groupings are very clearly discriminated while amongst the water-based population a strong variability is observed. Both discrimination and classification were tested on the pure material and extracted material with very good results. As per the DRIFTS–FTIR results,[48] the discrimination and classification results were similar for pure and extracted material. These results outline the interesting and promising results of Raman spectroscopy for condom lubricants analysis, with further analysis required on human samples and blind validation of the models.

6.2.2.2 Separation Techniques Coupled to Chemometrics

Chromatography was found to be rarely used for condom evidence analysis, its major use being for the detection of nonoxynol-9,[54] which is the only spermicide used on condoms,[32–34] or as a confirmatory tool for petrolatum residues.[36,55] The use is determined by the target compound. Silicones, for example, are very difficult to analyse using GC/MS due to their high apolarity and non-volatility, not to forget that PDMS, the main target lubricant from condoms, is also the main constituent of chromatographic columns. Extremely high concentrations of PDMS are required for detection using GC/MS,[56] which is incompatible with casework research where concentrations are extremely low[28–30] and the best extraction rates should be selected.[36,52] Therefore, pyrolysis-GC/MS (py-GC/MS) is one of the solutions for the use of GC/MS for silicone lubricants or siliconised products. Because of the strong variability coming from the pyrolysis unit, chemometrics is the key point to allow the adequate development and applicability of the method. Chemometrics was used for the design of experiments to obtain the best analytical conditions:[57] it was coupled to ANOVA statistics to select analytical conditions, which were then tested on a small sample set containing five condom samples.[57] PCA was not much of a success, samples seemed to be randomly clustered together and visual observation did not distinguish between any of them. An effect of too small a sample set or discrimination based on systematic errors? This is the main hypothesis, as further observations on 70 silicone-based condoms, sexual lubricants, personal hygiene products and creams revealed different silicone patterns in the py-GC/MS profile (Figure 6.2). Further discrimination between the different types of methylation of PDMS was also observed. As per the NAS report[58] and as seen in Figure 6.2, visual interpretation of some of the data is subjective to bias, and over 50 variables were targeted. PCA coupled with visual statistical tools were of significant help for variable selection, limited to 10 relevant discriminant variables. Further coupled with quadratic discriminant analysis (QDA), PCA confirmed the observed visual classes but also created five subgroupings within the silicone category, with a 97% accurate classification rate. If the method allows the discrimination of silicone products based on the difference in viscosity of the silicone constituents, it still does not allow the discrimination of products according to their brand and/or model, which suggests that other analytical methods would be more suitable than py-GC/MS for this purpose.

Further discrimination could be achieved by being able to detect minor compounds (Table 6.1) or other types of lubricants in the same products. These compounds are usually highly volatile, and GC/MS can be used for this purpose field.[40,54,59,60] In the only study using chemometrics coupled to GC/MS analysis, Burnier *et al.* illustrated firstly that silicone lubricants are not detected by GC/MS, and secondly that the wide range of compositions detected by the instrument amongst the water-based products allowed the individualisation of each of the products analysed. In that regard, the use of

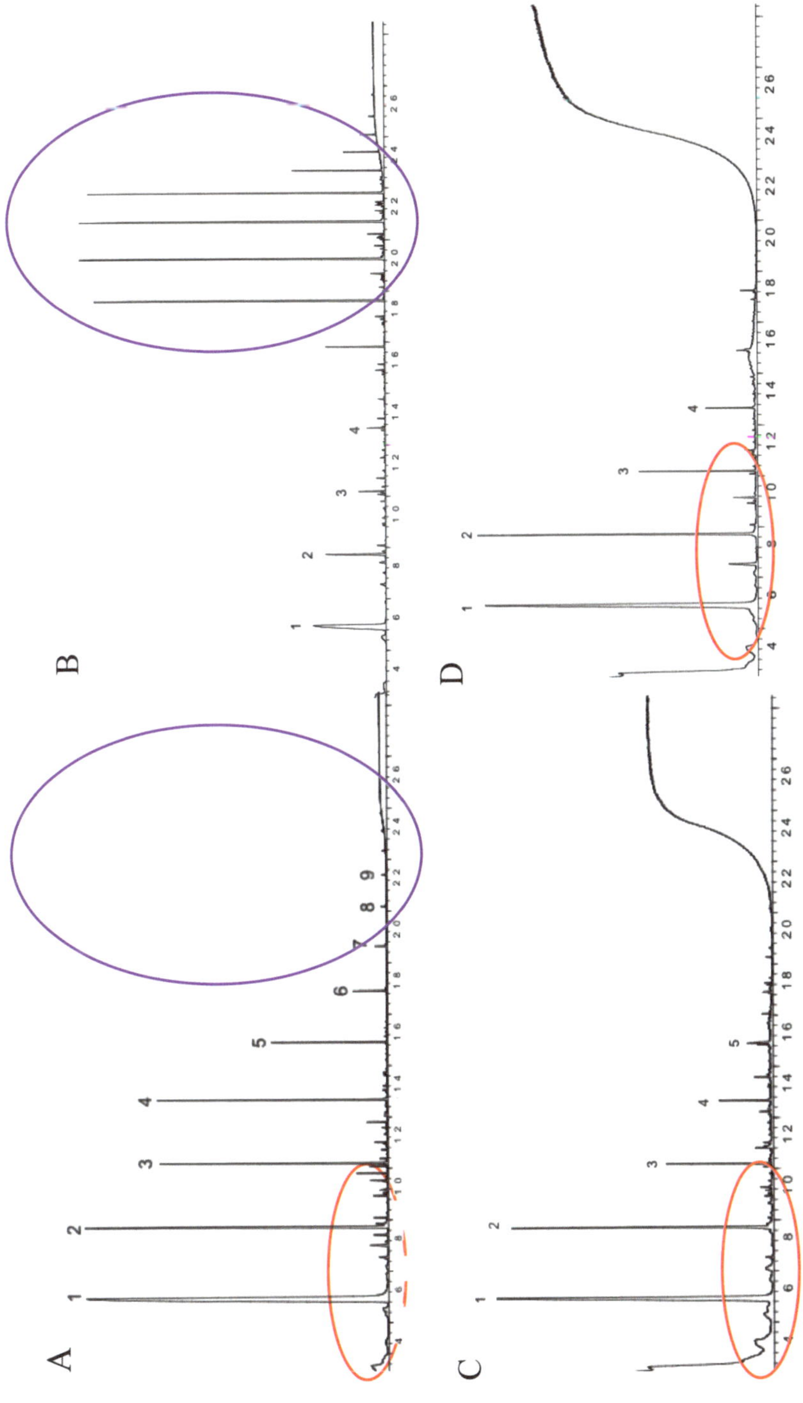

Figure 6.2 Illustration of the different patterns visually distinguishable after py-GC/MS analysis. (A) Silicone condom sample, (B) silicone gel lubricant sample, (C) silicone sexual lubricant sample, (D) silicone liquid for sexual use. Red and violet circles highlight the location of the differences between the different samples.

chemometrics for discrimination purposes seems rather unnecessary. Processed on the 166-sample dataset, chemometrics created 11 different groupings based on major composition and most abundant components; in this case, PPG, phenoxyethanol, benzocaine, glycerin and PEG. Samples that were not visually distinguishable were still not discriminated using PCA. Classification obtained a correct classification rate of 82%, which is lower than that of ATR–FTIR, but is questionable as the composition is different and, therefore, it is possible to individualise each product. Some silicone-lubricated condoms were also found to contain water-based products, highlighting the complementarity of the two methods. Both py-GC/MS and GC/MS were applied complementarily in casework,[25,27] with successful detection and classification of products when detected, allowing the expert to provide information on the initial question asked by the court, which was whether or not a condom was used.

Amongst other reported separation techniques, capillary electrophoresis coupled to UV-visible detection (CE/UV-Vis) and liquid chromatography coupled to mass spectrometry (LC/MS) are of high interest for the separation of various polymers. CE/UV-Vis was applied to 68 condom samples which were discriminated based on their chemical content, except for 12 which were clustered together.[18] On the other hand, LC/MS was used on nonoxynol-9 detection only[36,61–64] and has not been reported coupled to chemometrics for the discrimination of various samples. It highly depends on the solvent as some, like hexane, are inadequate as they dissolve the LC tubes. The use of DCM[17,24,52,55] or isopropanol[§§] should be considered for further analysis and discrimination of samples with promising possibilities.

6.2.2.3 *Mass Spectrometric Advanced Techniques Coupled to Chemometrics*

The development of extremely sensitive methods such mass spectrometric techniques offers chemometrics the possibility to show its highest potential: variable selection is enhanced allowing discrimination within the smallest groups. Amongst all the mass spectrometric methods, the most used for condom lubricants analysis and discrimination is DART–TOF/MS (direct analysis in real-time–time of flight mass spectrometry).[41–43,65,66] The other common methods are MALDI–MS (matrix-assisted laser desorption ionization mass spectrometry),[19,61] EASI–MS (electrospray ambient ionisation mass spectrometry),[67] and DESI–MS (desorption electrospray ionisation mass spectrometry).[68]

MALDI–TOF/MS coupled to the PCA algorithm allowed the separation of PEG-based lubricated products from PDMS-lubricated products, with a small variation within the silicone-containing cluster and a strong variability amongst the PEG-containing products field.[19] These results are similar to the results obtained with GC–MS[40] or py-GC/MS,[69] which illustrate that one

---

[§§]Personal communication, Dr Marc van Bochove, Netherlands Forensic Institute.

or the other of the methods can be used depending on the material available in the forensic laboratory, although tests on simulated casework samples should be processed first.

Two interesting methods due to the need for either minor or no sample preparation, EASI–MS and DESI–MS were tested and compared over 10 and 6 condoms respectively. Three out of six condoms analysed by DESI exhibited similar features in terms of composition but the relative abundance of the components varied, which provided discrimination potential. Of the samples analysed by EASI, one contained spermicide and PEG, one was found not to contain any lubricants, and the rest presented similar chemical features. Chemometrics discrimination was performed over 87 variables and 11 principal components, which is a significant amount of data to extract. EASI–MS was found to present a more accurate classification rate using an LDA algorithm and is therefore better adapted than DESI–MS for sample discrimination.

DART–TOF/MS is the most used technique, and probably one of the most powerful techniques for the discrimination of condoms and personal hygiene products, due to its high sensitivity. The first investigations were reported by Maric *et al.*[41] who discriminated 33 water-based lubricants. These were clustered into six groups and a classification accuracy of 100% was reached,[41] similar to the result obtained using ATR–FTIR by Burnier *et al.*,[39] with multiple samples in each group. In 2017, the method was applied to a wider range of products, covering 90 products including silicone-based and water-based lubricants and condoms.[42] Visual analysis grouped the 90 samples into 12 major groups based on their chemical content, and HCA and PCA were used complementarily to confirm the groupings, with a higher specificity seen using HCA. The classification accuracy reached 98%, which suggested that the bigger the sample set, the more difficult it was to classify them accurately. Interested in knowing whether modified DART–MS analysis would enhance the quality of the discrimination, a sample subset of 36 samples, covering the 12 groupings, were analysed using AccuTOF-DART.[65] Here again, HCA confirmed the 12 groups, which were not found to be clearly separated using PCA. There was a strong variability within the sample set, but silicone-lubricated products were under represented. These results are similar to those obtained by Burnier *et al.* using GC/MS data,[40] and the lack of investigation within each group to see whether subclusters are possible or not, highlights that the expensive DART–TOF/MS obtains similar results and classification rates to other more accessible instruments.[39,48]

When computing both silicone- and water-based lubricated products together, a strong variability is obtained, and classification issues are encountered. Therefore, Baumgarten *et al.* decided to investigate a classification scheme based on silicone-based lubricants exclusively.[43] Such studies are more relevant when considering condom investigation, as silicones, more specifically PDMS, cover more than 90% of the condoms on the international market.[17,35–37] A discrimination model was built based on

38 samples, comprising condoms, sexual lubricants and personal hygiene products, and another 18 were used to validate the model, which reinforces the value of the results. HCA created 11 classes based on the content, and PCA confirmed the classes. At this stage, the instrument was able to provide discrimination based on the different viscosities of the silicone lubricants used, with more information than the ones obtained with py-GC/MS.[69] These results also indicate that it is possible to detect differences within a sample population, such as within the condom population; a conclusion that had not been drawn with other instrumentations. Further investigations were led exclusively on condoms, with 110 samples, covering 16 different brands analysed.[66] The results are extremely powerful: using PLS-DA and sparse discriminant analysis (SDA), almost all the condoms were individualised according to their brand, and the classification model offers a 95% accurate classification rate.[66] However, when considering the prediction, classification was found to be highly dependent on the brands, with some brands presenting 100% correct classification, and others ranging between 85% and 98%. The model was then validated using an 85% confidence level for prediction.

All the aforementioned studies were based on reference or known materials purchased from the market. However, the detection of condom or personal hygiene product traces coming from casework requires an understanding of the interaction between the target compounds and a biological living matrix, as well as the lifetime in these matrices.[35,66] Therefore, it is crucial to understand what chemical markers can be recovered from casework and from different matrices (background), and how the evidence transfers and persists, to be able to adequately interpret the forensic evidence.

6.2.3 Identification of Chemical Markers for Casework

Condoms and sexual lubricants are investigated in sexual assault cases, which provides a wide range of matrices for sample collection, and therefore an interesting field for chemometrics to highlight differences or similarities within samples. Similar to DNA investigation in sexual assaults, the most common matrices for lubricant collection are the vaginal matrix, endocervix, vulva, anus, penis, skin or mouth.[26,28–30,35] Sexual assault kits contain swabs which are used by the forensic examiners/forensic nurse to collect trace evidence from the matrices. While the samples received by the forensic laboratories are mostly collected from the victim (vaginal, anal, endocervical or vulvar swabs), other trace media may also be encountered, such as the victim's underwear, as illustrated in the Malkinson case.[70] Identification of markers should be conducted on material recovered from casework, in this case, sampling swabs and/or textiles.

Two main approaches, including the use of chemometrics, are reported in the forensic literature for the identification of chemical markers. The first one is a chemical approach and consists of looking for the target chemical compounds on the different trace supports. The second one is a metabolomic approach, which aims to detect a reaction between the matrix and the chemical or a given activity. Both approaches are discussed hereafter.

6.2.3.1 Chemical Approach

6.2.3.1.1 Sampling Swabs. Based on DNA evidence sampling, swabs are used for condom/sexual lubricant evidence collection in casework.[25] The type and the number of swabs collected may vary according to the country, but cotton and nylon swabs are usually used and available in sexual assault kits.[25,71] The identification of chemical markers on the swabs requires investigation of multiple steps of the sample collection, including an understanding of the background at different steps of the sampling and analysis procedure, as well as contamination sources. Background investigation includes solvent blanks (if extraction is required), blank swabs and blanks of the matrix to ensure none of them contains the target compounds, before moving on to real samples.[28–30] Even though the procedures utilise sterile material, all blanks should be tested to ensure the absence of contamination, which may happen, as illustrated in the Malkinson case.

Condom residues, being an important source of evidence, should benefit from the use of specific swabs for sampling. However, in some protocols, the swabs used for condom investigation are the same as those used for DNA collection, and the two evidence types would be extracted from the same swab. To this extent, chemometrics was used to highlight whether the extraction of DNA prior to the condom extraction had a statistical impact on the discrimination of the samples.[71] It was found that DNA extraction generated an increase in the variability in the chemical profiles recovered from highly concentrated samples and a strong silicone loss at lower concentrations, leading to instrumental detection limits being reached. Therefore, condom evidence should be sampled with a dedicated swab and collected prior to other evidence and also prior to the insertion of a lubricated instrument for further investigation.[16,25,71]

Considering the investigation of extraction, swab and matrix interactions on the swabs, chemometrics in the form of PCA was used to compare the chemical profiles obtained from silicone-based condom lubricants solubilised in hexane, silicone lubricants deposited on a cotton swab and silicone lubricants sampled from the vaginal matrix, after a protected sexual intercourse.[30] Results illustrated that all the analysed samples created one big cluster with no possible subgroupings, therefore outlining the specificity and selectivity of the extraction. Chemical differences between samples collected from the skin, vaginal matrix, mouth and other contact matrices have never been reported to date.

If the vaginal matrix, endocervix, vulva, anus, penis, skin or mouth[26,28–30,35] were investigated as trace support, the only one which was further investigated with the help of chemometrics is the vaginal matrix.[29] PCA was used on background samples collected from human female volunteers to investigate firstly if silicones were present in the vaginal matrix for legitimate reasons, and secondly, the variability within the vaginal matrices, given that they are evolutive and constantly changing matrices.[29] Results have illustrated a strong variability on a very reduced spacing, highlighting that vaginal matrices are not as diverse as was initially thought. Notably, samples collected from the same volunteer were not clustered together specifically, revealing an interesting variability, the source of which could not be identified. Possibilities include that it came from the volunteer, the self-sampling procedure used and/or the instrumental variability. Such results underline the limitations of chemometrics applied to living samples, as it is still unclear what the most important factor is that generates discrimination of the samples.

6.2.3.1.2 Textiles. Cotton swabs are the most investigated support of lubricants evidence,[26,36,38] but textiles can also be of interest in forensic casework. For example, in the event a victim has washed, limiting the possibilities of recovering traces from the vaginal and/or endocervical swabs,[26,28] or an aggressor has wiped his fingers covered with lubricants on the couch. Presently, based on existing DNA research, the most interesting source of textiles to be investigated in sexual assault cases is the victim's underwear. There is only one preliminary study reporting the investigation of condom residues in textile underwear.[72] One of the main difficulties outlined is the fact that the manufacturing process uses silicone to protect the textile.[72] Assessing whether the silicone source is a condom or derived from the manufacturing process is challenging as it is unclear whether the chemometrics tool discriminates samples based on a concentration difference or real differences. Loading plots only show discrimination based on silicone peaks, which suggests that concentration is the main discrimination factor. The opportunity to explore the chemical profile variations amongst a bigger set of different underwear purchased on the market, and to investigate the possibility of detecting condom residues accurately on any of that underwear, would generate valuable and relevant data.

6.2.3.2 *Metabolomic Approach*

The vaginal matrix is an evolutive matrix with incredible extension properties and is subject to change induced by hormones secreted during the menstrual cycle, as well as during sexual intercourse.[73] The friction induced by the condom on the vaginal surface may induce acute alterations of the matrix and secretions produced by the vaginal epithelium and glands. Inflammatory cells or cell type changes may also be observed as a reaction of the matrix. Hence, condom-induced metabolomic changes may constitute additional evidence. It has unfortunately not yet been reported for condom evidence, but metabolomics is applied in forensic toxicology and various other forensic areas[74] and, if coupled with chemometrics, would provide interesting results for sexual assault investigation.

6.2.4 Factors Affecting Condom Evidence Interpretation

Condom evidence analysis is a real challenge worldwide, but the interpretation may well be the trickiest point of the framework,[25–30] especially with the currently limited number of reliable and valid protocols and the absence of interpretative models. Interpretation of the evidence considers parameters such as background, transfer and persistence of residues in a given situation.[25,75,76] Background considers the prevalence of the given profile in the population for legitimate purposes, whereas transfer and persistence take into account whether or not the compounds can be transferred to the matrix and how long after intercourse they might be possibly detected. All these parameters are anchored in forensic science as the way to understand the trace, how it got there and how it has evolved over time until detected by crime scene investigators.[75,76]

Comprehension and modelling of background, transfer and persistence are investigated using the five following parameters: donor, receiver, contact, elapsed time and activity of the victim.[35] Transfer requires an understanding of the donor properties (*i.e.*, the condom/sexual lubricant), the receiver properties (*e.g.*, the vaginal matrix) and the contact. Persistence adds to donor, receiver and contact the notions of elapsed time between the alleged assault and the moment of sample collection, as well as the activity of the victim during this time. Due to ethical constraints, the contact cannot be easily investigated, but the four other parameters have been partly investigated and are described hereafter.

Donor. The first parameter of interest, the donor, is the product itself. In this context, that equates to condoms and sexual lubricants. Two female volunteers had sexual intercourse with their partners, using condom samples of different types (lubricated with silicones, lubricated with water-based lubricants, non-lubricated, latex-based or latex-free) for a total of 10 protected intercourses and self-sampled straight after the intercourse.[30] Cotton swabs were extracted and analysed using DRIFTS–FTIR. Chemometrics applied to the collected DRIFTS–FTIR data revealed that the impact of the donor on the transfer was dependent on the type of condom used and that the discrimination patterns were similar to the ones obtained in the market studies.[39,48] These results are interesting as they highlight that the patterns obtained with ATR–FTIR are also applicable and transferable to data collected on humans and that the extraction procedure is extremely selective. The variability observed within the PCA scores plot presented is highly similar to the ones obtained from the DRIFTS–FTIR analysis, suggesting a limited effect of self-sampling.

Receiver. The receiver was investigated in the literature from the vaginal matrix and on textiles. The textile study is preliminary and does not outline the use of any chemometrics for discrimination purposes,[72] but the results obtained on the vaginal matrix were statistically analysed using PCA, after a visual inspection of the spectra.[30]

Ten volunteers had protected sexual intercourse with silicone-lubricated condoms and self-sampled immediately thereafter. Visual observation of the data revealed absorbance variations between the volunteers, which are likely to come from different contacts, different matrices and different self-sampling techniques. Interestingly, the PCA tool did not highlight such important differences with the samples being all clustered together. Within sample variability was found to vary between 0.1 and 0.3 on the first principal component, and around 0.1 on the second principal component. These are very small intervals considering the investigation of living biological samples. Between sample variability was also reported to be very small thus not allowing discrimination between volunteers, similar to the results that were already observed when investigating the background in the population.

Elapsed Time. Investigation into the effect of time on silicone residue persistence was first reported in 2018. The research focused on understanding for how long the detection of silicone and water-based lubricants was possible on different matrices, *e.g.* skin, mouth or vaginal matrix[26] after deposition. Multivariate models, namely PCA and PLSR, were used to model the elapsed time of samples collected from human donors at various post-coital intervals, between 30 min and 36 h, to define the relative stability of aging parameters (depending on the various influence factors), and construct a stable and reproducible model base for further interpretation of the evidence.[28] Observation of the PCA scores plot illustrated a decrease in the concentration of silicone lubricant over time, but no clear clusters of similar post-coital time intervals were highlighted. A strong variability was revealed in the PCA plot, variability which was also illustrated during the visual observation of the collected data. The calculated coefficient of variations was reported between 30% and 90%, which indicates errors as high as the measurement itself. Classification of samples was therefore not indicated. Samples at 18 h and 24 h were found to be very close to the background samples, outlining the real challenges that can be observed in casework, depending on the time of sampling after the aggression.

Activity of the Claimant. The last parameter of interest – the activity of the claimant – is one of the most difficult parameters to monitor. Specific activities such as showering are known to affect the recovery of the evidence. When comparing a passive (*i.e.*, laying), an active (*i.e.*, walking, sport) and a showering activity, showering was found to affect the results more significantly than the other two activities. However, the study led by Fischer *et al.*[28] did not report the use of chemometrics for the activity of the claimant. The use of discrimination patterns was not relevant as the comparison between the different activities would have required one volunteer to be assigned to the different activities for each elapsed time to be able to compare the data. This was further outlined by results illustrating the strong variation existing at the transfer moment for each different woman involved in the study.

6.2.5 Other Considerations on Condom Lubricants and Chemometrics

6.2.5.1 Application to Casework

Amongst all the different existing models for the discrimination of condom lubricants, sexual lubricants and personal hygiene products presented in the previous sections, very few applications to casework were outlined and are available in the literature, mostly due to the complexity of acquiring data from human volunteers.

Discrimination for current casework used py-GC/MS[69] and GC/MS[40] models developed to infer the source of a chemical profile to confirm or refute the victim's allegation.[25] Prior to casework application, the models were tested on proficiency trials to ensure the suitability of the models.[25] Once the models were validated, the casework samples were analysed and those that gave a chemical profile were projected into the discrimination models. They were found to cluster with the condom group, confirming the allegation of the parties. If the results were successful in one case, most of the other cases encountered in the forensic practice did not allow the detection of any chemical profile. This highlights the need for case-based research, as developing models is interesting but is of no use if evidence cannot be detected or is not collected.

6.2.5.2 Other Research

The previous sections have focused on the detection and classification of condom lubricants, sexual lubricants and personal hygiene products in the context of a sexual assault, with swabs being collected during the medical examination of the claimant. However, in the literature, other applications of chemometrics towards the analysis of condom lubricants can be found. For example, MALDI–MS, Raman and ATR–FTIR spectroscopy were used for the classification of fingermarks contaminated with different condom lubricants.[77] The model allowed the classification of four out of the five lubricants used while further discriminant analysis successfully distinguished the different brands or models of condoms that were contaminating the fingermarks.

PCA coupled to LDA was also used for the classification of condom residues detected using DESI–MS on fingermarks revealed using cyanoacrylate fuming.[78] The method was found to be sensitive enough to detect silicone-contaminated fingermarks under the cyanoacrylate layer. If the correct classification rate obtained for a model built on 32 bulk condoms was 99%, the accuracy achieved when plotting data coming from revealed fingerprints was found to be 90.9%, due to the cyanoacrylate layer deposited during the development procedure. The correct classification rate is still very high and applicable to casework.

6.3 Cosmetic Products

The term "cosmetics" covers an extraordinarily wide array of products, many of which are commonly used every day, such as toothpaste, handwash and deodorant. Perhaps the most frequently associated type of cosmetic is makeup; this was formally known as decorative cosmetics and accounts for around 30% of the global billion-dollar industry.[79] Traditionally, the analysis of cosmetics has been based either within the pharmaceutical or archaeological contexts; however, cosmetic products are also of significant interest in forensic casework. They have the potential to add valuable evidence to the most serious of criminal cases, such as murder, rape, sexual assault, child abuse, trafficking and kidnapping.[80,81] As is the case with condom traces, cosmetic traces would aid the understanding and reconstruction of the criminal activity by providing the link between suspect and victim, or suspect and scene. This would be particularly beneficial in cases where there is a lack of standard physical evidence, the type of cosmetic residue is uncommon, or cosmetics have been layered and therefore provide unique combinations. If we consider which sort of cosmetic products are likely to become trace evidence within a forensic context, it is reasonable to assume that decorative cosmetics would be the most relevant. For this chapter, therefore, the focus will be on decorative cosmetics. A short overview of various other cosmetic types will be presented in Section 6.3.5.

6.3.1 Composition and Components of Forensic Interest for Discrimination

Decorative cosmetics include products such as foundation, powder, blush, lipstick and lip gloss, lip and eyeliner, mascara, eyeshadow, and nail polish.[79] They tend to be mass-produced and the ingredients contained within them are subject to highly regulated laws and restrictions, to ensure consumer safety.[82,83] Alternatively, they can comprise local or homemade items; some consumers may even choose to use common household cooking ingredients such as cocoa powder in place of manufactured blusher or bronzer.[6] This results in a huge variety of potential decorative cosmetic traces that an examiner may find at a scene. Furthermore, the scope of different brands, series, formulations and shades, coupled with geographical location manufacture, inter/intra-batch variations, variability in particle sizes/shapes/layers/coatings, differing ages of used products, differing application methods and differing modes of trace deposition, are likely to result in a hypothetically unique form of trace evidence. Whilst the composition may vary, some components are commonly found in most of the decorative cosmetic types. Table 6.2 reports the major and minor components for a small selection of decorative cosmetics.

Nail polishes (not mentioned in Table 6.2) contain mostly nitrocellulose (as a film former), resins, plasticisers and solvents.[92] In addition to the

Table 6.2 Composition of foundations, lipsticks, mascaras, and eyeliners, based on the compositions obtained in the literature.[84–91] Columns are separated based on the major compounds commonly found in all the samples and minor compounds specific to different types of samples. Abr: PDMS = polydimethylsiloxane, PMMA = polymethylmethacrylate, PVP = polyvinylpyrrolidone, PEG = polyethylene glycol, PPG = polypropylene glycol.

Sample type	Major compounds	Minor compounds
Foundations	PDMS 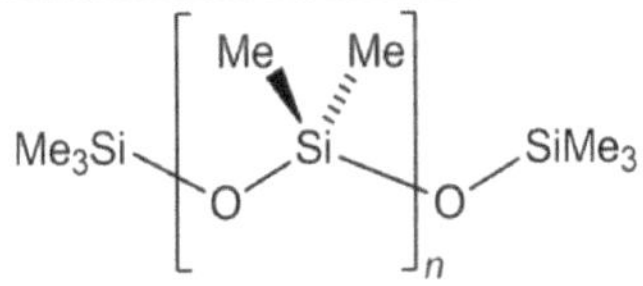	Cetyl palmitate, lecithin, PMMA, propylene/butylene glycol, silica, glyceryl stearate, isopropyl myristate, crosspolymer, acrylate copolymer, squalene, nylon-12, capric triglyceride
Lipsticks	Glycerin 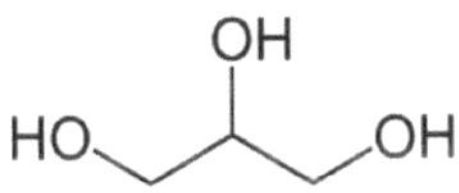 Castor oil (mixture of ricinoleic, oleic, and linoleic acids)	Beeswax, jojoba oil, lanolin alcohol, carnauba wax, candelilla wax, shea/cocoa butter, butylene glycol, paraffin, ozokerite, cetyl acetate, citric acid, stearalkonium bentonite
Mascaras	PVP 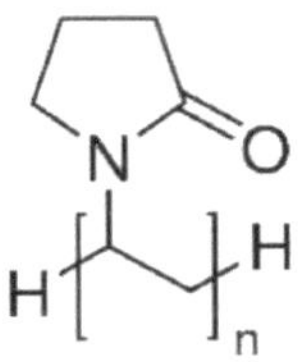	TEA-stearate, isopropyl myristate, carnauba wax, lanolin, ammonium acrylates copolymer, stearalkonium hectorite, paraffin, PEG, PPG
Eyeliners	PEG 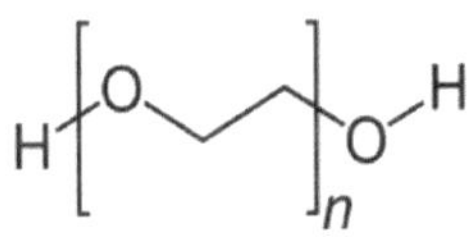	Ammonium acrylates copolymer, propylene glycol, mineral oil, carbomer, polysorbate 20, sorbitan oleate, PPG, PDMS, paraffin, hydroxypropylmethylcellulose

tabulated compounds, decorative cosmetics also include significant amounts of colourants (pigments or dyes), opacifying agents, and moisture absorbents. The most common ones are talc, titanium dioxide, mica, bismuth oxychloride, zinc oxide, magnesium aluminium silicate, kaolin and iron oxides.[86–88,92–94] Components systematically observed in many of the decorative cosmetics that are not reported in Table 6.2 are parabens and isothiazolinones (Figure 6.3), which are used as preservatives to avoid the development of microbes in cosmetics. However, both preservatives are known to be skin irritants and parabens specifically are suspected to act as endocrine (or hormone) disruptor chemicals and, therefore, their use is strictly limited in decorative cosmetics. None of the research in this area detected preservatives and this is likely due to a concentration below the instrumental detection limit. Nonetheless, for decorative cosmetics, it is the whole product composition that is of forensic interest, as the various

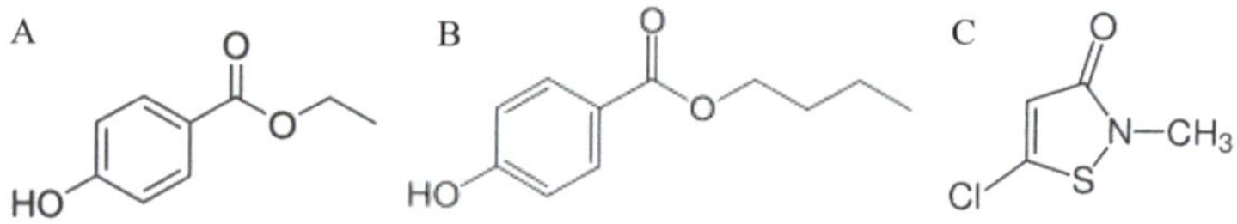

Figure 6.3 Chemical structures of (A) ethylparaben, (B) butylparaben and (C) methylchloroisothiazolinone.

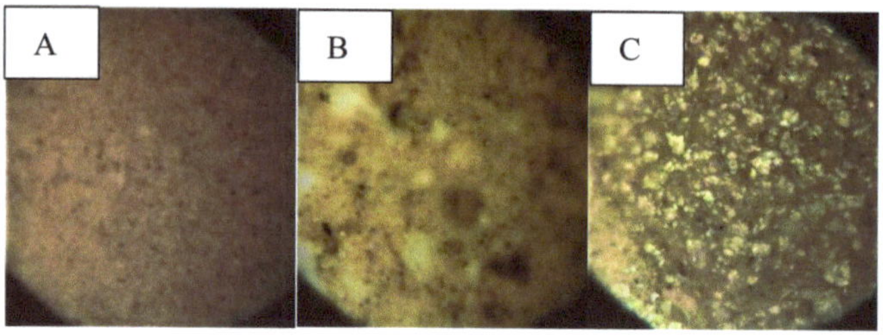

Figure 6.4 Colour photomicrograph images taken at 50× magnification of three different cosmetic foundation samples: (A) Chanel Vitalumière in shade 50 naturel, (B) Clarins Teint Éclat Lissant in shade 3.5 sandy beige, (C) Bare Minerals Original Foundation in shade 03 fairly light. Images taken from the poster entitled "The makeup of makeup", presented by Katy Bruce at the University of Kent Researchers' Showcase 2019, Canterbury, UK.

proportions of the components in the mixture confer a specific chemical pattern.

Initial visual discrimination between samples entails simply observing the colour of the sample.[85–87,95] However, discriminating between cosmetic samples based on visual comparison alone is a risky approach, as the thickness of the cosmetic trace may alter its perceived colour.[96] Furthermore, Figure 6.4 illustrates how cosmetic foundation samples that may appear visually similar by eye, can look very different when viewed on a microscopic scale. Incidentally, this hypothesis is refuted by Cho *et al.* when considering nail polishes.[92] The variation between foundation shades is often minor and, owing to the thin layer applied, this makes visual identification difficult and unreliable.[87] Additionally, it has been experimentally demonstrated that the apparent observed colour of transparent or translucent glitter particles changes depending on the orientation of the sample illumination, making visual discrimination highly subjective and circumstantial.[97,98]

Visual discrimination is also performed for the discrimination of samples based on the observation of the acquired chemical profile or spectrum, yet these discriminations are highly dependent on the analytical methods used (see Section 6.3.2). Visual discrimination of chemical profiles is mostly derived from the presence and/or absence of peaks, and differences in the relative intensities of those peaks, indicating that the samples are not identical. Some chemical profile differences may appear obvious but minor differences will be affected by operator interpretation and may also be extremely time consuming. Consequently, it is necessary to employ chemometric techniques for objective and unbiased data analysis of these forms of trace evidence.[99,100]

6.3.2 Analysis and Discrimination of Decorative Cosmetics

As previously mentioned, the discrimination between decorative cosmetic samples is dependent upon which compounds are detected; accordingly, it is dependent on the analytical method used. However, some sample types preclude the use of certain techniques, for example, the incompatibility of liquid samples with photoacoustic spectroscopy.[101] Consequently, the sample itself may dictate the analytical method used. Spectroscopic techniques are the most employed, with FTIR being the selected method in approximately half of the published papers, and Raman in around a fifth of the research papers included for the purposes of this chapter. Separation techniques and mass spectrometric methods are rarely used. The next sections will present the analytical and discrimination results obtained for each category.

6.3.2.1 Spectroscopic Techniques

Easily accessible in most forensic laboratories, spectroscopic techniques – the most popular being ATR–FTIR – are usually selected for their advantages of being non-destructive and requiring little to no sample preparation, thereby minimising the risk of sample alteration or contamination. Indeed, Raman spectroscopy was found to be effective at analysing samples through plastic evidence bags, further mitigating any sample interference or degradation as well as reducing any risk to the operator.[102]

The largest category of decorative cosmetics investigated using spectroscopic techniques is lipsticks, representing over half of the literature focused on these types of samples. Analysis is performed equally with ATR–FTIR and Raman spectroscopy. The populations of samples are also fairly similar: the use of around 40–50 samples is reported when running ATR–FTIR analysis,[85,95,103–105] whereas between 50 and 80 samples are used for Raman analysis.[102,106–108] The sample pools are relatively different, with products being sourced all over the world; nevertheless, the conclusions are all congruent.

With regards to ATR–FTIR analysis, the samples are found to primarily cluster by their colour. The most effective use of PCA is to first separate the samples into groups based on visual observation of their colour.[85,103] Wax and liquid lipsticks generate different chemical patterns based on their silicone and organic content, leading to an additional level of discrimination.[85] The variability between samples collected from the same manufacturer and series (or range) is found to be less than between samples originating from the same manufacturer but from a different series.[104]

All the studies report the use of PCA and LDA for the discrimination and classification of lipsticks[80,85,103–105,109] and, within these datasets, all the lipsticks are differentiated based on either the brands[80,103] or the brands and series.[85,104] Correct classification rates are found to vary depending on the colour of the sample,[103] suggesting that the homogeneity and composition of the samples are important factors. Nonetheless, most researchers report accurate classification rates, ranging between 98% and 100%.[80,85,103,104,109]

Raman analysis does not provide any further valuable information for the discrimination of lipsticks than that achieved *via* ATR–FTIR, except for one publication,[106] which highlights similar discrimination and classification results to those gained using ATR–FTIR. Research led using Raman analysis is rarely coupled to chemometrics, but mainly focuses on the enhancement of the spectral acquisition.[110,111] Visual spectral discrimination based on the presence or absence of peaks is the furthest reported discrimination method, resulting in a discrimination power of 99%.[107] Colourants allow discrimination of lipstick samples from the same product line but minimally pigmented lip cosmetics (such as lip glosses or lip balms) are found to yield very similar spectral profiles, inhibiting discrimination within brands.[112] However, Raman interest is found in the analysis of decorative cosmetic traces on a support, the results of which will be presented in Section 6.3.3.

Nail polishes as trace evidence have been the subject of attention for several research groups.[113,114] Confocal Raman spectroscopy was used for the analysis of 77 nail polishes but encountered undesirable effects when using the 532 nm laser of sample burning and/or fluorescence.[113] The use of a 780 nm laser circumvented some of these problems, yet, regrettably, no chemometric techniques for discrimination or classification were employed.[113] Visual comparison of the Raman spectra revealed that only 64% of the samples yielded a characteristic spectrum, further highlighting the need for chemometrics when analysing complex, multivariate data. Guided by this work, Chophi *et al.* trialled the analysis of 73 nail polishes using ATR–FTIR.[114] As predicted, sample burning and fluorescence were avoided, and all samples gave good-quality spectra. Interpretation of the spectra using PCA and correlation coefficient provided discriminating powers of 99.5% and 99.3% respectively.

Significant interest is reported for the analysis of eyeliners, under different forms such as kohl,[81] kajal[95,115] or more classic eyeliners,[116] and mascaras.[88] The first papers reporting eyeliner analysis used ATR–FTIR on around 40 samples.[81,95] The use of chemometrics confirms the conclusions reached *via* visual discrimination of the spectra, and their correct classification rates are 100%.

Mixing decorative cosmetic types likely to be encountered together in casework into a sample dataset is extremely interesting. Work by Chophi *et al.* demonstrates this by analysing a selection of eyeliners and eyeshadows.[117] PCA of the eyeliners initially showed good separation of the samples, but when triplicates of some of the eyeliners were plotted, they were not in close enough proximity to one another to permit interpretation. The PCA scores plot for the eyeshadows showed clustering for the majority of the samples but also some overlapping results. Therefore, LDA was applied to report classification accuracy. The LDA model resulted in only one misclassification for both the calibration and validation datasets, giving accuracy values of 98% and 96.87% respectively. Combining eyeliners and mascaras for discrimination is also interesting, given their similar composition.[88] In this case, PCA applied to ATR–FTIR spectra allowed the discrimination of samples based on their composition: paraffin, phosphates, alcohols and amides on one side, mica, esters, pyrophosphates and talc on the other. The result is a discrimination model with a specificity equal to 1 and a sensitivity of 0.9.

The analysis of kajal samples using ATR–FTIR and Raman spectroscopy is further questioned by Asri *et al.* They highlight that when the techniques are applied to the same set of samples, the classification accuracy of the Raman spectra is greater than the ATR–FTIR spectra, with 100% correct classification against 86% respectively.[115] This difference is likely due to the better suitability of Raman spectroscopy for the analysis of pigments.[118–121]

Glitter particles are often found in decorative cosmetics; they are small pieces of plastic or metal cut into various shapes (usually hexagonal).[122] ATR–FTIR analysis is particularly suitable in this case due to the composition and different thicknesses of various glitter particles.[90] Based on the discrimination of other particle types such as paint,[120,123] one can assume that glitter particles will also be easily discriminated. This is confirmed by the analysis undertaken by Najjar *et al.* which demonstrated that glitter classes are separated based on their acrylate, PET and PBT (polybutylene terephthalate) composition. A correct classification rate of 100% was obtained when coupling PCA to a *k*-nearest neighbour (*k*NN) algorithm, compared to 99% for LDA and 93% for SVM.[90]

Finally, ATR–FTIR has been employed for the differentiation of 31 cosmetic foundation creams of similar hues. Visual comparison of the spectra resulted in a discrimination power of 98%; however, once PCA was applied to the data the discriminating power rose to 99.3%, and increased further to 100% upon combination with LDA.[124] Cosmetic foundation creams were also investigated by Skobeeva and colleagues, who chose to combine near-infrared spectroscopy (NIR) with PCA-LDA to interrogate their samples.[86] They noted that the lighter-shaded foundations yielded a more accurate classification rate (96.7%) than those of the darker-shaded foundations (90%) and proposed this could be due to the optical effects of the white titanium dioxide pigments and black iron oxide pigments respectively. Surprisingly, the standard normal variate (SNV) spectral pre-processing technique did not improve the classification rates, yet it did enhance the spectra visually. Whilst NIR appeared to be a promising technique, the authors concluded that the use of ATR–FTIR would likely result in higher classification rates.[86]

6.3.2.2 Separation Methods

One of the most appealing separation methods in forensic science generally, is gas chromatography (GC), yet this technique has drawbacks when applied to the analysis of some decorative cosmetics. Coupled with flame ionization detection (FID), GC has been used for the analysis of facial foundations,[87] although its limitation for identifying this type of sample has been revealed, with half of the samples providing no chemical profile. The discrimination power was calculated at 82%, which is significantly lower than the results obtained *via* ATR–FTIR and SEM–EDX (scanning electron microscopy coupled to energy-dispersive X-ray spectroscopy). Furthermore, when coupled with both techniques, GC–FID only added an advantage of 0.2% to the discrimination of the samples.

GC/MS was more successful when applied to the discrimination of 43 lipsticks.[125] Based on a set of samples previously analysed using ATR–FTIR,[104] the discriminating power of GC/MS was superlative to that obtained with ATR–FTIR. Visual discrimination of the GC/MS data created six sample groupings, which were then confirmed using multidimensional scaling (MDS). Although correlations were observed between samples of the same manufacturer and series, no conclusive trends were identified. Focusing on enhancing the quality of the discrimination of the samples, the same set of 43 lipsticks was analysed using micellar electrokinetic capillary chromatography (MEKC).[126] No definitive trends could be observed among manufacturer, series, or index number and misclassifications were evident for samples originating from the same manufacturer and series. It should be noted though, that this method permitted the discrimination of one pair of samples that could not be achieved using GC/MS. An alternative version of capillary electrophoresis, microemulsion electrokinetic capillary chromatography (MEEKC), was employed for the analysis of red lipsticks. Regrettably, the sample set only comprised six lipsticks and analyses were performed *via* visual interpretation of the electropherograms, rather than based on chemometric techniques. The focus of the research was instead directed towards optimisation of the technique, and MEEKC proved to be an effective tool for the forensic analysis of lipstick traces.[127]

The lack of research using chromatographic techniques for decorative cosmetics analysis suggests that there are significant limitations when applied in a forensic context. It is not only the destructivity of the technique that must be considered but also the nature and the chemistry of the different cosmetics that render them unsuitable for GC analysis, *e.g.*, the lack of volatile components. This is a motivation to use pyrolysis-GC/MS (py-GC/MS), similar to condom lubricants (see Section 6.2.2.2). Py-GC/MS has been applied to lipglosses[128] and allowed the successful visual discrimination of 15 of the 21 samples based on variations in styrene and cyclohexane concentrations. The other samples could not be clearly visually distinguished; however, a subsequent blind validation procedure correctly attributed all randomly selected samples to their respective sources, revealing the potential of py-GC/MS for sample analysis and discrimination in forensic casework.[128]

6.3.2.3 Elemental Analysis Methods

Elemental analysis of decorative cosmetics has been performed using SEM–EDX,[87,89] atomic absorption spectroscopy (AAS),[94,129–131] X-ray fluorescence (XRF)[93,99,132] and inductively coupled plasma mass spectrometry (ICP–MS)[133] or coupled to optical emission spectrometry (ICP–OES)[134] for the detection of metals.

Eyeshadows have been investigated using AAS. Several metals were detected in the composition of the samples; the discrimination between those samples was defined by variations in the relative concentrations. Eyeshadows of a golden colour, for example, contain more cadmium than blue coloured eyeshadows.[130,131] No further discrimination on eyeshadows has been reported in the current literature.

Nail polishes have been investigated using portable XRF.[99,132] The discrimination of 42 nail polishes from five different brands was based on the elements phosphorus, sulfur, potassium, calcium, titanium and iron, with calcium and titanium content being the main discriminators. Using PCA, five groupings were successfully created, and these correlated directly back to the sample brand. A combination of PCA and PLS-DA was used to successfully identify sulfur content within a sample set of 100 nail polishes.[132]

XRF has also been used for the discrimination of 39 traditional and mineral foundation samples.[93] Applying HCA and then PCA, the majority of the mineral-based foundations were separated from the traditional formulations and correctly clustered into their respective manufacturer groupings. The main indicators of a traditional foundation were the magnesium and calcium concentrations, presumably due to the presence of talc and chalk in the formulations.[93] Differentiation within the mineral-based foundations was achieved by looking at the pigmentation profiles of titanium, iron and zinc, the addition of bismuth, and aluminium-containing clays.

Another set of 53 foundation samples, was subjected to a full sequence analysis of FTIR, SEM–EDX and GC–FID.[87] Similar elemental profiles were observed across all samples; however, the relative intensities of those elements were highly variable and provided a discrimination potential of 93.8%. Comparatively, SEM–EDX proved inferior to FTIR but superior to GC–FID. When coupled to FTIR, SEM–EDX analysis elevated the discriminating power to 99.5%, compared to 98.3% for FTIR analysis alone.[87]

SEM–EDX has been employed for the discrimination of 36 glitter and 40 shimmer particles.[89] The distinction between the two cosmetic sample types is that glitter comprises man-made multilayered foil and/or plastic pieces cut into regular shapes such as hexagons, whereas shimmer is a generic term that encompasses naturally occurring mineral flakes (or their synthetic analogues), such as mica, which are randomly shaped and sized. HCA and PCA yielded similar groupings for the glitter and shimmer samples; the common elements found to generate the discrimination were iron, aluminium, sulfur, silicon, and titanium. Among the results obtained, the ability to discriminate between natural and synthetic versions of mica was demonstrated.[89] Nevertheless, observations of the clusters revealed that samples of different colours were clustered together. This highlights one of the limitations of exclusively using elemental analysis and the necessity of combining instrumental techniques with the operator's visual assessment for the accurate discrimination of samples.

6.3.3 Identification of Chemical Markers for Casework

Given that cosmetic trace evidence is a relatively new research area, the current focus seems to be the discrimination of standard samples and furthering our understanding of how those samples compare to one another. Notwithstanding this, and bearing in mind the nature of evidence transfer during violent exchange, it is highly unlikely that any cosmetic sample retrieved from a scene will reliably compare with its original product. This is especially true if the original product's formulation was a liquid or gel as opposed to a loose powder.[¶¶] Ergo, by changing the substrate or performing ageing experiments, additional variables are studied and are likely to impact the initial models while being more representative of casework samples.

According to crime scene investigators, decorative cosmetic traces are most encountered on garments and bedding,[‖‖] thus indicating a need to investigate substrate interference. Recent studies have targeted substrates including cotton fabric, tissues, A4 paper sheets, skin, cigarette butts, drinking glasses and plastics as the main support likely to be recovered from casework samples.[80,88,90,99,103,104,106,135]

Interestingly, if the intensity of spectral peaks is universally reduced[104,108,136] by contact with an external matrix/substrate, visualisation of the major peaks is still possible despite the inference,[88] *i.e.*, only the minor peaks are affected, and good stability over time is observed.[95] Generally, very few interferences from the substrate are noticed[106] except when the substrate is fluorescent.[107] In this case, Raman analysis is affected, generating peak identification issues. Minor peaks are more easily disturbed and are found to either shift along the spectrum or disappear altogether.[88,107,108] The discrimination and classification of samples, when analysed upon various substrates, was still found to be efficient and, frequently, decorative cosmetic smears could be linked to their source,[86,104,106] predominantly *via* identification of the brand.

Despite garments and bedding being the most commonly encountered substrates for decorative cosmetic residues, fingermarks[137] and lip prints contaminated with cosmetics may also be found in casework. As a result, a ToF-SIMS (time of flight secondary ion mass spectrometry) method was

---

¶¶Personal communication, Robert Blackledge, Forensic Chemist Consultant.

‖‖Personal communications, Andrew Langley, University of Kent and Richard Hunt, Kent Police.

developed for the discrimination of lipsticks, various cosmetic creams, foundations and nail polishes that had been applied to a volunteer's hands and lips.[138] When coupled to PCA, the type of product could be identified from the deposited trace and linked to its source.[138] Another illustration can be seen by Ricci *et al.* who used ATR–FTIR spectroscopic imaging coupled to PCA to detect traces of face cream, foundation, body butter, and body lotion in deposited fingermarks on nonporous surfaces.[137] The contaminated fingermarks were recovered *via* gelatine lifters; the resultant spectra demonstrated that, despite the traces being invisible to the naked eye, they were successfully identified with this technique. Furthermore, fingermarks contaminated with traces of lip gloss were deposited upon a polycotton shirt (porous surface) and subjected to ATR–FTIR analysis. Reconstruction of the fingermark pattern was not possible, but the lip gloss traces were successfully located on the garment.

6.3.4 Factors Affecting Cosmetic Evidence Interpretation

Cosmetic evidence interpretation is principally considered as a source enquiry. That is to say, the forensic scientist wishes to know whether a trace detected on a substrate can be linked to either a reference sample from a database (investigative purpose) or to some material seized by/from a suspect (evaluative purpose). Comparison of samples at a source level is essential and these studies are conducted logically, starting with the samples in their most basic format and limiting the number of variables, so that models can be created.

However, the investigation of cosmetic trace evidence also requires comprehension of which cosmetic smears may be present on textiles for legitimate reasons (background considerations), how cosmetics transfer to various substrates, and their persistence on those substrates until collection. This framework is derived from the basic principles of forensic trace evidence and is used to understand the properties of the sample, how it got there, and how it has evolved until detected and collected by crime scene investigators.[75,76] It is worth mentioning at this point that the likelihood of incidentally finding decorative cosmetic traces in the environment is low,[97] and thus the discovery of these traces indicates a high probability that transfer has occurred.

The five parameters mentioned for condom evidence interpretation also apply here (see Section 6.2.4): donor, receiver, contact, elapsed time and activity of the victim.[35] In the case of cosmetics, donors are the different samples that are likely to be encountered in casework and their properties have been quite well investigated (see Section 6.3.2). The receivers are the substrates upon which the donor is found, and their interactions have also been well investigated, as reported in Section 6.3.3. The core missing elements and, hence, opportunities for further research, address cosmetic evidence interpretation with regards specifically to the contact, the background, the elapsed time effect and the activity before sampling for each of the cosmetic types presented. Research carried out in other forensic areas[139–141] could serve as a template for these investigations, particularly those pertaining to the investigation of perfumes as another type of cosmetic.[142,143] As an example, Gladysz *et al.* led an investigation into the different factors affecting lipstick transfer, and the effect of varying storage conditions and time on trace persistence.[144] A similar methodology was applied to the study of persistent lipsticks and lip prints.[145]

Further studies could entail examining the contact parameter by looking at the dynamics of cosmetic transfer during, say, a hug or a fight. The background parameter may be explored by collecting textiles within a given population and searching for cosmetic traces. The elapsed time effect may be modelled by performing ageing studies,[146] focusing on the chemical changes in the cosmetic at set intervals when applied to a given surface. Factors such as the loss of water-soluble components into the skin,[7,117] evaporation of solvents,[85] absorption of oils into the skin, UV degradation,[87] or alteration due to heat could all impact a deposited smear's composition. Finally, any changes induced in the cosmetic trace arising from the victim's activities after the moment of the aggression and up to the collection of the sample would need to be investigated. This may involve determining the effect of laundering a garment exhibiting a cosmetic trace, similar to other studies in microtraces.[72,139] The application of chemometric methods to model these changing parameters would allow analysts to make predictions based on the condition of the discovered samples in the context of the scene.

6.3.5 Other Considerations for Cosmetics and Chemometrics

6.3.5.1 Other Research

The preceding sections have focused on the detection and classification of decorative cosmetics for forensic purposes. However, in the literature, the application of chemometrics to the analysis of other cosmetics can be found. For instance, the discrimination of facial creams was investigated using ATR–FTIR[136] and Raman[147] coupled with chemometrics. ATR–FTIR was used to analyse a set of 57 cosmetic samples comprising herbal and non-herbal creams;[136] the PCA and LDA tools highlighted two clear groups of samples based on the sample type but the PLS-DA error rate was found to be 0.16. Conversely, the error rate was found to be diminished when using Raman analysis, as the results of PCA coupled to PLS-DA for 40 different cosmetic facial creams were found to be 0.07.[147] Enhanced and clearer discrimination than with ATR–FTIR was also observed.[147] Raman was also applied to the samples once deposited on various substrates, *e.g.*, textiles, and very little contribution of the substrate was observed. Correct classification was obtained for all but two of the 40 samples tested, those two samples coming from the same brand and same manufacturer.[147]

In an interesting case, the comparison of seized counterfeit perfumes with their genuine counterparts was examined using paper spray mass spectrometry (PS–MS). Trends in the data were observed using PCA, the genuine perfumes having more negative values along PC2. The classification methods PLS-DA and SIMCA were then employed to create models of the data; PLS-DA proved to be the superlative technique with no misclassifications observed. The authors note that, contrary to what might be predicted, the most abundant ions were not the most discriminating variables, and, furthermore, proposed which ions might be the best diagnostic predictors of counterfeit perfumes.[148]

Finally, chemometrics was used for the ATR–FTIR discrimination of 109 sunscreens.[149] In this study, the PCA algorithm confirmed the groupings obtained by visual spectral discrimination, enhancing the discrimination power of 0.1% to a final discrimination accuracy of 99.8%. Separation of the classes was found to be primarily due to the presence/absence of solvents, which are likely to evaporate and, secondly, based on the mineral composition (talc and mica) and paraffin concentrations.[149] Nevertheless, the solvent evaporation poses some ambiguity as it is unknown whether or not the solvent had already started to evaporate between the time of deposition and the subsequent analysis. This may render discrimination based on solvent concentration unreliable.

6.3.5.2 *Application to Casework*

Forensic research into the evidential value of cosmetic residues began in the 1980s.[22,150–153] Some of this research was case-based and was able to guide future research, for example by illustrating likely target substrates. Despite all the discrimination models discussed in the preceding sections, very few recent applications to casework are reported in the literature. The primary aim of these models is to provide information about the sample itself, *i.e.*, the extraction and analysis of the sample, without further consideration of the interpretation of the results. Assessing a sample at the source of the evidence is a valuable endeavour; however, comparing it to material seized from a suspect or a crime scene, and thus determining if the trace is linked to the criminal activity itself, would be far more beneficial to forensic practice and the criminal justice system.[***] To the best of the authors' knowledge, the application of chemometrics to the case-based analysis of decorative cosmetics has never yet been investigated. The only example of a study based on other forms of cosmetics involved the comparison of authentic and counterfeit perfumes, as discussed in Section 6.3.5.1 above.[148] Curiously, according to several reporting scientists, cosmetics analysis is performed in forensic laboratories, albeit on a minor scale, but it seems these analyses are not shared with the wider forensic community.[†††]

A current challenge facing the use of cosmetic evidence in forensic science is the collection of that evidence. Following several personal communications, it has been discovered that crime scene investigators – at both Kent Police and within the Metropolitan Police – do not routinely collect cosmetic

[***]Personal communication, Robert Blackledge.
[†††]Personal communication, Chris Brereton, Key Forensic Services.

evidence at scenes, even if those traces are immediately apparent, unless there is a particular indication to do so.[‡‡‡] Furthermore, no post-mortem cosmetic samples are taken from the cadaver, either as a reference for further investigations or as a contact-and-trace specimen.[§§§] This is a missed opportunity to add potentially valuable evidence to a case and demonstrates the urgent need for scientific validation of these types of traces. The implementation of cosmetic trace evidence into criminal cases would be hugely advantageous and the initial research in this area is encouraging. In addition to further research in the areas set out in the preceding sections, what also needs to change is peoples' acceptance of this type of evidence and their perception of its value. Every piece of evidence from a scene has the potential to provide the "missing piece of the jigsaw" and cosmetics are no exception.

6.4 Conclusion

Since the 1980s, cosmetics and personal hygiene products, such as condoms and lubricants, have become increasingly present as new types of evidence recovered in forensic casework. Although they are mass-produced products, the diversity of their chemical compositions may be exploited for the forensic identification of their source material and, consequently, aid in reconstructing the case scenario. Identification of the compounds likely to be detected in casework is a key starting point. However, given the high variability demonstrated among different sample types, discrimination of samples purely on visual assessment of the presence or absence of specific chemical pattern is not sufficient and statistical tools such as chemometrics are required to limit bias.

Chemometrics is used complementarily to these visual observations of the data, and at different stages over the course of the research. The most important focus is the discrimination of pure samples purchased from different markets, to build models dedicated to source identification and interpretation, within a Bayesian context of evidence interpretation. The most common algorithms used in cosmetic and personal hygiene product discrimination are HCA and PCA, coupled with classification algorithms such as LDA, SVM, QDA or *k*NN; the choice of technique is determined according to the highest discrimination levels and most accurate classification rates.

While source interpretation is the primary focus in casework, effective interpretation of that evidence cannot be made without certain prerequisite knowledge: the prevalence of the compounds of interest within the population, an understanding of how different substrates might impact the trace recovery, and how those substrate interactions will affect the subsequent trace chemical profile obtained. Chemometrics allows the discrimination of various samples collected for the background samples, and has illustrated for example the incredible variability due to human samples, making interpretation more complex. In addition to background considerations, an

---

[‡‡‡]Personal communications, Robert Green, University of Kent and Phill Pemble, Kent Police.
[§§§]Personal communication, Andrew Langley, University of Kent.

appreciation of the mechanisms of transfer and persistence of these kinds of traces will make research more applicable to casework, at both investigative and evaluative levels. Here again, chemometrics has been found to be of valuable help to observe patterns due to elapsed time before sample collection as well as to highlight a strong variability within the samples.

Future directions include raising awareness of the potential evidential value of cosmetics and personal hygiene products, among police forces and forensic laboratories, evaluating challenges and orienting research accordingly, and determining the most appropriate collection methods and analyses for this particular type of trace evidence. Ultimately, evidence that is not collected at the scene is evidence that is lost.

Abbreviations

AAS	Atomic absorption spectroscopy
AccuTOF-DART	Accurate time of flight direct analysis in real-time
ANOVA	Analysis of variance
ATR	Attenuated total reflectance
CE-UV-Vis	Capillary electrophoresis coupled to an ultraviolet-visible detector
CURML	Centre Universitaire Romand de Médecine Légale
DART–TOF/MS	Direct analysis in real-time–time of flight mass spectrometry
DCM	Dichloromethane
DESI–MS	Desorption electrospray ionisation mass spectrometry
DNA	Deoxyribonucleic acid
DRIFTS	Diffuse reflectance infrared Fourier transform spectroscopy
EASI–MS	Electrospray ambient ionisation mass spectrometry
FID	Flame ionization detector
FTIR	Fourier transform infrared spectroscopy
GC	Gas chromatography
GC/FID	Gas chromatography coupled to a flame ionization detector
GC/MS	Gas chromatography coupled to mass spectrometry
HCA	Hierarchical cluster analysis
ICP–MS	Inductively coupled plasma mass spectrometry
ICP–OES	Inductively coupled plasma optical emission spectrometry
ISO	International Standardisation Organisation
*k*NN	*k*-nearest neighbours
LC	Liquid chromatography
LC/MS	Liquid chromatography coupled to mass spectrometry
LDA	Linear discriminant analysis
MALDI–MS	Matrix-assisted laser desorption ionisation–mass spectrometry

MALDI–TOF	Matrix-assisted laser desorption ionisation–time of flight
MEKC	Micellar electrokinetic capillary chromatography
MEEKC	Microemulsion electrokinetic capillary chromatography
MeOH	Methanol
MDS	Multidimensional scaling
NAS	National Academy of Science
NIR	Near-infrared spectroscopy
PBT	Polybutylene terephthalate
PCA	Principal component analysis
PCAST	President's Council of Advisors on Science and Technology
PCR	Principal component regression
PDMS	Polydimethylsiloxane
PEG	Polyethylene glycol
PET	Polyethylene terephthalate
PLS-DA	Partial least squares discriminant analysis
PLSR	Partial least squares regression
PPG	Polypropylene glycol
PS–MS	Paper spray mass spectrometry
py-GC/MS	Pyrolysis gas chromatography coupled to mass spectrometry
QDA	Quadratic discriminant analysis
SARC	Sexual Assault Resource Center
SDA	Sparse discriminant analysis
SEM–EDX	Scanning electron microscopy coupled to energy-dispersive X-Ray spectroscopy
SIMCA	Soft independent modelling of class analogies
SNV	Standard normal variate
STD	Sexually transmitted diseases
SVM	Support vector machine
ToF-SIMS	Time of flight secondary ion mass spectrometry
XRF	X-Ray fluorescence

Acknowledgements

The author would like to acknowledge Prof. Patrick Buzzini (Sam Houston State University, USA), Dr Donna Arnold (University of Kent, UK), and Dr Amanda Frick (Curtin University, Perth, WA) for English proofreading and suggestions on the content of the present book chapter.

References

1. *Crime scene to court: the essentials of forensic science*, ed. P. White and Royal Society of Chemistry (Great Britain), Royal Society of Chemistry, Cambridge, 4th edn, 2016.

2. A. R. W. Jackson and J. M. Jackson, *Forensic science*, Pearson Prentice Hall, Harlow, 2008.
3. R. Saferstein, *Criminalistics: an introduction to forensic science*, Pearson Education, NY, 12th edn, 2018.
4. K. Rayment, *J. Law Inf. Sci.*, 2010, **20**, 238.
5. F. Gunn, *The artificial face: a history of cosmetics*, David & Charles, Newton Abbot, 1973.
6. L. Eldridge, *Face paint: the story of makeup*, Abrams Image, New York, 2015.
7. R. Chophi, S. Sharma, S. Sharma and R. Singh, *Forensic Chem.*, 2019, **14**, 100165.
8. International Organization for Standardization, *Condoms – Determination of nitrosamines migrating from natural rubber latex condoms*, 2010.
9. M. Potoczny, F. Truchetet and G. Gratier de Saint Louis, *Rev. Sage-Femme*, 2017, **16**, 119–123.
10. K. Tanfer, W. R. Grady, D. H. Klepinger and J. O. Billy, *Fam. Plann. Perspect.*, 1993, **25**, 61–66.
11. A. Raj, E. Reed, E. Miller, M. R. Decker, E. F. Rothman and J. G. Silverman, *AIDS Care*, 2007, **19**, 970–973.
12. National Center for HIV/AIDS, Viral Hepatitis, STD, and TB Prevention (U.S.), Division of STD Prevention, J. Braxton, D. W. Davis, B. Emerson, E. W. Flagg, J. Grey, L. Grier, A. Harvey, S. Kidd, J. Kim, K. Kreisel, E. Llata, K. Mauk, R. J. Presley, V. Ramirez, S. Shapiro, S. St. Cyr, M. Stenger, E. Torrone, H. Weinstock, E. J. Weston and N. Womack, *Sexually transmitted disease surveillance 2017*, Centers for Disease Control and Prevention, 2018.
13. E. N. O'Neal, S. H. Decker, C. Spohn and K. Tellis, *J. Forensic Leg. Med.*, 2013, **20**, 605–609.
14. K. Kelder and B. Burton, *Can. Soc. Forensic Sci. J.*, 2002, **35**, 165–175.
15. *Forensic analysis on the cutting edge: new methods for trace evidence analysis*, ed. R. D. Blackledge, Wiley, John Wiley [distributor], Hoboken, NJ, Chichester, 2007.
16. R. D. Blackledge, *Crime Lab. Dig.*, 1994, **21**, 57.
17. R. D. Blackledge and M. Vincenti, *J. Forensic Sci. Soc.*, 1994, **34**, 245–256.
18. F. Burger, M. Dawson, C. Roux, P. Maynard, P. Doble and P. Kirkbride, *Talanta*, 2005, **67**, 368–376.
19. S. E. Spencer, S. Y. Kim, S. B. Kim and K. A. Schug, *Forensic Sci. Int.*, 2011, **207**, 19–26.
20. E. J. Wagner, *The science of Sherlock Holmes: from Baskerville Hall to the Valley of Fear, the real forensics behind the great detective's greatest cases*, Wiley, Hoboken, NJ, 2006.
21. W. Farrukh and D. van der Haar, *IET Biometrics*, 2023, **12**, 1–12.
22. R. L. Keagy, *J. Forensic Sci.*, 1983, **28**, 11558J.
23. EWCA Crim., *Regina v. Andrew Nicholas Malkinson*, 2006.
24. R. D. Blackledge, *J. Forensic Sci.*, 1995, **40**, 467–469.
25. C. Burnier, M. Kelly, D. DeTata and K. Pitts, *Forensic Sci. Int. Rep.*, 2021, **4**, 100221.

26. L. S. Tottey, S. A. Coulson, G. E. Wevers, L. Fabian, H. McClelland and M. Dustin, *J. Forensic Sci.*, 2018, **64**, 207–217.
27. P. Tozzo, P. Nespeca, E. Ponzano and L. Caeazzo, *Forensic Sci. Int. Genet. Suppl. Ser.*, 2019, **7**, 85–86.
28. J. Fischer, N. Saric, L. Fabien, A. Hermelin, G. Massonnet and C. Burnier, *Forensic Chem.*, 2021, **26**, 100357.
29. A. Hermelin, L. Fabien, J. Fischer, N. Saric, G. Massonnet and C. Burnier, *Forensic Sci. Int.*, 2021, **321**, 110724.
30. N. Saric, L. Fabien, J. Fischer, A. Hermelin, G. Massonnet and C. Burnier, *Forensic Sci. Int.*, 2021, **325**, 110847.
31. T. Coyle and N. Anwar, *Sci. Justice*, 2009, **49**, 32–40.
32. International Organization for Standardization, *Natural rubber latex male condoms – Requirements and test methods*, 2015.
33. International Organization for Standardization, *Préservatifs en caoutchouc – Directives sur l'utilisation de l'ISO 4074 dans le management de la qualité des préservatifs en latex de caoutchouc naturel*, 2005.
34. International Organization for Standardization, *Lubrifiants additionnels et péparations médicamenteuses ou non, destinés à ou susceptibles d'être mis en contact avec des préservatifs masculins*, 2007.
35. C. Burnier and G. Massonnet, *Forensic Sci. Int.*, 2019, **302**, 109861.
36. P. Maynard, K. Allwell, C. Roux, M. Dawson and D. Royds, *Forensic Sci. Int.*, 2001, **124**, 140–156.
37. C. Burnier and G. Massonnet, *Forensic Sci. Int.*, 2020, **310**, 110255.
38. M. Tonkin, L. F. Yeap, E. K. Bartle and A. Reeder, *J. Forensic Sci.*, 2013, **58**, S26–S33.
39. C. Burnier, S. Coulson, G. Massonnet, K. Pitts, G. Sauzier and S. W. Lewis, *Sci. Justice*, 2021, **61**, 235–248.
40. C. Burnier, G. Massonnet, S. Coulson, D. DeTata and K. Pitts, *Forensic Sci. Int.*, 2020, **317**, 110513.
41. M. Maric and C. Bridge, *Forensic Sci. Int.*, 2016, **266**, 73–79.
42. M. Maric, L. Harvey, M. Tomcsak, A. Solano and C. Bridge, *Rapid Commun. Mass Spectrom.*, 2017, **31**, 1014–1022.
43. B. Baumgarten, M. Marić, L. Harvey and C. M. Bridge, *Forensic Chem.*, 2018, **8**, 28–39.
44. National Research Council (U.S.), *Strengthening forensic science in the United States: a path forward*, National Academies Press, Washington, D.C., 2009.
45. S. Bell, S. Sah, T. D. Albright, S. J. Gates, M. B. Denton and A. Casadevall, *Proc. Natl. Acad. Sci. U. S. A.*, 2018, **115**, 4541–4544.
46. President's Council of Advisors on Science and Technology, *Forensic Science in Criminal Courts: Ensuring Scientific Validity of Feature-Comparison Methods*, Washington, D.C., 2016.
47. L.-L. Cho and K.-B. Huang, *Forensic Sci. Int.*, 2012, **11**, 33–40.
48. C. Burnier, V. Favre and G. Massonnet, *Spectrochim. Acta, Part A*, 2021, **261**, 120025.

49. S. A. L. Thomas, N. Andersen, M. Marić and C. Bridge, *Forensic Chem.*, 2021, **24**, 100329.
50. C. A. Burnier, W. van Bronswijk and G. Massonnet, *Anal. Methods*, 2020, **12**, 657–665.
51. B. Baumgarten, S. Thomas, N. Flynn, M. Marić, N. T. Nguyen, L. Tang and C. Bridge, *Forensic Chem.*, 2022, **27**, 100403.
52. J. N. Lee, C. Park and G. Whitesides, *Anal. Chem.*, 2003, **75**, 6544–6554.
53. C. Burnier, M. Monzò, G. Sauzier and S. W. Lewis, *Forensic Sci. Int. Rep.*, 2022, 100283.
54. R. A. Musah, A. L. Vuong, C. Henck and J. R. E. Shepard, *J. Am. Soc. Mass Spectrom.*, 2012, **23**, 996–999.
55. R. D. Blackledge and L. R. Cabiness, *J. Forensic Sci.*, 1983, **28**, 451–462.
56. S. A. L. Thomas, B. Baumgarten, D. Green and C. Bridge, *Forensic Sci. Int.*, 2022, **335**, 111273.
57. J. Maurer, K. Buffaz, G. Massonnet, C. Roussel and C. Burnier, *J. Anal. Appl. Pyrolysis*, 2020, **149**, 104861.
58. S. Bell, S. Sah, T. D. Albright, S. J. Gates, M. B. Denton and A. Casadevall, *Proc. Natl. Acad. Sci. U. S. A.*, 2018, **115**, 4541–4544.
59. A. O. Onigbinde, G. Nicol and B. Munson, *Eur. J. Mass Spectrom.*, 2001, **7**, 279–291.
60. A. O. Onigbinde, B. Munson and B. M. W. Amos-Tautua, *Res. J. Appl. Sci. Eng. Technol.*, 2013, **5**, 2332–2339.
61. T. P. E. Hollenbeck, G. Siuzdak and R. D. Blackledge, *J. Forensic Sci.*, 1999, **44**, 783–788.
62. Y. S. Chhonker, H. Chandasana, V. Bala, L. Kumar, V. L. Sharma, G. Gupta and R. S. Bhatta, *J. Chromatogr. B: Anal. Technol. Biomed. Life Sci.*, 2014, **965**, 127–132.
63. B. A. Walter and G. A. Digenis, *Pharm. Res.*, 1991, **08**, 409–411.
64. M. J. Shifflet, M. Shapiro, C. Levin and R. DeNisco, *LCGC*, 2001, **19**, 312–317.
65. Y. Moustafa and C. M. Bridge, *Forensic Chem.*, 2017, **5**, 58–71.
66. A. M. Coon, S. Beyramysoltan and R. A. Musah, *Talanta*, 2019, **194**, 563–575.
67. M. F. Mirabelli, D. R. Ifa, G. Sindona and A. Tagarelli, *J. Mass Spectrom.*, 2015, **50**, 749–755.
68. M. F. Mirabelli, A. Chramow, E. C. Cabral and D. R. Ifa, *J. Mass Spectrom.*, 2013, **48**, 774–778.
69. C. Burnier, G. Massonnet, S. Coulson, D. DeTata and K. Pitts, *Forensic Sci. Int.*, 2021, **324**, 110793.
70. B. Woffinden, *The Nicholas cases: casualties of justice*, Bojangles Books, Great Britian, 2016.
71. C. Mbo, F. Honoré, M.-P. Milon, G. Massonnet and C. Burnier, *Forensic Sci. Int.*, 2022, **331**, 111141.
72. C. Mbo and C. Burnier, *Forensic Sci. Int.*, 2021, **329**, 111077.
73. G. Wagner and R. J. Levin, *Reproduction*, 1980, **60**, 17–27.

74. M. Szeremeta, K. Pietrowska, A. Niemcunowicz-Janica, A. Kretowski and M. Ciborowski, *Int. J. Mol. Sci.*, 2021, **22**, 3010.
75. C. Roux, R. Bucht, F. Crispino, P. De Forest, C. Lennard, P. Margot, M. D. Miranda, N. NicDaeid, O. Ribaux, A. Ross and S. Willis, *Forensic Sci. Int.*, 2022, **332**, 111182.
76. C. Roux, S. Willis and C. Weyermann, *Sci. Justice*, 2021, **61**, 678–686.
77. R. Bradshaw, R. Wolstenholme, L. S. Ferguson, C. Sammon, K. Mader, E. Claude, R. D. Blackledge, M. R. Clench and S. Francese, *Analyst*, 2013, **138**, 2546.
78. W. van Helmond, M. P. V. Begieneman, R. Kniest and M. de Puit, *Forensic Sci. Int.*, 2019, **305**, 110005.
79. S. Antczak and G. Antczak, *Cosmetics unmasked: your family guide to safe cosmetics and toiletries*, Thorsons, London, 2001.
80. V. Sharma, A. Bharti and R. Kumar, *Spectrochim. Acta, Part A*, 2019, **215**, 48–57.
81. V. Sharma, S. Bhardwaj and R. Kumar, *Vib. Spectrosc.*, 2019, **101**, 81–91.
82. European Parliament and the Council of the European Union, *Regulation (EC) no. 1223/2009 of the European Parliament and of the Council of 30 November 2009 on cosmetic products*, European Parliament and the Council of the European Union, 2009.
83. European Parliament, *Regulation (EC) No. 1907/2006 of the European Parliament and of the Council, concerning the Registration, Evaluation, Authorisation and Restriction of Chemicals (REACH)*, 2006.
84. M. Gładysz, M. Król and P. Kościelniak, *Microchem. J.*, 2021, **164**, 106002.
85. J. X. W. Wong, G. Sauzier and S. W. Lewis, *Forensic Sci. Int.*, 2019, **298**, 88–96.
86. S. Skobeeva, A. Banyard, B. Rooney, R. Thatti, B. Thatti and J. Fletcher, *Sci. Justice*, 2022, **62**, 327–335.
87. A. L. Gordon and S. Coulson, *J. Forensic Sci.*, 2004, **49**, 1244–1252.
88. T. Arora, R. Verma, R. Kumar, R. Chauhan, B. Kumar and V. Sharma, *Microchem. J.*, 2021, **164**, 106080.
89. K. Najjar and C. M. Bridge, *Forensic Sci. Int.*, 2020, **317**, 110527.
90. K. Najjar and C. Bridge, *Spectroscopy*, 2018, **33**, 30–37.
91. D. N. de Oliveira, M. Siqueira, S. Sartor and R. Catharino, *Int. J. Cosmet. Sci.*, 2013, **35**, 467–471.
92. L.-L. Cho and I.-C. Lee, *Forensic Sci. J.*, 2020, **19**, 39–48.
93. E. Kulikov, K. Latham and M. J. Adams, *X-Ray Spectrom.*, 2012, **41**, 410–415.
94. É. F. Batista, A. dos, S. Augusto and E. R. Pereira-Filho, *Talanta*, 2016, **150**, 206–212.
95. S. Sharma, R. Chophi, R. Kumar, V. Sharma and R. Singh, *Forensic Sci. Int.*, 2019, **303**, 109930.
96. R. Goodacre, N. Kaderbhai, A. C. McGovern and E. A. Goodacre, in *Kohonen Maps*, Elsevier, 1999, pp. 335–347.

97. K. Aardahl, S. Kirkowski and R. D. Blackledge, *Sci. Justice*, 2005, **45**, 7–12.
98. S. Gross, K. Igowsky and E. Pangerl, *JASTEE*, 2010, **1**, 62–72.
99. G. G. Shimamoto, J. Terra and M. I. M. S. Bueno, *J. Braz. Chem. Soc.*, 2013, **24**, 731–735.
100. Y. Ehara and Y. Marumo, *Forensic Sci. Int.*, 1998, **96**, 1–10.
101. S. Pasieczna-Patkowska and T. Olejnik, *Ann. UMCS Chem.*, 2014, **68**, 95–106.
102. F. Salahioglu and M. J. Went, *Forensic Sci. Int.*, 2012, **223**, 148–152.
103. K. Kaur, P. K. Yadav, G. S. Bumbrah and R. M. Sharma, *Vib. Spectrosc.*, 2020, **110**, 103146.
104. M. Gładysz, M. Król and P. Kościelniak, *Forensic Sci. Int.*, 2017, **280**, 130–138.
105. R. Chophi, S. Sharma and R. Singh, *Forensic Chem.*, 2020, **17**, 100209.
106. F. Salahioglu, M. J. Went and S. J. Gibson, *Anal. Methods*, 2013, **5**, 5392.
107. P. Gardner, M. F. Bertino, R. Weimer and E. Hazelrigg, *Forensic Sci. Int.*, 2013, **232**, 67–72.
108. M. López-López, N. Özbek and C. García-Ruiz, *Talanta*, 2014, **123**, 135–139.
109. F. M. Ghazali and D. Ismail, *Int. J. Med. Sci.*, 2018, **3**, 48–52.
110. C. Rodger and D. Broughton, *The Analyst*, 1998, **123**, 1823–1826.
111. M. Gładysz, M. Król, A. Karoly, R. Szalai and P. Kościelniak, *J. Forensic Sci.*, 2022, **67**, 494–504.
112. P. Gardner, M. F. Bertino and R. Weimer, *JASTEE*, 2015, **6**, 42–57.
113. M. López-López, J. Vaz and C. García-Ruiz, *Talanta*, 2015, **138**, 155–162.
114. R. Chophi, S. Sharma and R. Singh, *Aust. J. Forensic Sci.*, 2021, **53**, 325–336.
115. M. N. M. Asri, R. Verma, M. H. Ibrahim, V. Sharma and N. A. M. Nor, *Microchem. J.*, 2021, **169**, 106556.
116. K. Jallad and H. Hedderich, *J. Hazard. Mater.*, 2005, **124**, 236–240.
117. R. Chophi, S. Sharma, J. K. Jossan and R. Singh, *Forensic Sci. Int.*, 2021, **329**, 111062.
118. P. Buzzini, G. Massonnet and F. Monard Sermier, *J. Raman Spectrosc.*, 2006, **37**, 922–931.
119. R. Chen, J. Lv and J. Feng, *Anal. Lett.*, 2015, **48**, 1502–1510.
120. D. Lambert, C. Muehlethaler, L. Gueissaz and G. Massonnet, *J. Raman Spectrosc.*, 2014, **45**, 1285–1292.
121. J. De Gelder, P. Vandenabeele, F. Govaert and L. Moens, *J. Raman Spectrosc.*, 2005, **36**, 1059–1067.
122. L. Vernoud, H. A. Bechtel, M. C. Martin, J. A. Reffner and R. D. Blackledge, *Forensic Sci. Int.*, 2011, **210**, 47–51.
123. C. Muehlethaler, PhD thesis, University of Lausanne, 2015.
124. S. Sharma, R. Chophi, H. Kaur and R. Singh, *J. Forensic Sci.*, 2020, **65**, 751–761.
125. M. Gładysz, M. Król, P. Własiuk, M. Piwowar, G. Zadora and P. Kościelniak, *J. Chromatogr. A*, 2018, **1577**, 92–100.

126. M. Gładysz, M. Król, K. Mystek and P. Kościelniak, *Forensic Sci. Int.*, 2019, **299**, 49–58.
127. M. Król, M. Nowak, M. Gładysz and P. Kościelniak, *Microchem. J.*, 2020, **155**, 104735.
128. M. Zellner and L. Quarino, *J. Forensic Sci.*, 2009, **54**, 1022–1028.
129. A. Sani, M. B. Gaya and F. A. Abubakar, *Toxicol. Rep.*, 2016, **3**, 866–869.
130. L. Świerczek, B. Cieślik, A. Matysiak and P. Konieczka, *Monatshefte Für Chem. - Chem. Mon.*, 2019, **150**, 1675–1680.
131. Z. Mousavi, P. Ziarati and A. Shariatdoost, *J. Environ. Anal. Toxicol.*, 2013, **3**, 1000193.
132. F. L. Melquiades and A. M. Alves da Silva, *Anal. Methods*, 2016, **8**, 3920–3926.
133. I. de la Calle, M. Menta, M. Klein and F. Séby, *Talanta*, 2017, **171**, 291–306.
134. F. A. Ababneh and I. F. Al-Momani, *Environ. Forensics*, 2018, **19**, 134–142.
135. J. Lee and Y. Lee, *J. Anal. Chem.*, 2021, **76**, 854–867.
136. A. Sharma, R. Chauhan, R. Kumar, P. Mankotia, R. Verma and V. Sharma, *Spectrochim. Acta, Part A*, 2021, **258**, 119803.
137. C. Ricci and S. G. Kazarian, *Surf. Interface Anal.*, 2010, **42**, 386–392.
138. T. Terlier, K.-B. Lee and Y. Lee, *Microchem. J.*, 2020, **159**, 105446.
139. T. Hicks, R. Vanina and P. Margot, *Sci. Justice*, 1996, **36**, 101–107.
140. L. Kitto, G. P. Campbell, A. L. Gordon and S. Coulson, *Clearwater Beach*, Florida, 2009.
141. R. Palmer and H. J. Burch, *Sci. Justice*, 2009, **49**, 259–264.
142. S. Gherghel, R. M. Morgan, J. F. Arrebola-Liébanas, C. S. Blackman, A. Garrido-Frenich and I. P. Parkin, *Sci. Justice*, 2020, **60**, 53–62.
143. S. Gherghel, R. M. Morgan, J. F. Arrebola-Liébanas, C. S. Blackman and I. P. Parkin, *Sci. Justice*, 2019, **59**, 256–267.
144. M. Gładysz, M. Król, A. Chudecka and P. Kościelniak, *Forensic Sci. Int.*, 2020, **309**, 110230.
145. M. A. Seguí, M. M. Feucht, A. C. Ponce and F. A. V. Pascual, *Forensic Sci. Int.*, 2000, **112**, 41–47.
146. S. Sharma, S. S. Sahota and R. K. Garg, *Int. J. Med. Toxicol. Leg. Med.*, 2016, **19**, 36–41.
147. M. N. M. Asri, R. Verma, M. H. Ibrahim, N. A. M. Nor, V. Sharma and D. Ismail, *Sci. Justice*, 2021, **61**, 687–696.
148. J. A. R. Teodoro, H. V. Pereira, D. N. Correia, M. M. Sena, E. Piccin and R. Augusti, *Anal. Methods*, 2017, **9**, 4979–4987.
149. A. Angrish, R. Kumar, R. Chauhan and V. Sharma, *Spectrochim. Acta, Part A*, 2020, **242**, 118708.
150. L. W. Russell and A. E. Welch, *Forensic Sci. Int.*, 1984, **25**, 105–116.
151. M. C. Grieve, *J. Forensic Sci. Soc.*, 1987, **27**, 405–412.
152. J. Andrasko, *Forensic Sci. Int.*, 1981, **17**, 235–251.
153. D. J. Reuland and W. A. Trinler, *J. Forensic Sci. Soc.*, 1980, **20**, 111–120.

CHAPTER 7

Trace Evidence

G. SAUZIER* AND S. W. LEWIS*

School of Molecular and Life Sciences, Curtin University, Australia
*Emails: Georgina.Sauzier@curtin.edu.au; S.Lewis@curtin.edu.au

7.1 Introduction

A key underlying tenet of forensic science is Locard's exchange principle, named for the forensic pioneer Edmond Locard. This principle states that during a criminal act, a cross-transfer of physical material (however minute) can occur:

> "The truth is that none can act with the intensity induced without leaving multiple traces of his passing.... Sometimes the criminal leaves traces at a scene by his actions; sometimes, alternatively, he picked up on his clothes or his body traces of his location or presence." (E. Locard, L'enquête criminelle et les méthodes scientifiques, translated F. Crispino 1920, cited in Houck.[1])

Locard's key focus in establishing this principle was the transfer of dust. This would be recognised today as a form of trace evidence; defined as minute quantities of material recovered as evidence that can be used to help solve criminal cases.[2]

The key value of trace evidence in criminal investigations is that, due to its small size, it is highly likely to be inadvertently transferred during criminal activity. The recovery of these traces leads to the possibility of establishing associations between people, locations, and objects of interest. Beyond this evidence of association, it also has the potential to provide information concerning the activity surrounding the exchange, thus aiding in the reconstruction of events.

Chemometric Methods in Forensic Science
Edited by Vishal Sharma, Georgina Sauzier and Simon W. Lewis

Published by the Royal Society of Chemistry, www.rsc.org

Many types of evidence can exist in trace form; however, the focus of this chapter will be those forms of physical trace evidence that are frequently encountered by the forensic scientist.† Most of these are sourced primarily from manufactured items (*e.g.* paint, fibres, and glass) rather than having a biological origin. Although many such items are mass produced, they may possess distinctive properties that lend them probative value as forensic evidence.

7.1.1 Forensic Analysis of Trace Evidence

Forensic analysis of trace evidence consists of characterisation of its chemical and physical properties to determine its identity. It is important to note that this step will only categorise an item of evidence as a member of a group (class). The analysis workflow will typically involve a sequence of techniques, as shown in Figure 7.1. This sequence will begin with various forms of microscopic analysis, followed by spectroscopic analysis. Depending on the nature of the evidence, this will be supported by chromatographic separations and/or elemental analysis. The essential rule is that destructive forms of analysis are only applied at the end of the sequence to maximise preservation of the evidence.

Samples collected for analysis can be divided into known and questioned specimens. A known sample is intentionally collected to represent the item population from which it was taken. Examples might include the paint from a car involved in a hit-and-run incident, or the fibres from the clothing of

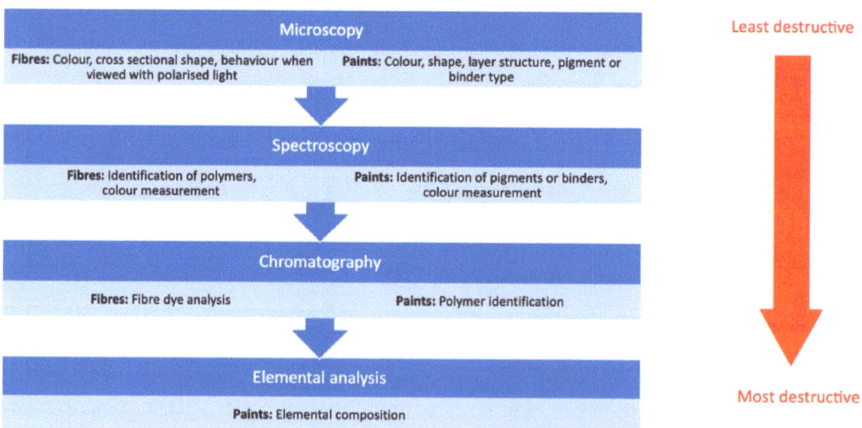

Figure 7.1 Summary of typical analytical sequences for trace evidence and the properties that may be characterised, using the examples of fibre and paint evidence.

---

†For brevity, selected forms of trace evidence are discussed separately in other chapters of this volume; notably Chapter 4 (Fire and Explosions Investigation) and Chapter 6 (Cosmetics and Personal Hygiene Products).

the victim. A questioned sample is incidentally produced by the interaction of item(s) during a crime and is often of undetermined origin. Using the example of the hit-and-run, this could include smears of paint on the victim's clothing, or any fibres from their clothing transferred by the impact to the damaged area of the car.

In cases where both questioned and known samples are available, the analytical results for both will be directly compared, *i.e.* a "questioned *versus* known" (or Q *vs.* K) comparison. The aim is to determine whether they share sufficient physical and/or chemical characteristics to suggest that an association exists between them; that is, whether it can be concluded that they share a common source. In scenarios where there are no known samples available for comparison, the chemical and physical properties of the questioned sample may instead provide investigative leads.

When an item of evidence is categorised as belonging to a class, it shares certain characteristics with other members of that class. These classes are scalable in membership, and in principle once a class has a membership of one, then the item has been individualised. In other words, that item has been associated with a specific single source (Figure 7.2).[3]

However, considering the size of a potential population of a particular type of evidence in the world, it cannot be considered realistic to truly individualise such forms of evidence. In the absence of individualisation, it follows that the value and interpretation of trace evidence is highly dependent on the rarity of class features and the ability to reliably distinguish these for determining provenance. This applies both with respect to Q *vs.* K comparisons and generating investigative leads or forensic intelligence. The challenge lies in how to appropriately determine and convey such information. While this issue has had a raised profile in recent

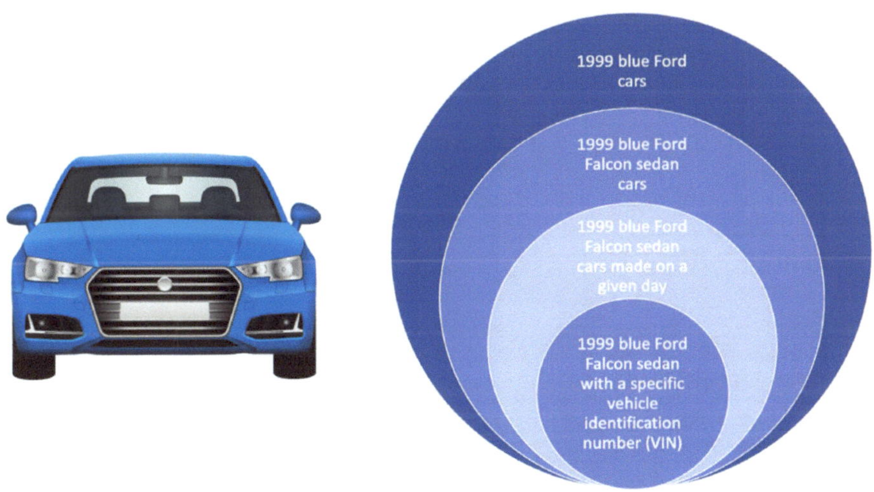

Figure 7.2 Schematic demonstrating the scalability of class characteristics for trace evidence.

years, it has long been recognised as an issue with physical evidence. As noted by chemical criminalist Paul Kirk:

> "...it is possible to identify all types of textile fiber with exactness, through a variety of methods... However, the probabilistic value of a fiber transfer between two sets of clothing, as in a crime, is still a matter of controversy."[4]

7.2 Chemometric Techniques for Trace Evidence

Chemometric methods are an emerging approach to addressing the challenges outlined above, as they can rapidly extract or generate data from large populations to statistically determine feature rarity or sample similarity. These methods also offer a powerful tool for investigating contextual factors affecting analysis or interpretation of traces, supporting the increasing requirements for activity-level interpretation. To date, however, most studies applying chemometrics to trace evidence have focussed on answering questions related to Q *vs.* K comparisons, *e.g.* "are these items similar/could they have originated from the same source?". Such questions can be answered through the use of pattern recognition (see Figure 7.3).

Unsupervised (exploratory analysis) techniques model sources of variation between samples without assuming that this variation is due to different sample classes. These methods are ideal for objectively assessing the extent and (in some cases) underlying sources of diversity between samples. This can establish whether traces show detectable variability based on certain properties, and hence whether discrimination based on those properties could be achieved. If enough samples are available then it may be possible to draw general inferences about a certain type of trace material, but in other cases, analysis could be limited to just two samples under direct comparison.

Conversely, supervised techniques are used to build classification models for known sample classes. Depending on how significant the variation between

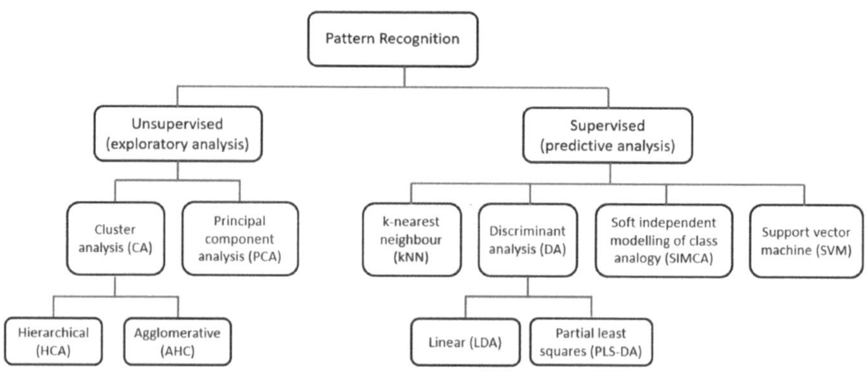

Figure 7.3 Common pattern recognition methods used in research studies examining trace evidence of manufactured origins.

samples is, these classes may correspond to a broad set of sample properties, or to more specific characteristics such as sample origin (*e.g.* brand or manufacturer). Supervised techniques are thus suited to classification or source prediction of questioned traces. They are often applied in sequence after an initial unsupervised analysis, which identifies groupings and major sources of variation for supervised model building.

Less commonly, experimental design and regression techniques can be used to answer questions relating to the suitability of methods used to examine trace materials, or to quantify sample properties.

7.3 Polymers

Polymer (plastic) items make up a sizeable proportion of everyday consumer products, ranging from acrylate paints to polyethylene food wraps. Although these items can act as substrates for other traces such as fingermarks and DNA, their potential value as a form of trace evidence themselves should not be dismissed. Although plastic items are mass-produced, most raw polymers require the addition of co-polymers or additives to create the desired properties for common end uses. This means that variations in polymer composition and structure can exist due to different combinations of raw materials, additives, or production processes.

As with all forms of trace evidence, polymers are only of strong evidential value if they can be classified within a limited class of items based on distinguishable features. Several studies have hence used chemometrics to assess the chemical diversity amongst polymer products including paints, adhesives tapes, or food wraps to better exploit them as forms of evidence.

7.3.1 Paint

Paints and varnishes are routinely applied to manufactured items as a means of protection or aesthetic modification. Significant chemical variation exists amongst paint formulations due to factors including:

- Pigments used to create specific colours or visual effects.
- Additives used to impart properties suited to intended end-uses, such as reflective glass beads in paint designed for road markings.
- Environmental and health regulations regarding solvent use, which has recently resulted in a trend towards waterborne formulations.

The prevalence and diversity of paints makes them potentially valuable evidence in criminal cases involving architectural damage, automotive vehicles, or artworks. A high-profile example is the use of microscopic traces of automotive paints to link the suspect of the Green River murders to his victims.[5] Significant research has used chemometrics to profile various paint types for purposes including prediction of the paint manufacturer, identification of key components, or investigation of degradation effects.

Table 7.1 Purposes of research studies applying chemometric methods to the forensic characterisation of architectural, automotive, artist, and spray paints. NA = not applicable.

Paint type	CA, PCA and DA	SIMCA, SVM, and PLS-DA	Other
Architectural	Exploratory analysis[6–8]	N/A	Exploratory analysis[9]
Automotive	Exploratory analysis[10–18] Classification[19] Manufacturer prediction[11–13] Degradation studies[19–21]	Classification[19] Degradation studies[19]	Database searching[22–25] Manufacturer prediction[26] Optimisation[14,27]
Artist	Component identification[28–30] Degradation studies[31] Layer analysis[32,33]	N/A	Dating[34]
Spray	Exploratory analysis[35] Classification[35]	Classification[35,36]	N/A

Table 7.1 summarises several of these studies, a selection of which are further discussed below.

Muehlethaler *et al.* applied Fourier transform infrared (FTIR) and Raman spectroscopy with PCA to analyse 34 red household paints.[8] This approach was able to distinguish samples according to their binder type (alkyd or acrylic), calcium carbonate content, or pigment composition. The same authors later applied FTIR to 74 red, green and blue spray paints.[35] Iterative PCA, in which successive models are constructed for overlapping samples to try and improve separation, was able to distinguish over 90% of samples in each colour category. Several supervised methods were used to construct models for classifying spray paints into chemical categories, with SIMCA giving the best results (*ca.* 95% correct classification rate) using a separate test set.

More recently, He *et al.* demonstrated the ability of microscopic laser Raman spectroscopy in combination with Bayes discriminant analysis to differentiate 252 white architectural paint samples representing seven separate manufacturers.[9] This study also showed the importance of appropriate data pre-processing prior to chemometric analysis, with several polynomial corrections required to combat the effects of baseline drift and high frequency noise.

Automotive paint is frequently encountered due to incidents such as hit-and-run cases. It may be particularly distinctive due to its multi-layered structure and the wide variety of colours or finishes available. A series of studies by Maric *et al.* applied PCA and LDA to the vibrational spectral profiles of clear (top) coats or primer layers of over 100 automotive vehicles.[11–13] This ultimately revealed 19 classes associated with the vehicle origin, such as manufacturer, model, and in some instances specific production plants or year ranges. Further investigations found that classifications based on the clear coat layer could be affected by the application of post-manufacture coatings or long-term weathering.[20,21] This has implications for paint sampling from known vehicles and interpretation of comparative paint analyses.

Lambert *et al.* similarly demonstrated the importance of sample preparation for automotive paints analysed by Raman spectroscopy.[27] Using a fractional factorial design, it was determined that layered paints prepared as cross-sections showed greater spectral variability than non-sectioned samples, likely due to sample surface texture. This variability was also noticeable when the spectra were subjected to PCA, implying that this could affect the ability to reliably distinguish similar paints or to match a paint trace to its potential source.

Lavine *et al.* trained a series of genetic algorithms (optimisation tools modelled on natural selection to continually "evolve" better solutions) to match automotive paints against the spectral library of the paint data query database.[23–25] These algorithms were able to distinguish the model, line, or manufacturing year for a range of vehicles from six manufacturers (General Motors, Chrysler, Ford, Honda, Nissan, and Toyota) produced between 2000 and 2006. However, as the data drawn from the database consisted solely of North American-made vehicles, the methodology has not yet been validated for international jurisdictions.

Chemometrics has also been shown to generate useful information in the analysis of artworks. Of particular interest, Ortiz-Herrero *et al.* built orthogonal PLS dating models for two brands of artist paints based on pyrolysis gas chromatography–mass spectrometry (py/GC–MS) and attenuated total reflectance (ATR)-FTIR data.[34] A significant correlation was found between natural and accelerated ageing for Liquitex® paints, with suggestions that the accelerated model could be applied to date naturally aged artworks up to at least 22 years old. Although developed from an art conservation and restoration standpoint, this approach could also be applicable in cases of suspected art forgery.

7.3.2 Adhesive Tapes

Pressure sensitive adhesive tapes are a common source of evidence in criminal investigations as they are routinely used in the packaging of controlled substances, as a makeshift restraint, or for the preparation of improvised explosive devices. Depending on their intended use, these tapes can vary in both their adhesive and backing material. Most studies on the chemical characterisation of tape have focussed on electrical (vinyl) tapes, as these are frequently submitted as evidence in cases of assault, kidnapping, or explosion incidents.

Goodpaster *et al.* used a PCA–DA approach to distinguish 67 black electrical tapes based on the surface texture and elemental composition of their backing materials, as determined *via* scanning electron microscopy with energy-dispersive X-ray spectrometry (SEM–EDX).[37] This approach identified 36 distinguishable classes, to which samples could be assigned with up to 94% accuracy. A follow-up study examined both the backing material and the adhesive material of 72 electrical tapes, this time using ATR–FTIR spectroscopy to profile the organic composition. DA models based on the combined data were able to achieve correct classification rates of up to 99%, as well as correctly assigning two fragments of tape recovered from a detonated explosive device to their retail brands.[38]

A similar recent study by Nimi *et al.* examined 25 electrical tapes using ATR–FTIR spectroscopy of both the adhesive and backing sides.[39] PCA–LDA models revealed that nearly 93% of the tapes could be distinguished based on their adhesive composition, compared to only 88% based on the backing material. However, the ability to correctly associate a tape to its source based on adhesive composition decreased dramatically for tapes that had been applied and removed from common substrates. This highlights the importance of validating models using samples realistic of forensic casework.

In a subsequent study, the same group used ATR–FTIR spectroscopy to analyse the adhesive, backing, and textile fabric reinforcement of 12 duct tapes.[40] The results similarly showed that each individual component could be used to discriminate the tapes, albeit with varying levels of success. The most successful approach was analysis of the adhesive layer, which discriminated approximately 97% of the samples. Comparatively, analysis of the backing, warp yarn or fill yarn achieved discrimination rates ranging from 66–72%. In a similar fashion to the previous study, the discriminatory power decreased after the tape had been applied and removed from a substrate.

7.3.3 Other Polymeric Products

Several other polymeric products are also of interest for forensic purposes. Food wrap (also known as cling or saran wrap) for example is commonly used in the packaging of illicit drugs, hence their analysis could potentially establish linkages between seizures. Telford *et al.* studied nine rolls of low-density polyethylene cling films using ATR–FTIR spectroscopy, with PCA and LDA distinguishing three main clusters associated with the manufacturing brands.[41] PCA factor loadings showed that discrimination of these classes was based on variations in the crystallinity, short-chain branching and oxidation of the low density polyethylene base.

While there has been an abundance of research on automotive paint, other automotive polymers can also be recovered as trace evidence. May and Watling showed that polycarbonate headlamp lenses from different vehicle types exhibited distinct trace element profiles when analysed using laser ablation–inductively coupled plasma–mass spectrometry (LA–ICP–MS).[42] Blind validation trials for each headlamp type gave correct assignment of simulated unknowns using PCA–LDA models, except for some concurrently produced lenses. More recently, Grant *et al.* examined the adhesives of 40 automotive tints using ATR–FTIR spectroscopy, uncovering substantial variability between different brands or specific products.[43] Iterative PCA–LDA models were able to correctly assign separate tint samples to their brand category, even when tints had been applied and removed from glass after up to five months of sun exposure.

Counterfeit polymer-based identity documents are of strong interest in forensic intelligence due to their role in facilitating organised crimes such as drug trafficking. McGann and Willans *et al.* highlighted SVM as a possible tool to support the identification of counterfeit identity cards.[44] Three issuing series of Western Australian driver's licences were distinguished based on their cross-sectional ATR–FTIR spectra, and test licences of each series correctly assigned. Although not yet validated using actual counterfeits, this approach could form the basis for a simple, non-destructive screening tool for security purposes.

7.4 Textile Fibres

Natural and synthetic textiles make up our clothing, upholstery, furnishings, linens, and more. Fibres from these textiles are amongst the most encountered forms of trace evidence due to their natural tendency to transfer through physical contact or proximity. The wide variety in potential sources (see Figure 7.4) or treatments such as dyeing, texturing, or coating makes many fibres highly distinctive. This was evident by the significant role played by fibre evidence in the successful cold case investigation into a sexual assault and two related homicides in Perth, Western Australia, where fibres accounted for the majority of trace evidence relied upon.[45]

Chemometric methods have been applied in several studies to distinguish and classify fibres based on their textile and/or dye composition, quantify this composition, or determine the effects of weathering (see Table 7.2).

Early research by Kokot *et al.* found that a combination of PCA with Fourier transform Raman spectroscopy could distinguish cotton fibres

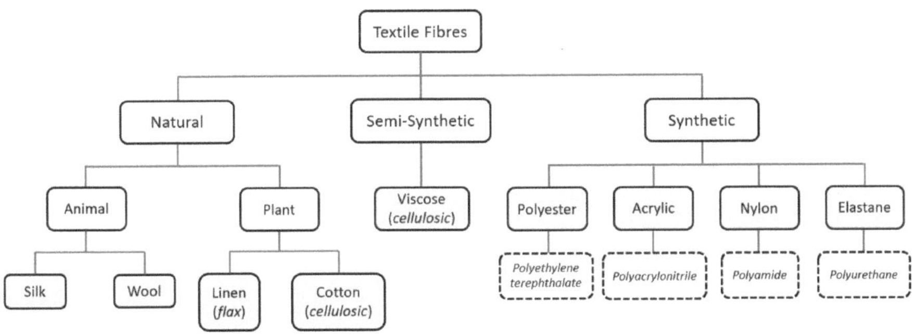

Figure 7.4 Classes of frequently encountered textile fibres and their composition according to source.

Table 7.2 Summary of selected research studies applying chemometric methods to the forensic characterisation of textile fibres. NIP = no information provided.

Ref.	Fibre type(s)[a]	Colour status[b]	Chemometric methods
46	Cotton, modal, viscose, linen	Both	PCA, SIMCA
47	Cotton, wool	Coloured	PCA
48	Cotton	Coloured	PCA, PLSR
49	Cotton	Uncoloured	PCA
50	Cotton	Uncoloured	PCA
51	Acrylic	Uncoloured	PCA
52	Cotton, polyester	NIP	PCA, PLSR
53	Cotton, polyester, acrylic, nylon	Coloured	PCA, LDA
54	Cotton	Uncoloured	HCA, PCA
55	Cotton	Uncoloured	PCA
56	Nylon, viscose, polyester, acrylic	Coloured	Parallel factor analysis, LDA, PLS-DA
57	Acrylic	Coloured	PCA, DA
58	Wool, silk, cotton, viscose, acetate, linen, polyester, nylon, acrylic, elastane	NIP	PCA
59	Polyester	Coloured	AHC, PCA, DA
60	Acrylic, nylon, cotton	Coloured	PLS-DA
61	Cotton, polyester, acrylic, nylon	Coloured	Logistic regression, PCA, DA
62	Cotton, polyester, acrylic, nylon, silk	Both	PCA, multi-variate curve resolution
63	Cotton, polyester, polypropylene, polycotton, viscose	NIP	Random forest

[a] "Cotton" includes muslin and poplin; "viscose" includes rayon; "polycotton" in this instance refers to a blend of cotton and polyester.
[b] "Uncoloured" may refer to both raw (unprocessed) fibres and treated white fibres.

containing different forms of the same reactive dye.[48] Partial least squares regression (PLSR) models were then built to predict dye concentration and fixation times. Low values for the standard error of estimation suggested that such predictions were feasible, although the limited sample size and lack of a separate validation set made this difficult to definitively assess.

Causin *et al.* combined py/GC–MS with PCA to distinguish 36 undyed, morphologically similar acrylic-based fibres from 11 manufacturers.[51] Given the lack of dyes, these fibres primarily differed in their specific polymer composition. Fibres from one manufacturer were exclusively distinguishable based on the presence of methyl methacrylate as a co-monomer, whilst the remainder formed two broad clusters according to the presence or absence of vinyl acetate.

Morgan *et al.* applied PCA and LDA to a database of over 5000 microspectrophotometry (MSP) spectra representing *ca.* 500 dyed fibres.[53] Ultraviolet-visible (UV-Vis) absorbance and fluorescence spectra each gave discriminating information, though UV-Vis absorbance was determined to be the most effective single technique with a nearly 80% discrimination accuracy. This may in fact be an under-estimated measure of performance given that all fibres of a particular colour were compared simultaneously, rather than performing pairwise comparisons.

A challenge when comparing dyed fibres is potential degradation due to weathering. Mujumdar *et al.* tested the effects of desert and humid climates in the United States using dyed acrylic, nylon, and cotton fibres exposed for up to 12 months.[60] PLS discriminant analysis was able to distinguish fibres exposed to dry *versus* humid weather conditions, as well as those analysed following different periods of exposure. Information of this nature is essential in validating methods used by forensic experts to evaluate fibre evidence where the weathering history is unknown.

More recently, Zapata *et al.* applied Raman microspectroscopy with PCA to distinguish fabrics and fibres based on both the fabric material and colouring agents.[62] Interestingly, some discrimination was achieved even amongst white fibres of different cotton materials, potentially due to the use of different bleaching agents or whitening dyes. This is a significant result given that such fibres are often considered to have weak forensic value due to their commonness.

An interesting recent development has been the development of portable instrumentation capable of performing forensic analysis at or close to the incident scene. Whilst this has largely been restricted to relatively simple measurements of identification over the past two decades, Rashed and co-workers demonstrated that hand-held near infrared spectrometers in combination with random forest classification could be used to distinguish six different textile types.[63] The specificity of this approach varied in the range 78–100% for binary classification and 65–100% for multi-class classification. Although this demonstration was not applied to individual fibres, it does show that at-scene analysis of trace evidence with more advanced interpretive capability is a realistic future possibility. However, deployment of such approaches will require close attention to quality-centred protocols to deal with issues of potential contamination.

7.5 Soils

Soils have a long history of use in forensic investigations due to their ability to link individuals and objects to a given location. They often exhibit significant heterogeneity between locations and can be highly complex due to the combination of natural and man-made components. Although soils are mostly made up of natural materials, the composition of "urban soils" in populated areas are heavily influenced by human activities such as mining, building, or planting. Soils have historically been under-utilised as forensic evidence; however, there has been a recent growth in the prevalence of soil evidence in recent investigations. This has consequently led to an increased volume of literature aiming to discriminate or classify soils of different origins using chemometrics, as shown in Table 7.3.

Table 7.3 Summary of selected research studies applying chemometric methods to the forensic characterisation of urban soils. NIP = no information provided.

Ref.	Sampling region(s) (# of locations)	Soil layer[a]	Analysed component	Chemometric methods
64	Epirus, Greece (5)	Surface	Organic	CA, PCA, DA
65	Serbia, Montenegro (15)	Surface	Inorganic	PCA
66	Lincoln, UK (12)	Topsoil	Organic	LDA, PLS-DA
67	South Korea (NIP)	NIP	Organic	PCA, DA
68	Mississippi, US (4)	Surface	Inorganic	PCA, DA
69	New Jersey, US (12)	Surface	Organic	PCA, LDA
70	India (36)	Surface, topsoil	Both	PCA, LDA
71	Paraná, Brazil (5)	Surface	Inorganic	Factor analysis
72	India (18)	Surface, topsoil	Both	CA, PCA, LDA
73	China (15)	NIP	Inorganic	PCA
74	United States (NIP)	NIP	Inorganic	CA, SVM
75	China (8)	Surface	Inorganic	HCA, PCA, *k*NN
76	Perth, Australia (9)	Surface	Both	PCA, LDA
77	Canberra, Australia (268)	Topsoil	Both	Pearson's correlation
78	Victoria, Australia (4)	Surface	Organic	PCA, LDA

[a] "Surface" soil is defined as that collected from a depth less than 5 cm; "topsoil" is defined as that collected from a depth of 5–30 cm.

Thanasoulias *et al.* were able to distinguish soils from five different areas within a single region of Greece based on UV-Vis measurements, despite these soils being visually similar.[64] This was attributed to variations in UV absorbance caused by different proportions of aromatic groups in the fulvic and humic acid fractions. A cluster analysis followed by PCA and DA resulted in 85.0% discrimination accuracy, highlighting the capability of chemometric methods to detect heterogeneity amongst urban soils within a narrow geographical region.

A later study by Baron *et al.* examined topsoils from the Lincoln, UK area using ATR–FTIR spectroscopy. Factor analysis showed clear separation of soils according to their land use type (woodland, riverbank, and garden), with greater variation seen amongst the woodland and riverbank soils. Separate examination of soils from each land use category also revealed some distinction according to specific sampling location. LDA and PLS-DA models were then compared as methods of classifying soils according to land use type or sampling site. LDA produced lower error rates, particularly for sampling sites, as PLS-DA is restricted to modelling two classes at a time.

De Caritat *et al.* used a multi-modal provenancing approach based on a compositional soil survey of North Canberra in the Australian Capital Territory.[77] The survey consisted of chemical data from 268 samples of topsoil analysed using FTIR spectroscopy, X-ray fluorescence and inductively coupled plasma–mass spectrometry (ICP–MS). Correlation coefficients derived from this data were used to generate an interpolated "heat map", allowing the identification of regions in a search area that were most likely to be the source of a questioned sample. When tested using three blind samples, the generated maps were able to exclude at least 43% of the search area, which would allow investigators to prioritise resources more efficiently. However, the authors noted that poor results were obtained if the questioned sample was collected in a non-representative location.

Recently, Newland *et al.* used a multimodal spectroscopic approach based on MSP, IR spectroscopy, and X-ray diffraction (XRD) to distinguish heavily leached soils from the Swan Coastal Plain in Western Australia, based on the quartz-recovered fine fraction.[76] PCA on each dataset found IR spectra to be the most discriminating single technique, although a combination of the methods in sequence maximised the ability to distinguish all soil types according to sampling location. This approach has particularly significance for jurisdictions with dry, sandy soils exhibiting similar mineral compositions.

The previously discussed approaches have focussed on spectroscopic techniques in combination with chemometrics. In contrast, Hay *et al.* used ultra-high performance liquid chromatography with PCA to discriminate soils from the Geelong/Surf Coast region of Victoria, Australia.[78] The developed method was not only able to discriminate four sampling locations, but was used to investigate changes in soil chemistry with time, revealing a cyclic seasonal pattern in compositional variation.

In addition to studies of simulated evidentiary samples, the applicability of chemometric methods to soil analysis has been proven in reported casework. Guo *et al.* reported the use of HCA carried out on mineralogical (XRD) data collected in a murder investigation in Hunan Province, Southern China.[79] In this case, the victim's body had been found in an outdoor irrigation ditch. Comparison of soils collected from the crime scene and the suspect's clothes established a high probability of a common provenance between them. This result, along with complementary pollen and diatom analysis, contributed to the suspect's confession.

7.6 Glass

Glass fragments are often recovered from crime scenes involving vehicle collisions, automotive theft, or burglaries. Although many forms of glass appear visually identical, differences in the raw materials and production method can produce substantial differences in both physical and chemical properties (see Figure 7.5). Unlike many other forms of trace evidence, glass is also a relatively stable substance that is less prone to degradation over time. While substantial literature exists on the evaluation of glass using probabilistic (likelihood ratio) approaches, less work has been carried out using chemometric methods (Table 7.4). These studies have mostly focussed on annealed glasses used in containers and building windows, or tempered and laminated glasses used in automotive vehicles.

Initial research by Almirall *et al.* used LDA to discriminate four glass types according to their refractive index and elemental composition measured using ICP atomic emission spectroscopy (ICP–AES).[80] Headlamp and the

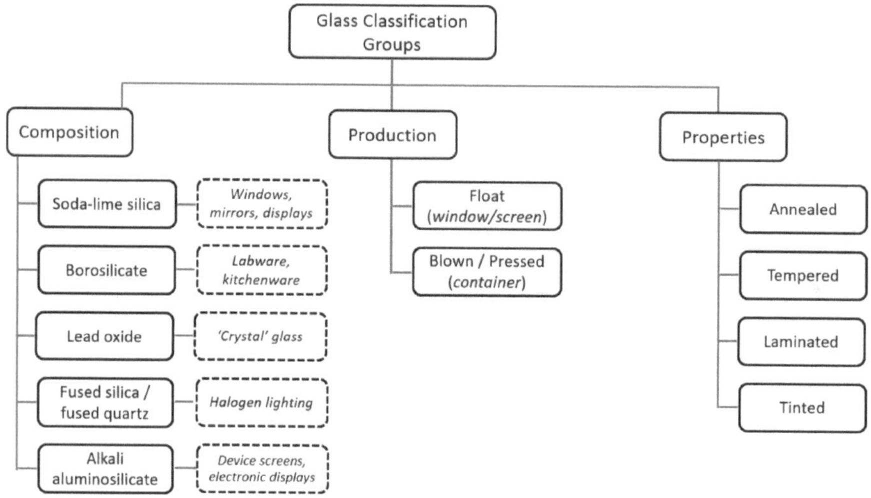

Figure 7.5 Overview of glass type classifications.

Table 7.4 Summary of selected research studies applying chemometric methods to the forensic characterisation of glass.

Ref.	Glass source(s)	Chemometric method
80	Headlamp, container, building and auto window	LDA
81	Headlamp, auto windscreen	CA
82	Building and auto window, lighting, headlamp	SVM
83	Container, lighting, mirror, labware	PLSR
84	Auto windscreen, lighting	PCA
85	Auto window, lighting	PCA
86	Auto window, screen protector, touch screen	PCA
87	Container	PCA, LDA

majority of container glass fragments were clearly distinguished both from each other and from float glass, although some misclassifications occurred between architectural and automotive float glass. The overall model gave 92% discrimination accuracy between the glass types, although the authors acknowledged that this could be an over-estimation due to lack of an independent test set.

Zadora similarly developed predictive models to classify glass fragments into end-use categories, this time comparing naïve Bayes classifiers with SVM.[82] Unlike the previous study by Almirall *et al.*, this research applied a two-step approach in which glass from headlamps or light bulbs were first distinguished from float glass, followed by discrimination of float glass from architectural and automotive sources. Both approaches gave comparable results in the first classification stage, although SVM out-performed naïve Bayes classifiers in distinguishing architectural and automotive float glass. SVM was ultimately recommended as an approach, particularly as there is no need for *a priori* data, although the authors warned that the results should be interpreted carefully given the potential legal consequences of any misclassifications.

Later work by Teklemariam and Gotera compared assorted container glass with that used in laboratory equipment, electronic displays, and other miscellaneous sources using laser induced breakdown spectroscopy (LIBS) with HCA.[83] Most container glasses possessed distinct elemental profiles to other glass types. This could be associated with the raw materials used, as containers are typically made of soda-lime glass whereas laboratory glass for example is usually borosilicate-based. Some variations were also found between container glass of different brands, indicative of manufacturing variations. However, no supervised models were used to assess whether glass fragments could be reliably classified according to end-use type or brand.

Most recently, Merk *et al.* suggested an approach for discriminating automotive glass samples using a data fusion approach, combining PCs derived from LIBS and Raman spectroscopy data.[85] Of 120 possible sample pairs, 119 were deemed separable using the combined LIBS-Raman data. Analysis based on each individual technique also gave effective discrimination, but with greater potential for the model to be affected by sources of chemical modification over time.

Counterfeiting is a significant issue for a variety of consumer products, including high-cost perfumes. Von Wuthenau *et al.* used laser ablation (LA)–ICP–MS to analyse 105 glass perfume bottles.[87] PCA in combination with statistical testing was used to identify 15 marker elements (including aluminium, titanium, barium and lead) that could be used to differentiate the origins of the glass. An LDA model constructed from this data was able to fully distinguish glass from six different manufacturing facilities in four countries. It was proposed that this approach of analysing the packaging could be used as an indirect approach to the identification of counterfeit perfumes, rather than analysis of the fragrance itself.

7.7 Impressions

Although not strictly "trace" material, impressions can be considered a manufactured form of evidence as they are usually created through contact of manufactured items (shoes, tyres, or tools) with a pliant surface. Despite being formed from mass-produced items, the presence of distinct wear patterns or physical defects can make impressions ideal candidates for discrimination. The primary challenge in applying chemometrics to this form of evidence is converting the pattern-based features of impressions into quantitative data that can be processed through statistical algorithms.

Petraco *et al.* approached this by converting accidental mark or wear patterns in shoe impressions to feature vectors.[88] Analysis was performed on 116 partial impressions left by five shoe pairs of the same brand and style, worn by a single individual for 30 days. PCA and DA were then to predict the shoe pair responsible for each impression, producing correct identifications in 81% to 100% of cases. The authors contended that this information could assist in validating footwear impression analysis (addressing concerns of admissibility raised in the 2009 NAS report) by providing statistical credibility to the distinctiveness of shoe impressions.

A similar methodology was later used to distinguish toolmark impressions from identical model screwdrivers.[89] An image processing program was used to convert the striation marks from nine screwdrivers into binary feature vectors, with PLS-DA and PCA-SVM then used to predict the screwdriver responsible for each impression. The performance of each model was assessed through multiple modes of validation, producing 97% or greater classification accuracies for both approaches.

7.8 Chemical and Biological Agents

The use of a Novichok nerve agent in the 2018 poisonings of several individuals in the United Kingdom highlights the continuing threat of chemical and biological warfare (CBW) agents. Mounting an effective response to CBW threats requires rapid but reliable methods for profiling these agents or their precursor materials. Chemometrics has emerged as a means of differentiating between CBW agents and (for chemical agents) potentially distinguishing their synthetic routes for purposes of forensic intelligence.

Munson *et al.* evaluated the use of PCA and SIMCA to distinguish the LIBS spectra of three bacterial spores and five organophosphorus compounds simulating CBW agents, along with potential interferents such as pollens or dust.[90] Weighted-variable PCA and SIMCA were able to distinguish the three biological simulants from each other and from biological interferents. However, resolving the chemical compounds proved challenging due to variability between the single-shot LIBS spectra.

Hoggard *et al.* instead used two-dimensional GC with time of flight–mass spectrometry (TOF–MS) to reveal the chemical impurity profiles from commercial dimethyl methylphosphonate as a chemical agent simulant. Parallel factor analysis, a derivative of PCA, was used to resolve overlapping peaks to enable spectral matching against a chemical database. Amongst six samples tested, five were found to have distinct impurity profiles, showing promise for impurity profiling using analytical and chemometric approaches for investigations of chemical attacks. However, no supervised models were built for classification purposes.

Galan-Freyle *et al.* demonstrated the use of Raman spectroscopy coupled to PLS-DA and ANN for the detection, identification and quantification of the chemical agent simulant triethyl phosphate when mixed with the contents of commercial beverage containers.[91] The approach made use of a purpose-built fibre optic probe enabling analysis of the container contents without the need to sample directly from the container, thus reducing the risk of exposure. PLS-DA performed well in recognising containers with or without triethyl phosphate provided that the container material of the test sample was the same as that used in model construction; however, ANN provided more robust performance (with an accurate classification rate of 90%) when dealing with large datasets. Although the addition of further data to the PLS-DA matrix was able to improve predictions, ANN analysis remained the more efficient approach.

A study by Włodarski *et al.* subjected various dry "white powder" samples to stroboscopic pulses at different excitation wavelengths, followed by HCA and PCA to distinguish the powders based on their fluorescence decay.[92] Bacterial endospores and flours were generally distinguished from other biological materials (vegetative bacteria, fungi, and protein powder), although, as with the previous study, no classification models were constructed. Nonetheless, this non-destructive approach could form the foundation of a rapid screening technique for field use.

Recent work by Lu *et al.* aimed to establish a method for identifying production of the nerve agent tabun based on profiling of the precursor compound *N*,*N*-dimethylphosphoramidic dichloride.[93] GC–MS was used to analyse 27 samples of the compound using three synthetic pathways, followed by HCA, PCA and PLS-DA to establish whether samples could be distinguished according to their production method or batch. Both HCA and PCA indicated that the three synthetic pathways produced samples of distinct chemical attribution signatures. The samples were subsequently divided into a training and test set for PLS-DA, which successfully identified the synthetic route for all except one test sample.

Höjer Holmgren *et al.* used a similar approach for the route determination of sulfur mustard synthesised using 11 production methods.[94] Unlike in previous studies, the synthesised samples were analysed both in their "raw" form and after being spiked onto five soils and textiles overnight before re-extraction. PCA was used to gain an overall view of the data or identify outliers, followed by PLS-DA and RF modelling to classify samples according to the synthetic route. Given the large number of potential classes, each algorithm was used in a hierarchical approach with a total of five iterative models. Both methods gave similar performance, with better classification rates for raw samples (>80%) than those spiked onto a matrix (between 67% and 73%).

7.9 Conclusions and Future Directions

Trace evidence is an extremely useful tool in criminal investigations for establishing associations between people, objects, and locales. However, the exploitation of trace materials requires a baseline understanding of their variability. In this chapter we have revealed how chemometrics can be used to study the inherent chemical variability of different traces, build objective models to rapidly generate investigative leads from their analysis, and develop statistically valid metrics to communicate the strength of these results to the legal system. At the centre of these approaches is an impetus to support, not supplant, the knowledge and training of the forensic expert.

What, then, is the next frontier in the exploration of chemometrics for trace evidence analysis? Many studies outlined in this chapter can be described as "market surveys" focussed on collecting in-depth data on target forms of evidence. The continuing importance of such studies is not to be dismissed, as information regarding the population distribution and background abundance of physical traces is needed for source-level interpretations. Despite a high volume of literature for materials such as fibres or paints, there is arguably a need for broader research on traces such as glass that leverage the power of chemometrics. Additionally, new casework and changes in manufacturing mean there are always trace materials of emerging interest. One example is building materials, as seen in a 2007 Western Australian homicide where small brick fragments were analysed in an attempt to identify the primary crime scene.[95] Another example is so-called "eco-friendly" plastics made from biodegradable or compostable materials, which have been investigated as a vector for latent fingermarks[96] but not yet as a form of evidence in itself. It should also be noted that forensically relevant datasets must be generated across different geographical contexts. This not only ensures that fit-for-purpose information is available to each jurisdiction but also reflects the increased likelihood of materials from international markets being encountered locally due to globalisation.

Moving beyond a foundational understanding of a trace's commonality, chemometric tools are also ideal for understanding the effect of contextual factors such as the environment on analytical results. Several studies have used trained classification models to subsequently predict samples exposed to light, humidity, explosive detonation, *etc.* to assess the impact on chemical composition.[19,43,60] The information derived from these studies can help establish measures of confidence and limits of performance in forensic analysis, contributing to quality forensic service provision. In some instances, it may even be possible to model these effects for the purpose of generating further information of use to investigators, such as through the use of dating curves.[34]

Additionally, there remains an ongoing need for research into the nature of forensic transfer and persistence to inform the interpretation of trace evidence. One area of trace evidence research likely to benefit from the use of chemometrics is that of activity-level studies. There is an increasing focus on evaluative reporting that goes beyond source determination and into the circumstances that resulted in a transfer of a trace.[97,98] de Ronde and coworkers have illustrated how chemometrics can be applied to investigate specific activity-level propositions. They conducted a study in which the pattern of fingermarks recovered from pillowcases differed between the pillowcase being changed or used in a simulated smothering.[99] A model based on quadratic discriminant analysis was successful in distinguishing between the two activities with an evaluated accuracy of 98.8%. They note that this approach could be used for interpretating other forms of two-dimensional traces at the activity level.

Although the primary focus of this chapter has been the use of pattern recognition and regression techniques, there is an unutilised potential for experimental design to be applied to trace evidence research. In particular, approaches such as factorial or central composite designs may be used to optimise the initial recovery of trace evidence or the analytical conditions for its examination. Jones *et al.* took the approach of a factorial design to examine how fibre recovery from multiple surfaces using tape-lifting was impacted by the tape type, taping approach and tape storage temperature.[100] A subsequent four-way factorial ANOVA established that not only were all of the investigated variables significant in affecting fibre recovery, but an interaction existed between them. In essence, the optimal recovery approach varied between surfaces, reinforcing the need to establish flexible protocols considering different possible scenarios. Such studies will make substantial progress towards the establishment of validated "best practices".

Finally, future research should consider how to effectively translate chemometric outputs into a form readily understood by external stakeholders and accepted by the courts. This may require the setting of decision criteria (for example, threshold similarity values for inclusion or exclusion in comparative analyses) or integration into existing interpretation frameworks. Readers interested in a more in-depth analysis of this subject are directed to Chapter 11: Practitioner Aspects.

Expanded research into the role of chemometrics in trace evidence examination is likely to underpin ongoing scientific advances in the field, providing a pathway towards more robust criminal investigations.

Acknowledgements

Portions of this chapter were adapted from the author's previous work,[101] reproduced with permission of the Royal Society of Chemistry. Professor Wilhelm van Bronswijk (Curtin University) is thanked for assistance with reviewing this material.

References

1. M. M. Houck, PhD thesis, Curtin University, 2010.
2. *Trace Evidence Analysis: More Cases in Forensic Microscopy and Mute Witnesses*, ed. M. Grieve and M. Houck, Academic Press, San Diego, 2003, pp. 1–26.
3. *Fundamentals of Forensic Science*, ed. M. M. Houck and J. A. Siegel, Academic Press, San Diego, 3rd edn, 2015.
4. P. Kirk, *Science*, 1963, **140**, 367–370.
5. S. Palenik, *Microsc. Microanal.*, 2009, **15**, 62–63.
6. M. Falardeau, V. Moran and C. Muehlethaler, *Forensic Sci. Int.*, 2019, **297**, 72–80.
7. D. Lambert, C. Muehlethaler, P. Esseiva and G. Massonnet, *Forensic Sci. Int.*, 2016, **263**, 39–47.
8. C. Muehlethaler, G. Massonnet and P. Esseiva, *Forensic Sci. Int.*, 2011, **209**, 173–182.

9. X. He, J. Wang, C. Gao, Y. Liu, Z. Li, N. Li and J. Xia, *Spectrochim. Acta, Part A*, 2021, **248**, 119284.
10. K. B. Ferreira, A. G. G. Oliveira, A. S. Gonçalves and J. A. Gomes, *Forensic Chem.*, 2017, **5**, 46–52.
11. M. Maric, W. van Bronswijk, S. W. Lewis and K. Pitts, *Anal. Methods*, 2012, **4**, 2687–2693.
12. M. Maric, W. van Bronswijk, S. W. Lewis and K. Pitts, *Talanta*, 2014, **118**, 156–161.
13. M. Maric, W. van Bronswijk, K. Pitts and S. W. Lewis, *J. Raman Spectrosc.*, 2016, **47**, 948–955.
14. B. Kochanowski, *J. Chromatogr. Sci.*, 2000, **38**, 100–108.
15. E. A. Liszewski, S. W. Lewis, J. A. Siegel and J. V. Goodpaster, *Appl. Spectrosc.*, 2010, **64**, 1122–1125.
16. A. N. Mendlein, J. A. Siegel and J. V. Goodpaster, 42nd Cent. Reg. Meet. Am. Chem. Soc., 2011.
17. S. Muramoto, G. Gillen and E. S. Windsor, *Surf. Interface Anal.*, 2018, **50**, 889–896.
18. S.-H. Yang, J. Y. Shen, M. S. Chang and G. J. Wu, *Anal. Methods*, 2015, **7**, 1527–1534.
19. J. M. Duarte, N. G. S. Sales, J. W. B. Braga, C. Bridge, M. Maric, M. H. Sousa and J. de A. Gomes, *Talanta*, 2022, **240**, 123154.
20. G. Sauzier, M. Maric, W. van Bronswijk and S. W. Lewis, *Anal. Methods*, 2013, **5**, 4984–4990.
21. K. J. van der Pal, G. Sauzier, M. Maric, W. van Bronswijk, K. Pitts and S. W. Lewis, *Talanta*, 2016, **148**, 715–720.
22. A. Fasasi, N. Mirjankar, R.-I. Stoian, C. White, M. Allen, M. P. Sandercock and B. K. Lavine, *Appl. Spectrosc.*, 2015, **69**, 84–94.
23. B. K. Lavine, N. Mirjankar, S. Ryland and M. Sandercock, *Talanta*, 2011, **87**, 46–52.
24. B. K. Lavine, K. Nuguru, N. Mirjankar and J. J. Workman, *Appl. Spectrosc.*, 2012, **66**, 917–925.
25. B. K. Lavine, A. Fasasi, N. Mirjankar, M. Sandercock and S. D. Brown, *J. Chemom.*, 2014, **28**, 385–394.
26. F. Kwofie, N. U. D. Perera, K. S. Dahal, G. P. Affadu-Danful, K. Nishikida and B. K. Lavine, *Appl. Spectrosc.*, 2022, **76**, 118–131.
27. D. Lambert, C. Muehlethaler, L. Gueissaz and G. Massonnet, *J. Raman Spectrosc.*, 2014, **45**, 1285–1292.
28. C. Defeyt, J. Van Pevenage, L. Moens, D. Strivay and P. Vandenabeele, *Spectrochim. Acta, Part A*, 2013, **115**, 636–640.
29. T. Gatta, L. Campanella, C. Coluzza, V. Mambro, P. Postorino, M. Tomassetti and G. Visco, *Chem. Cent. J.*, 2012, **6**, S2.
30. C. Miguel, J. A. Lopes, M. Clarke and M. J. Melo, *Chemom. Intell. Lab. Syst.*, 2012, **119**, 32–38.
31. F. Pereira and M. Bueno, *J. Coat. Technol. Res.*, 2009, **6**, 445–455.
32. F. Rosi, A. Federici, B. Brunetti, A. Sgamellotti, S. Clementi and C. Miliani, *Anal. Bioanal. Chem.*, 2011, **399**, 3133–3145.

33. G. Sciutto, P. Oliveri, S. Prati, M. Quaranta, S. Lanteri and R. Mazzeo, *Anal. Bioanal. Chem.*, 2013, **405**, 625–633.
34. L. Ortiz-Herrero, I. Cardaba, S. Setien, L. Bartolomé, M. L. Alonso and M. I. Maguregui, *Talanta*, 2019, **205**, 120114.
35. C. Muehlethaler, G. Massonnet and P. Esseiva, *Forensic Sci. Int.*, 2014, **244**, 170–178.
36. R. Vitale, G. Spinaci, F. Marini, P. Marion, M. Delcroix, A. Vieillard, F. Coudon, O. Devos and C. Ruckebusch, *Talanta*, 2022, **243**, 123360.
37. J. V. Goodpaster, A. B. Sturdevant, K. L. Andrews and L. Brun-Conti, *J. Forensic Sci.*, 2007, **52**, 610–629.
38. J. V. Goodpaster, A. B. Sturdevant, K. L. Andrews, E. M. Briley and L. Brun-Conti, *J. Forensic Sci.*, 2009, **54**, 328–338.
39. C. Nimi, R. Chophi and R. Singh, *J. Forensic Sci.*, 2022, **67**, 911–926.
40. C. Nimi, R. Chophi and R. Singh, *Forensic Chem.*, 2022, 100461.
41. C. J. Telford, B. A. Burrows, G. Sauzier, W. van Bronswijk, M. M. Houck, M. Maric and S. W. Lewis, *Anal. Methods*, 2017, **9**, 192–197.
42. C. D. May and R. J. Watling, *J. Forensic Sci.*, 2011, **56**, S47–S57.
43. B. Grant, G. Sauzier and S. W. Lewis, *Forensic Sci. Int.*, 2019, **313**, 110338.
44. J. McGann, M. Willans, G. Sauzier, M. J. Hackett, S. W. Lewis, J. McGinn, T. Trubshoe and W. van Bronswijk, *Forensic Sci. Int.: Reports*, 2020, **2**, 100149.
45. R. Powell, P. Collins, G. Horsley, J. Coumbaros and W. van Bronswijk, *Forensic Sci. Int.*, 2021, **325**, 110894.
46. C. Gilbert and S. Kokot, *Vib. Spectrosc.*, 1995, **9**, 161–167.
47. C. Chen, C. W. Brown and M. J. Bide, *J. Soc. Dyers Colour.*, 1997, **113**, 51–56.
48. S. Kokot, N. Anh and T. Rintoul, *Appl. Spectrosc.*, 1997, **51**, 387–395.
49. Y. Liu, *Analyst*, 1998, **123**, 633–636.
50. Y. Liu, S. Kokot and T. J. Sambi, *Analyst*, 1998, **123**, 1725–1728.
51. V. Causin, C. Marega, S. Schiavone, V. Della Guardia and A. Marigo, *J. Anal. Appl. Pyrolysis*, 2006, **75**, 43–48.
52. C. Ruckebusch, F. Orhan, A. Durand, T. Boubellouta and J. P. Huvenne, *Appl. Spectrosc.*, 2006, **60**, 539–544.
53. S. L. Morgan, S. H. Hall, J. E. Hendrix and E. G. Bartick, in *2007 Trace Evidence Symposium*, Clearwater Beach, Florida, 2007.
54. N. Nic Daéid, W. Meier-Augenstein and H. F. Kemp, *Rapid Commun. mass Spectrom.*, 2011, **25**, 1812–1816.
55. C. Fortier and J. Rodgers, *Fibers*, 2014, **2**, 264–274.
56. A. Muñoz de la Peña, N. Mujumdar, E. C. Heider, H. C. Goicoechea, D. Muñoz de la Peña and A. D. Campiglia, *Anal. Chem.*, 2016, **88**, 2967–2975.
57. G. Sauzier, E. Reichard, W. van Bronswijk, S. W. Lewis and J. V. Goodpaster, *Forensic Chem.*, 2016, **2**, 15–21.
58. P. Peets, I. Leito, J. Pelt and S. Vahur, *Spectrochim. Acta, Part A*, 2017, **173**, 175–181.

59. E. J. Reichard, E. G. Bartick, S. L. Morgan and J. V. Goodpaster, *Forensic Chem.*, 2017, **3**, 21–27.
60. N. Mujumdar, A. M. de la Peña and A. D. Campiglia, *Forensic Chem.*, 2018, **12**, 25–32.
61. D. C. Rich, K. M. Livingston and S. L. Morgan, *Forensic Chem.*, 2020, **18**, 100213.
62. F. Zapata, F. E. Ortega-Ojeda and C. García-Ruiz, *Spectrochim. Acta, Part A*, 2022, **268**, 120695.
63. H. S. Rashed, P. Mishra, A. Nordon, D. S. Palmer and M. J. Baker, *Vib. Spectrosc.*, 2021, **113**, 103205.
64. N. C. Thanasoulias, E. T. Piliouris, M.-S. E. Kotti and N. P. Evmiridis, *Forensic Sci. Int.*, 2002, **130**, 73–82.
65. S. Dragović and A. Onjia, *J. Environ. Radioact.*, 2006, **89**, 150–158.
66. M. Baron, J. Gonzalez-Rodriguez, R. Croxton, R. Gonzalez and R. Jimenez-Perez, *Appl. Spectrosc.*, 2011, **65**, 1151–1161.
67. C. S. Lee, T. M. Sung, H. S. Kim and C. H. Jeon, *J. Anal. Appl. Pyrolysis*, 2012, **96**, 33–42.
68. L. Reidy, K. Bu, M. Godfrey and J. V. Cizdziel, *Forensic Sci. Int.*, 2013, **233**, 37–44.
69. J. Bonetti and L. Quarino, *J. Forensic Sci.*, 2014, **59**, 627–636.
70. R. Chauhan, R. Kumar and V. Sharma, *Microchem. J.*, 2018, **139**, 74–84.
71. V. F. Melo, J. M. L. Mazzetto, J. Dieckow and E. J. Bonfleur, *Forensic Sci. Int.*, 2018, **290**, 244–250.
72. R. Chauhan, R. Kumar, P. K. Diwan and V. Sharma, *Forensic Chem.*, 2020, **17**, 100191.
73. X. Xu, C. Du, F. Ma, Y. Shen and J. Zhou, *Forensic Sci. Int.*, 2020, **310**, 110222.
74. C. Maione, N. Luíza da Costa, F. Barbosa and R. M. Barbosa, *Front. Artif. Intell.*, 2021, 1–20.
75. H. Guo, P. Wang, Y. Li, C. Hu, J. Zheng, H. Mei, J. Zhu, S. Fan and Q. Zhong, *Sci. Justice*, 2022, **62**, 76–85.
76. T. G. Newland, K. Pitts and S. W. Lewis, *Forensic Chem.*, 2022, **28**, 100412.
77. P. de Caritat, B. Woods, T. Simpson, C. Nichols, L. Hoogenboom, A. Ilheo, M. G. Aberle and J. Hoogewerff, *J. Forensic Sci.*, 2022, **67**, 927–935.
78. J. G. P. Hay, A. P. A. Oxley, M. L. Wos-Oxley, R. Hayes, T. Pickles, K. Roberts and X. A. Conlan, *Talanta Open*, 2022, **6**, 100126.
79. H. Guo, Y. Yao, Y. Li, P. Wang, C. Hu, M. Yuan, H. Mei and J. Zhu, *J. Forensic Sci.*, 2022, **67**, 766–774.
80. J. R. Almirall, M. D. Cole, G. Gettinby and K. G. Furton, *Sci. Justice*, 1998, **38**, 93–100.
81. G. Zadora and Z. Brozek-Mucha, *Z Zagadnien Nauk Sadowych*, 1999, **40**, 33–71.
82. G. Zadora, *J. Chemom.*, 2007, **21**, 174–186.
83. T. A. Teklemariam and J. Gotera, *Spectrochim. Acta, Part B*, 2019, **155**, 34–43.

84. A. Gupta, R. Corzo, A. Akmeemana, K. Lambert, K. Jimenez, J. M. Curran and J. R. Almirall, *J. Chemom.*, 2020, e3298.
85. V. Merk, D. Huber, L. Pfeifer, S. Damaske, S. Merk, W. Werncke and M. Schuster, *Spectrochim. Acta, Part B*, 2021, **180**, 106198.
86. C. L. S. Costa, C. T. Prais and C. C. Nascentes, *Talanta*, 2022, **243**, 123354.
87. K. von Wuthenau, T. Segelke, A. Kuschnereit and M. Fischer, *Talanta*, 2021, **235**, 122686.
88. N. D. K. Petraco, C. Gambino, T. A. Kubic, D. Olivio and N. Petraco, *J. Forensic Sci.*, 2010, **55**, 34–41.
89. N. D. K. Petraco, P. Shenkin, J. Speir, P. Diaczuk, P. A. Pizzola, C. Gambino and N. Petraco, *J. Forensic Sci.*, 2012, **57**, 900–911.
90. C. A. Munson, F. C. De Lucia, T. Piehler, K. L. McNesby and A. W. Miziolek, *Spectrochim. Acta, Part B*, 2005, **60**, 1217–1224.
91. N. J. Galan-Freyle, A. M. Figueroa-Navedo, Y. C. Pacheco-Londoño, W. Ortiz-Rivera, L. C. Pacheco-Londoño and S. P. Hernández-Rivera, *Anal. Chem. Res.*, 2014, **2**, 15–22.
92. M. Włodarski, M. Kaliszewski, E. A. Trafny, M. Szpakowska, R. Lewandowski, A. Bombalska, M. Kwaśny, K. Kopczyński and M. Mularczyk-Oliwa, *Forensic Sci. Int.*, 2015, **248**, 71–77.
93. X. Lu, Z. Zhang, H. Liu, H. Tang, R. Gao, C. Pei, H. Wang and J. Xiao, *Talanta*, 2021, **232**, 122476.
94. K. Höjer Holmgren, L. Mörén, L. Ahlinder, A. Larsson, D. Wiktelius, R. Norlin and C. Åstot, *Anal. Chem.*, 2021, **93**, 4850–4858.
95. M. D. Raven, R. W. Fitzpatrick and P. G. Self, *Geol. Soc. Spec. Publ.*, 2021, **492**, 165–179.
96. G. Illston-Baggs, P. Deacon, J. Ivanova, L. Nichols-Drew, R. Armitage and K. J. Farrugia, *Forensic Chem.*, 2022, **29**, 100427.
97. A. Biedermann, C. Champod and S. Willis, *Int. J. Evidence Proof*, 2017, **21**, 14–29.
98. D. Catoggio, J. Bunford, D. Taylor, G. Wevers, K. Ballantyne and R. Morgan, *Aust. J. Forensic Sci.*, 2019, **51**, S247–S251.
99. A. de Ronde, M. van Aken, M. de Puit and C. de Poot, *Forensic Sci. Int.*, 2019, **295**, 113–120.
100. Z. V. Jones, C. Gwinnett and A. R. W. Jackson, *Sci. Justice*, 2019, **59**, 268–291.
101. G. Sauzier, W. van Bronswijk and S. W. Lewis, *Analyst*, 2021, **146**, 2415–2448.

CHAPTER 8

Questioned Document Examination

R. KUMAR* AND V. SHARMA*

Institute of Forensic Science and Criminology, Panjab University Chandigarh, India
*Emails: rajgoel7007@gmail.com; vsharma@pu.ac.in

8.1 Introduction

Forensic science is an integrative discipline that investigates criminal cases using the aspects of chemistry, physics, biology, and computer sciences.[1] The numbers of associated areas of forensic science are represented in Figure 8.1. Among all of these fields, the "question documents" is one of the significant area in the field of forensic science.

Different writing systems have existed in human civilizations since prehistoric times, either for documentation reasons or as a medium of communication. In the past, people used stone, wax, vellum, quills, leaves, and wooden materials as various writing materials and tools.[2] Other writing instruments, such as ballpoint pens, gel pens, and rollerball pens, *etc.* have supplanted them. The invention of writing was later followed by documentation, which was equally important to record.[3] To date, different documents have equivalent significance and are used at every stage of life, *i.e.*, from the birth certificate to the death certificate. Substances like ink and paper used in the documentation are helpful in the characterisation of such documents. Despite the intricate chemical formulations of ink, toner, and paper, enhanced preparations are being introduced on a daily basis, with inadequate information about the standards utilised throughout their manufacture and external attributes such as humidity, light, temperature, and storage environment of the document, creating issues in their identification and dating

Chemometric Methods in Forensic Science
Edited by Vishal Sharma, Georgina Sauzier and Simon W. Lewis

Published by the Royal Society of Chemistry, www.rsc.org

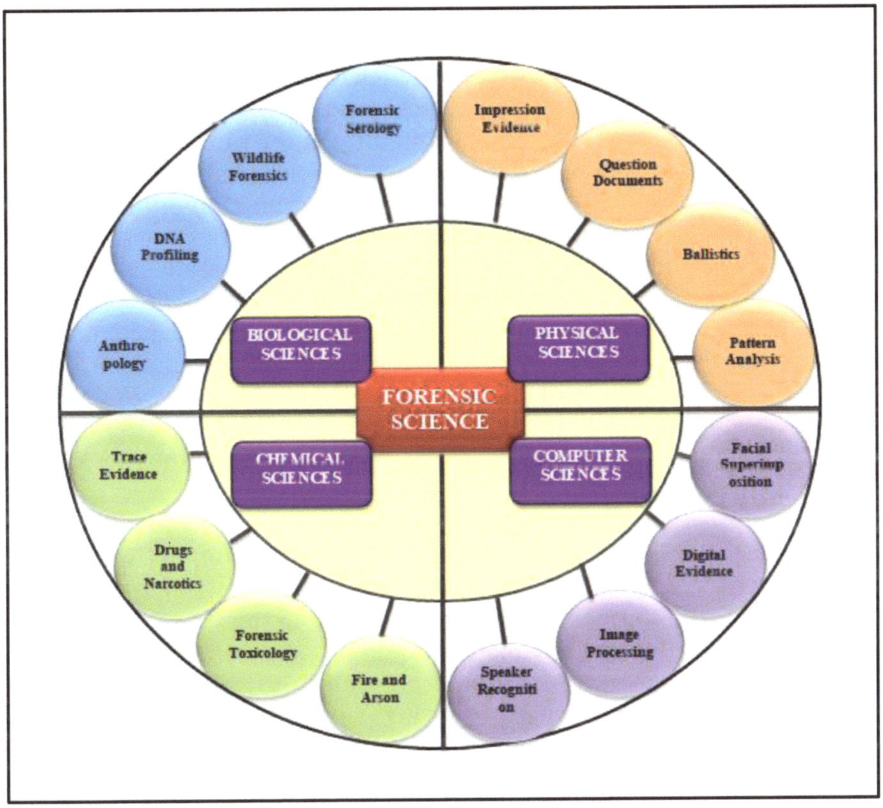

Figure 8.1 Different areas of forensic science.

of such documents. Substantial progress has also been made in the inspection of questioned documents[4] in the last few decades.

Forensic science laboratories (FSLs) evaluate questioned documents for legitimacy issues on a continuous basis. This includes checks, question papers, passport duplication, medical records, insurance claims, tax statements, and duplication of other documents related to personal identity such as Aadhar cards, voter cards, driver's licence, and permanent account number (PAN) cards, among others.[5] It is a constantly expanding field with an emphasis on innovation in order to obtain a reliable outcome for the enquiries of courts of law. With their frequent application, forensic scientists will need to establish specific strategies for determining the source, uniqueness, and antiquity of the questioned documents. As a result, the approaches applied in forensic settings must be scientific, reliable, unbiased, and methodical, as well as reproducible. In a recent report on the current incarnation of forensic science, the National Academy of Sciences, USA, highlighted the issues that forensic professionals confront as well as the essential remedies to overcome these issues.[6]

This book chapter emphasizes that, in an effort to acquire better results from the evaluation, the analytical approaches ought to be computationally efficient and associated with statistical reliability, with the objective of minimising the uncertainty to the greatest extent attainable. To investigate the degree of feature rarity or sample similarity in the sample under review, *i.e.*, ink, toner, and papers, quantitative approaches integrated with statistical analysis are employed for the accurate interpretation of the outcome. The age of the document through the degradation of ink or paper material is still an open ended problem in the discipline of questioned document examination. The rapid implementation of analytical methods by the forensic community is attributed to innovations in these analytical methods such as high sensitivity, resolution, reliable/reproducible results, minimal or no sample pre-treatment, and so on.

8.1.1 Questioned Document and Its Examination

The document is specifically described in Section 8.3 of the Indian Evidence Act (1872) as: "Any information expressed or labelled on any material by means of figures, letters, marks, or by any other means, intended to be used, or which may be used, for the purpose of recording that matter".[7] In other words, a document can be any material that contains some figures, marks, signs, or symbols and conveys meaning to someone.[2]

Since the introduction of paper and ink, the document has exercised a fundamental aspect in communication, education, graphical art forms, music, and literature, among other things. Despite the fact that the exchange of digital information is at an all-time high, the application of documents is still a preferred method in most of the applications, such as currency notes, tickets, advertisements, books, wills, letters, agreements, and other day to day work. Moreover, signatures are used in order to authenticate payment transactions, statements, and deeds like important contracts, declarations, or passports. No matter how far advancements are made in the fields of technology, the integration of ink and paper will remain a matter of concern for most forensic document examiners. In a nutshell, all forgery/fraud-related concerns fit within the umbrella of "questioned documents," a comprehensive discipline of forensic science.[8] The primary objective of investigating questioned documents is to furnish relevant information about the history and production of the documents, to support the criminal justice system.

A document expert investigates a wide variety of suspect documents located in a range of criminal prosecutions, such as falsified identity cards, bank cheques, letters, wills, licensing deals, compensation claims, travel documents, petitions, writs, suicide notes, threatening letters, lottery tickets, currency notes, and so on. As, over recent years, technological developments have rapidly expanded the range of ink and paper compositions used throughout the world, each exhibit, whether printed, handwritten, typescript, signature, or a composite of these documents, is rigorously inspected to ascertain the legitimacy of the document in question. The fundamental inquiries presented in a legal proceeding or by forensic professionals are:

- Whether the disputed documents, *i.e.*, identity documents (PAN card, licenses, voter card, passport, *etc.*), cheques, currency notes, letter head, *etc.* are original or fake.
- Who is the creator/writer of the document?
- What type of ink or pen or printing method (toner) is used in creating the document under question.
- What is the time period of its manufacture?

A thorough examination of standards (exemplars) and admitted writings or prints with the item whose legitimacy must be confirmed dictates the authorship of the document. The paper and ink samples have to be examined for characterisation, differentiation, and age estimation of the questioned document.[9] It is worth mentioning here that a questioned document is generally affected or self-degraded by the environmental/storage conditions and the time of recording/generation of the document. The following factors generally affect such a document:[10]

- **Temperature:** temperature is the major factor affecting the document; it can be temperature during storage, such as a document kept on an open shelf or in a closed cabinet, *etc.*
- **Humidity:** humidity also affects documents to some extent, and mainly affects the paper and ink composition.
- **Document age:** the age of the document, *i.e.*, the time since it was produced, also affects the document. With time, the paper and ink contained in the document deteriorate. The solvent in the ink evaporates with time.

For decades, analytical techniques such as chromatography and spectroscopic methods have been employed for the examination of questioned documents. The early works include the chemical analysis of inks/toners using thin layer chromatography (TLC),[11,12] pyrolysis gas chromatography mass spectrometry (Py-GC/MS),[13,14] ultra-violet-visible spectroscopy (UV-Vis),[15,16] Fourier transform infrared spectroscopy (FTIR),[16,17] laser induced breakdown spectroscopy (LIBS) and laser ablation inductively coupled plasma mass spectrometry (LA–ICP–MS),[18–20] *etc.* Although the majority of these methods are useful to classify forensic exhibits, no such technique has progressed as a routine mode of examination in forensic laboratories. The major disadvantage that is common to all the methods is the destructive mode of obtaining the sample. This may include direct scratching of inks or tones using a scalpel from the surface of the paper or using a heating method in order to obtain a sample for further analysis. The other drawbacks linked with the different techniques include partial or complete deterioration of documents in TLC, UV-Vis, and Py-GC/MS for sample extraction processes, along with a large amount of time required to generate the results. The tedious task of making KBr pellets with extracted toner in infrared spectrophotometry and the high cost of scanning electron microscopy with energy dispersive X-Ray analysis (SEM–EDS) for elemental analysis, and other factors like absence of trained operator/professionals in laboratories for interpretation of data, has made optical examination typically the method of choice so far. Modern techniques like LIBS and LA–ICP–MS also involve the destruction of documents from the area under observation. In this context, diffuse reflectance UV-Vis-NIR spectrophotometry, attenuated total reflectance (ATR)–FTIR spectroscopy, Raman spectroscopy, and other spectroscopic techniques are persuasive analytical tools, and have proven to be successful methods of analysis in the field of questioned document examination.[11–20]

8.2 Chemometric Methods

In the forensic realm, laboratories might state that an analysis was conducted; however, the spectra/chromatograms or any other evidence of the experiment is not submitted along with the reports. Similarly, the possible errors in sources and analytical data are still not reported in the final reports. However, this type of reporting style is often not adequate because it lacks enough detail to understand the nature of the sample as well as its sampling scheme, analysis, or interpretation. Again, because of the lack of statistical confidence, the examiners generally choose the probabilistic approach depending on the research expertise. However, a statistical approach that offers the quantitation of such claims is essential. Further, statistical interpretations, in terms of the probabilistic approach, strengthen the evidential values in a court of law. Statistical interpretations of multidimensional datasets are routinely undertaken in forensic science, either on unprocessed or log-transformed datasets. Data acquisition *via* a spectroscopic or chromatographic procedure is exceedingly troublesome. As a consequence, these datasets are first normalised to either mean 0 or 100 through various normalisation processes. Efficient and appropriate research relies heavily on the correct treatment of unprocessed data.[21,22]

Many international reports acknowledge the importance as well as requirements of objectiveness in the evidence interpretation, *e.g.* those issued by the Forensic Science Regulator (United Kingdom),[23] National Academy of Sciences (United States),[6] and President's Council of Advisors on Science and Technology (United States).[24] These reports highlighted the requirement for statistical validation in the output of an experiment that ultimately enhanced the validity of the outcomes of the experiment and better interpretation of the results. Further in the efforts in this direction, NIST (National Institute of Standards and Technology)[25] has suggested a new centre of excellence titled "The Centre for Statistics and Applications in Forensic Evidence". This centre will concentrate its research endeavours on enhancing the statistical strategies in various research areas of forensic sciences, including questioned document examination. Similarly, according to Morgan (2007), awards have been given to scientists for the advancement of tools and technologies that provide a quantitative measure or statistical evaluation of forensic comparisons and the identification or characterisation of new analytes of forensic importance.[26]

Pattern recognition statistical algorithms are employed to extract and customise elements within multi-elemental evidence or exhibits. Unsupervised pattern recognition is the initial stage in data processing, and it's used to identify features from the dataset. Principal component analysis (PCA) is a popularly exploited unsupervised pattern recognition approach.[27,28] This decreases the data's dimensionality without compromising any of the original data's integrity. In addition to PCA analysis, cluster analysis, such as hierarchical cluster analysis (HCA), is also employed as an unsupervised approach for feature representation.[29] Supervised approaches are presently employed for the interpretation of large analytical datasets for purposes including individualisation, classification, discrimination, and detection of impurities, *etc.*, of forensic exhibits under examination.[30] Frequently, there are different methodologies in the realm of questioned document examination, including linear discriminant analysis (LDA),[31] partial least square-discriminant analysis (PLS-DA),[32] support vector machine (SVM),[33] *k*-nearest neighbour (*k*NN),[34] and soft independent modelling of class analogy (SIMCA).[30]

In addition to pattern recognition, regression methods are also employed to measure the relationship between several independent or predictor variables and a dependent variable, *e.g.* in ink dating analysis. The multiple linear regression (MLR) model is built with descriptive variables using the least squares methods to minimise the residuals.[35] Another approach of partial least squares regression (PLSR), is a biased approach that employs factor analysis, and the most popular regression method in chemometrics. The purpose of this technique is to estimate the space of the original measures into one of reduced size. The model is built in a single step, in which the information from matrix "*X*" and values of interest are considered during the data decomposition and compression.[36]

Basic chemometric publications and reference books[37,38] provide additional details about the multivariate data processing methodologies (*e.g.*, the PCA algorithm, dimension reduction methods, and clustering algorithms).

8.2.1 Statistical Considerations

Despite the ample application of spectroscopic methods in chemical, biological, and physical sciences, the utilisation of these methods is less in use in document examination. According to the Scopus data search with "documents AND chemometrics OR statistical methods", there are about 150 publications which utilise chemometric methods for forensic document examination worldwide which includes research articles, book chapters, reviews, *etc.* (see Figure 8.2). Most of the articles published during the last decade have utilised spectroscopic and other methods. It is worth

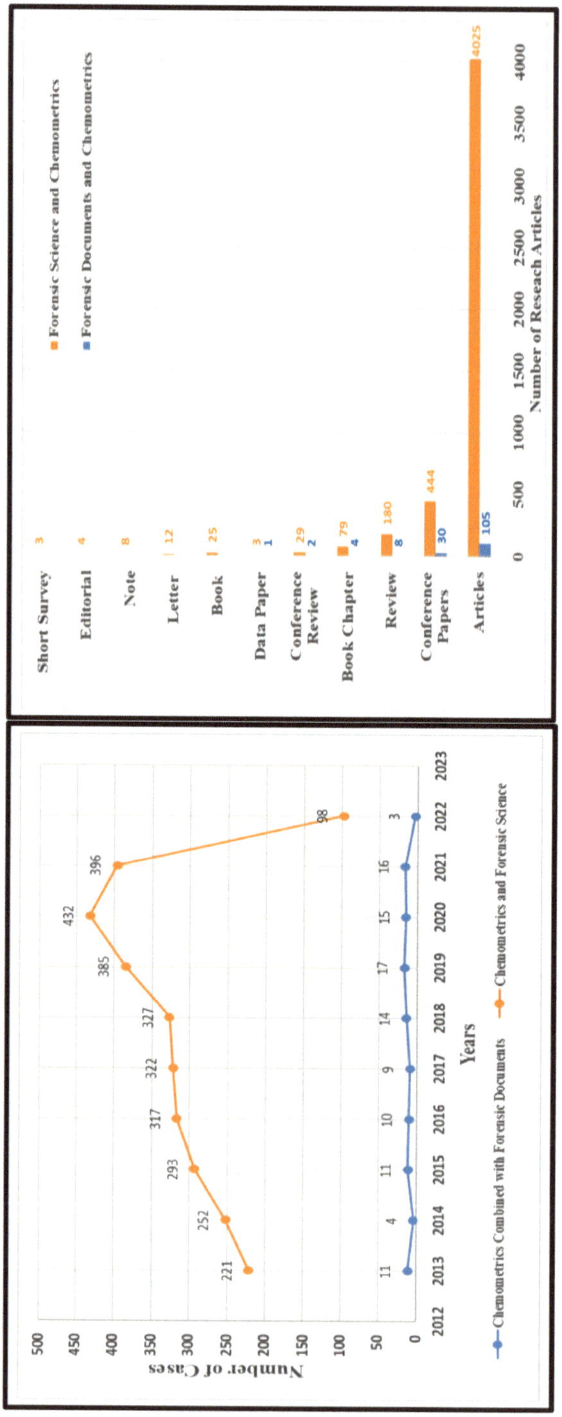

Figure 8.2 The work done on chemometrics combined with forensic documents and reported in the literature as per SCOPUS data.

mentioning here that when we search with "forensic AND chemometric OR statistical OR multivariate", in forensic science, overall 3043 articles that have utilised chemometric methods have been published in the last decade. However, among 3043 articles, only 110 research publications from forensic document examination have been published, which highlights the use of chemometric methods.

The majority of analytical techniques provide good results for document examination along with statistical confidence. The use of spectroscopic methods provides fast, non-destructive, and accurate results in the determination of chemical makeup of the ink, toners as well as paper samples. The most of the studies explain characterisation and discrimination of ink, paper and toner samples along with the studies related to ink dating. In spectroscopic methods, Fourier transform infrared (FTIR) and UV-Vis-NIR spectroscopy are the most commonly used spectroscopic techniques for the investigation of forensic questioned documents, rather than the others such as Raman spectroscopy, LA–CP–ICMS, HPTLC, mass spectrometry, *etc.*

8.3 Examination of Related Exhibits

8.3.1 Ink Analysis

The identification and comparison of writing ink involved in the making of the document is one of the crucial components of questioned document analysis. The first ink dates back 4500 years (Egypt), and was made by mixing animal or vegetable charcoal with glue.[39] Iron-gallotannate inks have been extensively used for writing and drawing purposes since the 12th century. Iron salts provide a blue colour to the ink. Carbon was used as a source of black ink. Later, the salts of vanadium and aniline were used until the late 19th century.[40] Today, inks are mostly either water-based (contains water as a solvent; fountain pens, fibre pens, *etc.*) or glycol-based (contains glycol as a solvent; ballpoint pens). Modern inks have complex chemical compositions (see Figure 8.3). From the chemistry point of view, the ink is a colloidal system of coloured or uncoloured pigment/dye particles dispersed in an organic or aqueous solvent. Colorants (*i.e.*, dyes or pigments), a solvent or solvent system, polymeric resins, pH controlling agents, antifoaming agents, surfactants, lubricants, emulsifiers, and other additives make up writing inks.[41,42] The proportion and composition of these components vary from one product to the next. The type of ink used in a pen varies from one pen to another depending on its application, such as whether it's meant for drawing, writing, artwork, *etc.* Each brand of pen has unique physical and chemical characteristics that set it apart from other types or classes. For instance, a ballpoint pen's ink is primarily composed of solvents (50%), resins (25%), and dyes/pigments (25%).[43] These discrepancies will be examined by a questioned document examiner in order to ascertain how many inks were used to generate a specific document. A comparison between the two ink samples is performed, with both similarities and differences highlighted.

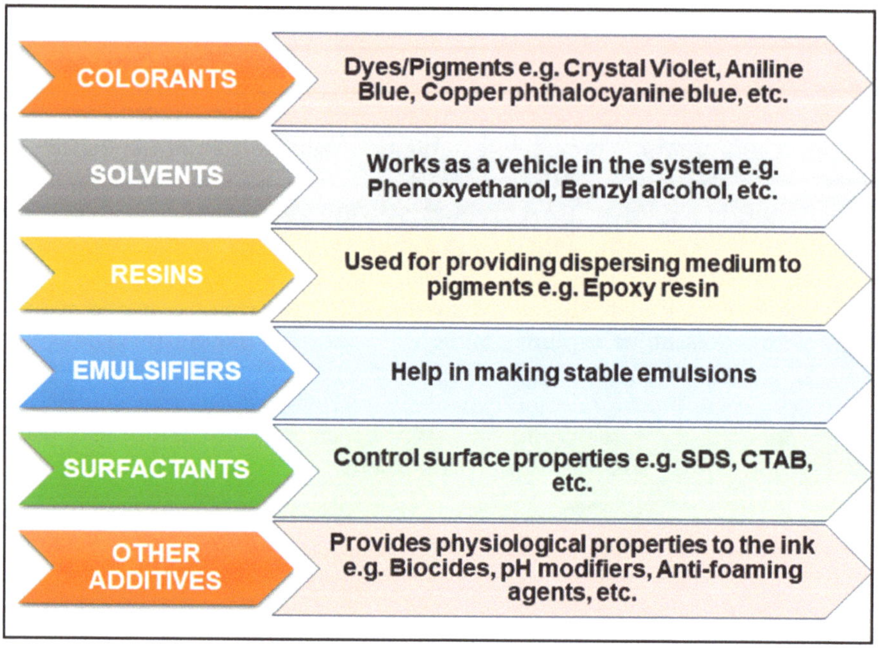

Figure 8.3 Modern ink compositions.

The ink analysis is performed for class characteristics and for chemical compositions to find out the similarities or differences among different inks. The quality of the ink also helps in its differentiation from other inks. The most significant issue for a document expert to address is whether two or more inks involved in the manufacturing of a document are from the same source or distinct. The second important question is about the date of the document, *i.e.*, whether the particular document was written on the claimed date or not. Sometimes, queries related to artificial ageing and manipulation of writing ink are also faced by document examiners. Experts may evaluate the composition of inks used in document generation at the molecular level in order to respond to such questions. In addition, the document specialist should be familiar with ink formulation, as each type of ink has different constituents.

In very old practices, the composition of writing inks is determined by various chemical methods, like the application of chemical reagents to writing ink, which generates different colour reactions for different types of ink. Moreover, the results obtained by using such chemical methods will not be conclusive. The same class of inks can be differentiated by comparing the two writing inks for their colour strength, contrast, tint, complimentary colour in oblique or UV/IR light, the diameter of strokes, reflection, gloss, *etc.* Similarly, microscopic analysis of inks may not always be able to distinguish between the same classes of ink because the ink's surface characteristics will be similar.

The procedure of ink examination should start with visualising the ink using the naked eye, then with a magnifying lamp and a microscope, and last but not least, with the analytical methods for objective results. A casual visual examination of two different writing inks can undoubtedly lead to an incorrect judgement. This is due to the fact that ink changes colour/shade with age, environmental circumstances, the paper used for writing, document storage, and so on. As a result, several considerations must be made before examining the document in question. It may also be possible that two pieces of writing, written with the same ink, are deliberately placed in two places to create the differences between them.

Document examiners now use a range of procedures that provide better resolution and consistency than microscopic, chemical, and other scanning methods. Non-destructive approaches such as diffuse reflectance UV-Vis-NIR (ultraviolet-visible-near infrared) spectroscopy, ATR–FTIR, hyper spectral imaging (HIS-NIR) and Raman spectroscopy are included. Destructive procedures such as high performance liquid chromatography (HPLC), gas chromatography mass spectrometry (GCMS), liquid chromatography mass spectrometry (LCMS), and scanning electron microscope–energy dispersive X-ray spectroscopy (SEM–EDS), among others, are also used. Table 8.1 shows how several of these methods are combined with multivariate analysis for fast, reliable, and correct findings interpretation. Some of the listed studies have been discussed in detail in this book chapter.

In forensic documents, one of the most common queries is whether a particular document was written with one or more pens/inks. Various approaches have been used in this direction, but some of the analytical methods require ink extraction and hence the integrity of the documents is destroyed. It is worth mentioning that some modern techniques do not require such destruction of documents. Gorziza *et al.*, for example, used PhotoMetrix PRO®, a mobile device app, to distinguish blue ballpoint pen inks. This app provides qualitative and colorimetric analysis tools that apply to univariate as well as multivariate analysis. The results revealed clear differentiation between colourful ballpoint pens, and there was a satisfactory tendency for separation for different brands of blue ballpoint pens, most commonly used in Brazil.[44]

Sauzier *et al.* analysed 35 blue ballpoint inks by using diffuse reflectance visible spectroscopy combined with PCA and LDA. PCA revealed the clear segregation of many pens based on their visible spectra, while some pens showed overlapping results. Further, the authors built an LDA model for classification of pen inks and cross validation of the model resulted in 71.7% accurate assignment of pens and 16.7% accurate assignment of pen suppliers. It was also concluded that the ink remains chemically stable for two months when stored in dark conditions, while heat and ultraviolet light play a crucial role in the ageing process.[47]

Sharma and Kumar discriminated between 57 blue ballpoint pen inks with the help of ATR–FTIR spectroscopy combined with chemometrics and high performance thin layer chromatography (HPTLC). A clear segregation

Table 8.1 Pen ink analysis by using analytical techniques coupled with chemometric approaches.

Ref.	Sample type(s)	Analytical techniques	Chemometric methods	Objective of study
3	Markers pen inks	UV-Vis and UV-NIR spectroscopy	PCA & DA	Discrimination of pen inks
44	Blue ballpoint pens	Mobile phone	PCA, HCA, & PLS-DA	
45	Blue ballpoint pens	VSC	PCA	
46	Blue ballpoint pens	Raman imaging	Mean-field approach independent component analysis (MF-ICA)	
47	Blue ballpoint pens	Diffuse reflectance visible spectroscopy	PCA & LDA	
48	Black ballpoint pens	FTIR spectroscopy	PCA & HCA	
49	Blue ballpoint pens	ATR–FTIR spectroscopy, HPTLC	PCA	
50	Ballpoint pens (blue and red)	Raman spectroscopy	PCA	
51	Black ballpoint pens	ATR–FTIR spectroscopy	PCA	
52	Blue ballpoint pens	UV-Vis-NIR spectroscopy	PCA	
53	Blue ballpoint pens	FTIR spectroscopy	PCA & HCA	
54	Blue ballpoint pens	Raman spectroscopy	PCA, HCA & PLS-DA	
55	Blue ballpoint pens	LA–ICP–MS spectroscopy	Multivariate analysis of variance (MANOVA)	
56	Black pens	VSC, LC/MS–TOF spectroscopy	PLS-DA	
57	Fibre-tip pens (black, red, green and blue)	ATR–FTIR	PCA & LDA	
58	Black ballpoint and gel pens	HSI–NIR spectroscopy	PCA, MCR-ALS & PLS-DA	Sequence of strokes
59	Black ballpoint pens	ToF–SIMS	PCA & MCR	
60	Blue ballpoints pen	VSC	MCR	
61	Blue and black gel pens	Raman spectroscopy	MCR-ALS & PLS-DA	
62	Blue tips pens	UV-Vis spectroscopy	PCA, LDA & PLSR	Ink dating
63	Blue ballpoint pens	Visible spectroscopy	PCA, HCA, & Orthogonal-partial least squares (OPLS)	
64	Blue and black ballpoint pens	FTIR spectroscopy	PCA & HCA	
65	Black and blue ballpoint pens	UV-Vis-NIR spectroscopy	PLS	
66	Blue ballpoint pen	UV-Vis spectroscopy	PCA & MLR	
67	Blue ballpoint pens	UV-Vis, IR spectroscopy, HPTLC	PCA	

among ink samples by using a PCA plot between PC3 and PC4 for 57 ink spectra, which divided the whole plot into four different zones according to their loading value, *i.e.*, positive or negative, is shown in Figure 8.4. This plot was obtained after multivariate analysis of ink spectra for discrimination purposes. The overall discrimination with non-destructive ATR–FTIR analysis came out to be 99.69%, while visual inspection of the spectra and destructive HPTLC chromatography revealed 97.93% and 93.80% discrimination, respectively.[49]

Borba *et al.* utilised Raman spectroscopy and chemometrics for the differentiation of 14 blue ballpoint inks. After suitable pre-processing methods, the authors utilised PCA and HCA analysis for the discrimination of ink samples. In the findings, almost all inks were differentiated through PCA and HCA. Further, correct classification of inks was achieved by using PLS-DA models.[54]

Yadav and Sharma discriminated fibre-tip pen inks based on ATR–FTIR spectroscopy and chemometrics methods. However, due to the subjectivity entailed in the visual examination, PCA and LDA were applied. The spectra resulted in 77.20%, 98.40%, 96.60%, and 96.60% discrimination of black, red, green, and blue fibre-tip pen inks, respectively.[57]

The determination of the sequence of strokes is a recurrent issue for forensic document examiners. In this direction, Brito *et al.* evaluated the potential of hyperspectral imaging in the near infrared range (HIS-NIR) combined with multivariate analysis for non-destructive analysis of crossing ink lines (black inks). The authors selected 21 black gel and ballpoint pens from different manufacturers to prepare a combination of crossing lines. They extracted the ink spectra by using a PCA score plot and applying PLS-DA to the most important variables. The multivariate curve resolution-alternating least squares (MCR-ALS) method resulted in the correct determination of the order of crossings in only one pair of gel–gel pen crossings. The rest of the crossing lines were either impossible to discriminate between or led to incorrect determinations of the sequences. The results demonstrate the difficulties of using this technology for application in real forensic cases.[58]

Sauzier *et al.* presented research utilising visible spectroscopy combined with chemometrics for the estimation of the age of blue ball tip inks. In the findings, the authors concluded that the ink kept in the dark could be matched with the source pen after 32 months of ageing while the light-exposed inks resulted in rapid spectral changes. Dispersion graphs for reference values (actual age) *versus* estimated values (predicted age) by PLSR models were obtained for the ink dating of six blue ball tip inks exposed to light over a 32 months period. Excellent agreement was observed between the actual and estimated values, with correlation coefficients greater than 0.97 obtained for all models (see Figure 8.5). PLSR was used to develop a model for light-exposure inks. The prediction accuracy of ink aged for 2–6 months was challenging. However, with an increase in the age of the ink, *i.e.*, more than 2 years, the prediction accuracy increased considerably.[62]

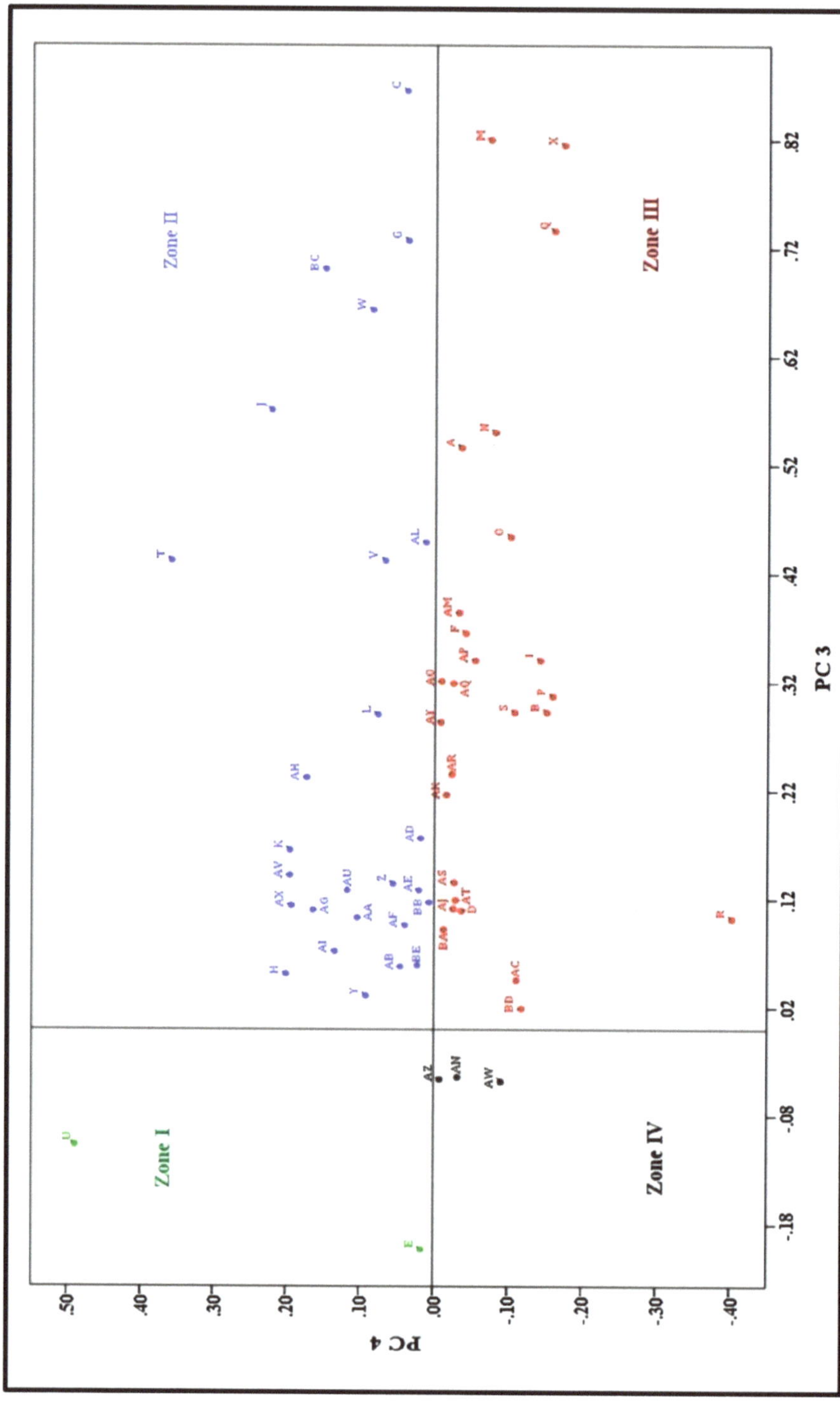

Figure 8.4 PCA scatter plot between PC3 and PC4 for 57 ink spectra, divided into four different zones according to their loading value; *i.e.*, positive or negative, obtained after multivariate analysis of ink spectra for discrimination purposes. Reproduced from ref. 49 with permission from Elsevier, Copyright 2017.

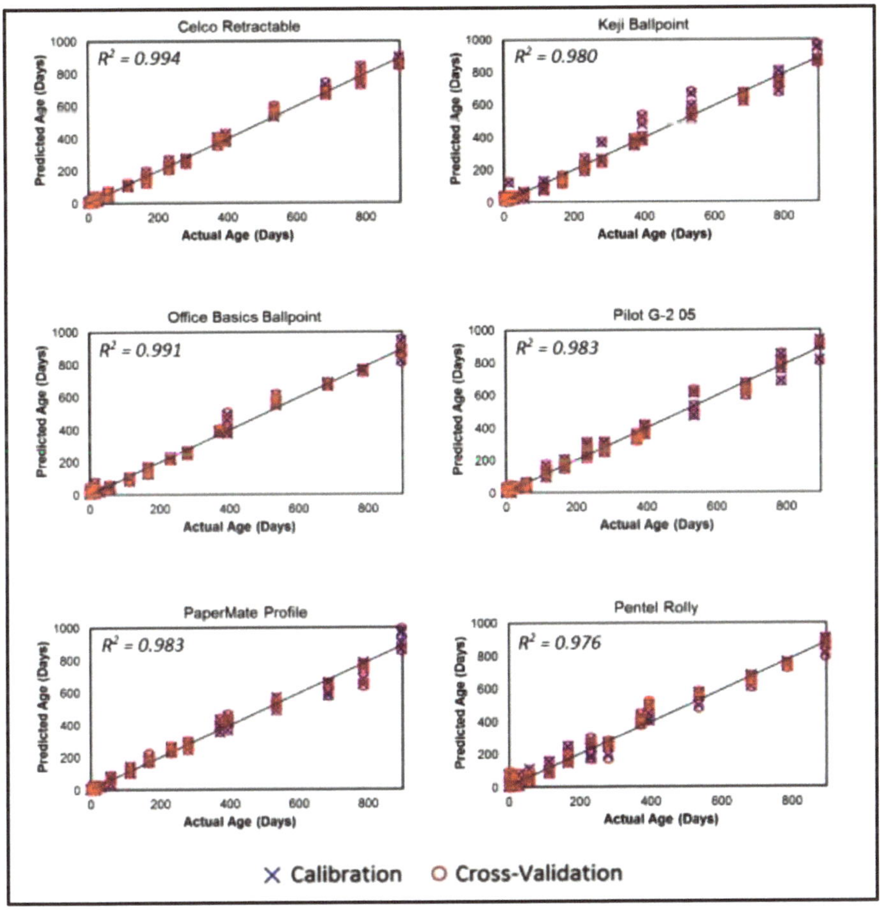

Figure 8.5 Dispersion graphs for reference values (actual age) *versus* estimated values (predicted age) by PLSR models for the ink dating of six blue ball tip inks exposed to light over a 32 months period. Excellent agreement was observed between the actual and estimated values, with correlation coefficients greater than 0.97 obtained for all models. Reproduced from ref. 62 with permission from the Royal Society of Chemistry.

Ortiz-Herrero *et al.* studied the feasibility of ink dating based on UV-Vis-NIR spectroscopy combined with the PLS model. The authors exposed the ink samples to artificial aging. The best model was obtained after standard normal variate (SNV) treatment. They stated that the inks sharing a common chromatographic profile proved to fit correctly in the predictive model, indicating that the methodology shows great potential for future applications in the field of questioned document dating.[65]

Sharma and Kumar proposed an MLR model for the estimation of blue ballpoint pen ink dating by measuring the fading of ink with respect to time using a UV-Vis spectrophotometer. The ink samples were analysed with a common method to avoid differences caused by the ink extraction procedure. The validation of the developed model is done by calculating the ageing of already known values of the same samples. The differences between expected and estimated dating were used to determine the average errors in the dating method.[66]

From Table 8.1 it is clear that most studies have utilized non-destructive analytical approaches such as video spectral comparator (VSC), ATR–FTIR, RAMAN, UV-Vis spectroscopy, HSI-IR, smart phones, *etc.*, but the most common multivariate method used for discrimination of pen inks is PCA. Further, most studies focused on the discrimination of pen inks because the maximum numbers of cases have queries related to writing ink, *i.e.*, whether the written matter has been executed by a single pen or by multiple pens. Various approaches have been used to predict the date of the pen inks. However, due to environmental conditions of the document, a reliable model for all the pen inks has not been developed up to now, as per the knowledge of the authors. Similarly, for sequence of stroke determination, no reliable method is available for the detection of sequence in the case where the same ink has been used in the execution of crossing lines.

8.3.2 Toner Analysis

Toner powders are essentially complex mixtures of a binder resin (60–80% of the whole mixture) with a particle size ranging from 8 to 10 μm. They contain, *e.g.*, low molecular weight polypropylenes, polyethylene, ethylene-vinyl acetate, styrene copolymers, *etc.* Other important ingredients like colorant for, *e.g.*, carbon black, furnace black, thermal black, magnetite, and nigrosine provide black colour, copper phthalocyanines are used for providing cyan colour, azo-pigments for proving yellow and quinacridones or rhodamines for the magenta colour. Charge control agents such as quaternary ammonium salts and nigrosine control the charge level, magnetic additives, essentially magnetite, carry the necessary electrostatic action, and surface additives like fumed silica, titanium dioxide, and organo-metallic salts control the flow properties. Some more constituents, like silicon oil or a low molecular weight polypropylene, polyethylene wax, microcrystalline wax, a paraffin wax or wax composed of fatty esters such as carnauba wax, metal nanoparticles or surfactants, which act as a dispersing agent and prevent adhesion of toner particles to the rollers during the process of fusing, are also added during the blending of ingredients to obtain toner powders with the desired set of properties.[68] Figure 8.6 depicts the toner particles with its major constituents in toner powders used in the process of electro photography. There are currently two types of toner manufactured with the help of different processes: the conventional toner (pulverized toner) and the chemically processed toner (CPT). These can be further classified as mono-components or dual-components. Bi-component toners use magnetic carriers, unlike mono-component toners, for the process of printing, without themselves being transferred to the paper surface.[69,70] Due to the differences

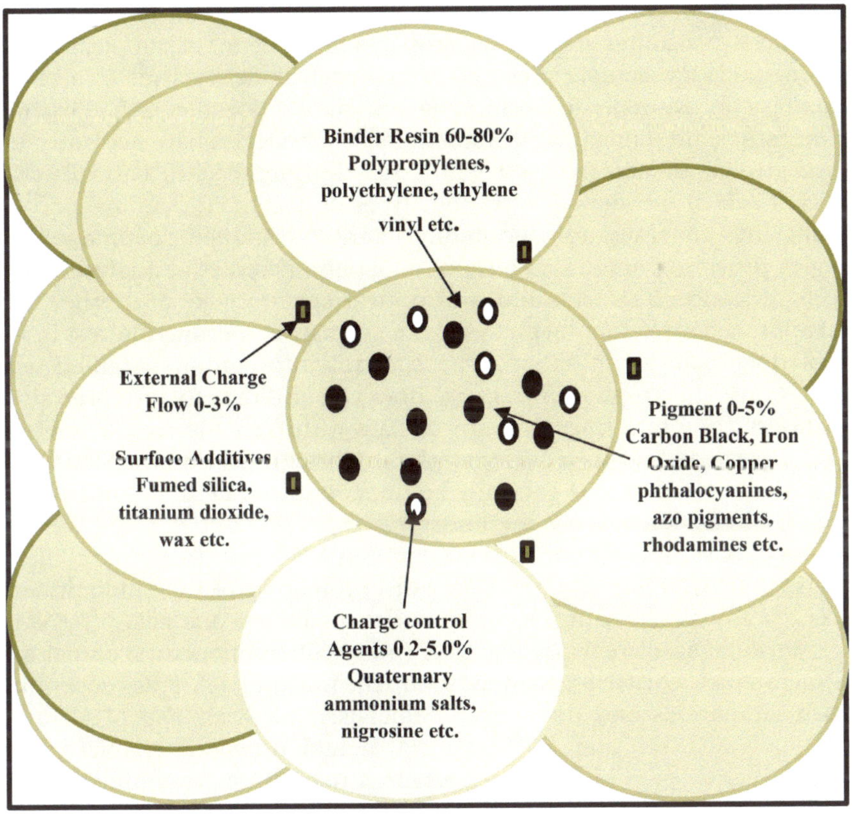

Figure 8.6 Image showing toner particles with their major constituents in toner powders used in the process of electrophotography.

in constituting materials and method of preparation, toner powders of varying emergence tend to show a variety of functionalities like mass, morphological features, electric charge and magnetic quality.[71]

Toner powders from various origins have varying physical and chemical properties[72] and forensic document examiners typically find it difficult and challenging to answer questions like: (a) is the questioned document received in the laboratory an original or a replica? (b) Whether the single document (build up) was created using one or more technologies. (c) Are there any signs of alteration, deletion, addition, or obliteration?

There have been a number of attempts in the past to look into the chemical makeup of toner powders using established methods for estimating and evaluating the authenticity of questioned documents. Microscopy, thin layer chromatography (TLC), infrared spectroscopy (FTIR), ultraviolet spectroscopy (UV-Vis), pyrolysis gas chromatography (Py-GC), laser ablation spectroscopy (LA–ICP–MS), and laser ablation spectroscopy (LIBS) have all been shown to be important in analysing toners. However, such assessments are primarily qualitative (in microscopic, TLC and other scanning methods) and hence rely on the forensic document examiners' (FDE) judgement. Furthermore, while most of these procedures are only sensitive to a tiny amount of sample, in some of the methods the sample preparation approach is destructive. As a result, developing effective tools for examining questioned documents has become both necessary and difficult, as document forgery strategies have become more sophisticated nowadays. In this regard, analytical approaches have played a significant role in the analysis of various types of document examination.[10]

Taking into consideration the nature of the questioned documents, the strategies providing non-destructive or quasi-non-destructive analysis are of great significance. The techniques of optical spectroscopy and other analytical approaches (elemental analysis) are extensively gaining interest in the field of document examination. The spectral information obtained with contemporary spectrophotometers is, however, difficult to interpret since the datasets contain a huge amount of data within themselves. Therefore, there is a perpetual need to detect all the important information that exists in large data sets, and this can only be done with the help of some robust techniques like chemometric methods.[73,74]

A number of statistical approaches are presently adopted in almost all fields of science to contribute towards better discrimination of the similarity and differences among the samples, with similar qualitative analysis.[75,76] For instance, unsupervised methods like PCA (principal component analysis) and HCA (hierarchical clustering analysis) group the individual samples depending on their similarities and differences. Supervised methods like LDA (linear discriminant analysis) and ANN (artificial neural networks) enable a large number of variables to be compared within a dataset. Multivariate statistical analysis for the extraction of important spectral features necessary for discrimination and classification of documents is very well presented with the analysis of paper, toners, and inks. Keeping this in mind, the present section of the book chapter focuses on the analysis of printed document samples of different sources of origin with the help of various analytical techniques and chemometrics.

With the improvements in the process of printing and composition of toner powders, the number of document crimes registered every year has increased, while the distinction between the documents produced from different sources of printers and photocopiers utilising only optical measures has been reduced to a large extent. In this view, a few researchers have proved the capability and potential of various analytical methods like separation techniques, spectroscopy studies, and other analytical methods as an application to discriminate against and characterise printed documents to relate them to their source of origin. The utility of multivariate analysis serves as an added advantage for handling large amounts of data and offers improvements in the interpretation and validation of methods used for the analysis. The study in this field is limited, but some of the important and relevant past works are reviewed as follows.

Materazzi *et al.* (2017) explored the use of FT-NIR (Fourier transform near-infrared spectroscopy) with chemometrics to analyse black toners in suspected documents. The author studied the toner samples with near-infrared spectroscopy (NIR) spectroscopy in reflectance mode from 10 000 to 4000 cm^{-1}. NIR spectra were recorded and treated with principal component analysis (PCA) to identify and predict the toner depending on various principal components. The PCA provided an overall explained variance of not less than 96.9% for the total set of toner samples. The methodology presented by the author is capable of analysing toner samples from the same manufacturers without interference from the paper substrate and printing process. Hence, the authors concluded that the combination of NIR with chemometrics was an encouraging technique to compare documents of unknown origins while achieving high discrimination between them.[15]

Johnson *et al.* (2017) studied the printing inks of various brands, colours, and chemical compositions with the help of Raman micro-spectroscopy at two wavelengths (532 and 785 nm). The toner samples from 76 different sources with different colours were examined at a 50× objective lens for the determination of peaks with individual spectral locations and their relative intensities. On the basis of spectral comparison, a discrimination power of 61% was achieved with 1738 discriminated pairs out of 2850 total pairs. However, the discrimination of black toners with Raman spectroscopy and statistical measures was poor (10%). Among different colours of varying sources, yellow-coloured toners could be best discriminated against, while magenta, cyan, and black could not be differentiated among different brands. The authors concluded that the capability of ATR–FTIR in discriminating toners is greater than that of Raman spectroscopy. Moreover, the author stated that the signals of Hewlett-Packard were similar to those of Canon, concluding that they share the same source of origin for toner powders.[77]

Corzo *et al.* (2016) made use of SEM–EDS (scanning electron microscopy–energy dispersive X-ray spectroscopy) and LA–ICP–MS (laser ablation–inductively coupled plasma–mass spectrometry) to develop the correct association of the toner samples from different sources with the individual techniques to discriminate them with significant results and high accuracy. The results generated by SEM–EDS were found to be complementary to LA–ICP–MS. SEM has the dual capability of discriminating samples depending on their particle morphology as well as on their elemental composition. The advantage of SEM is that the electron strikes only the layer of toner present on the surface of the paper, whereas in LA–ICP–MS the beam penetrates into the paper. Moreover, the elements with a polyatomic interface, for example, ${}^{16}O_2{}^{+}$, ${}^{14}N^{18}O^{+}$, ${}^{14}N^{16}O^{1}H_2{}^{+}$, ${}^{15}N^{18}O^{1}H^{+}$, and ${}^{16}O^{18}O^{+}$, were analysed more efficiently with SEM–EDS.[18]

Trejos *et al.* (2016) assessed the analytical activity of five major techniques used in the field of forensic science, *i.e.*, SEM–EDX, LA–ICP–MS, direct analysis in real time mass spectrometry (DART–MS), FTIR, and pyrolysis-gas chromatography–mass spectrometry (Py-GC–MS) to deduce the differences in the polymeric as well as chemical composition, mainly elements, in the toner samples originating from various sources. The authors tried to

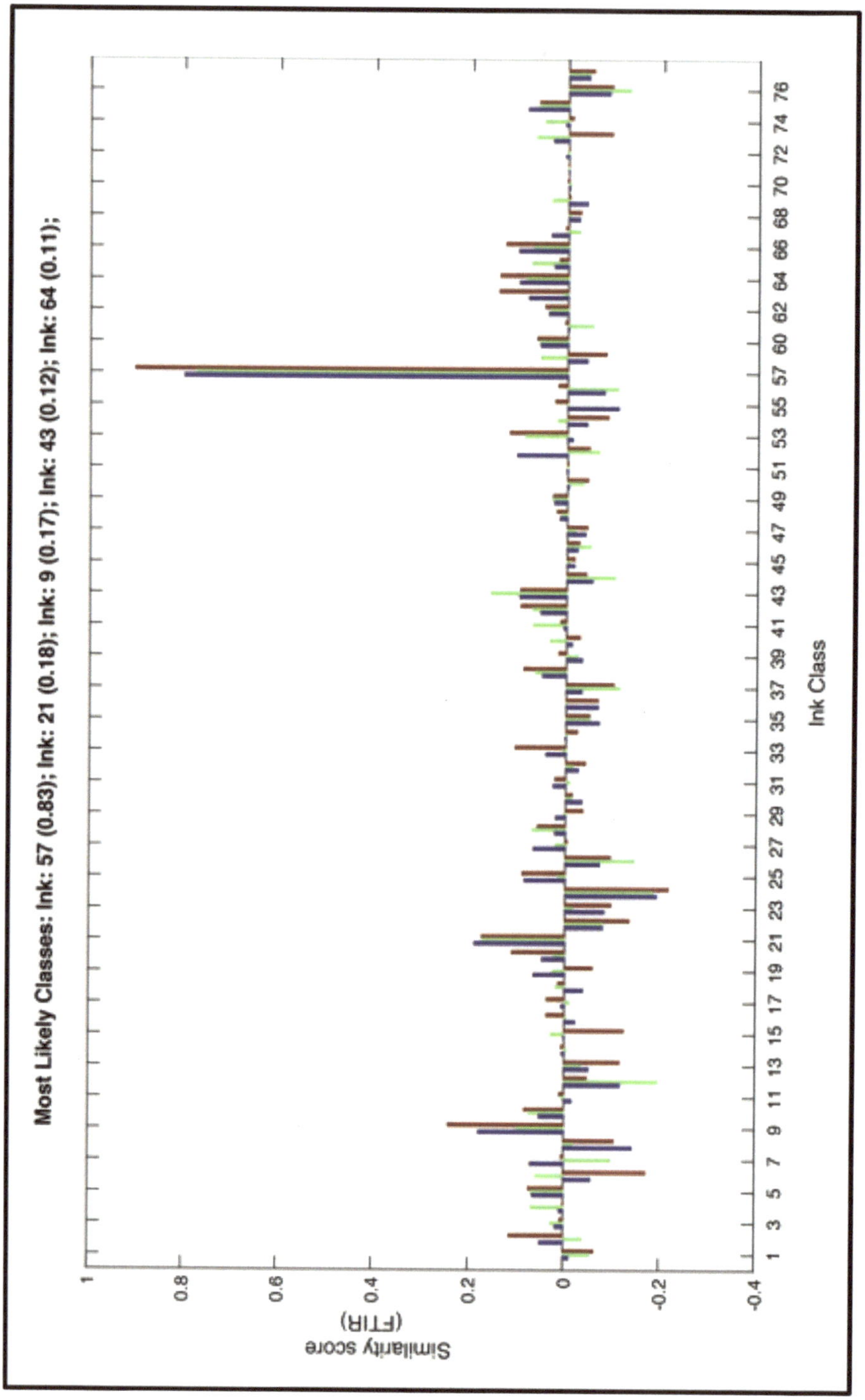
Most Likely Classes: Ink: 57 (0.83); Ink: 21 (0.18); Ink: 9 (0.17); Ink: 43 (0.12); Ink: 64 (0.11);
Similarity score (FTIR)
Ink Class

summarise the study by mentioning each technique with its advantages, disadvantages, limitations, and limits of detection in order to analyse a different range of materials. Thereafter, partial least square discriminant analysis (PLS-DA) and *k*-nearest neighbour (*k*NN) algorithms were performed on the data by calculating the search assessment from the created library to check the viability of the study (see Figure 8.7).[78]

Almirall *et al.* (2016) conducted experiments using LIBS and LA–ICP–MS to emphasise the capabilities of laser ablation methods in the field of forensic document examination. These techniques showed improved detection limits, superior sensitivity, and specificity with a simple experimental setup, generating high accuracy while preventing damage or destruction of the sample.[79]

Williamson *et al.* (2016) worked on toner samples from various sources with the objective of determining the capability of both ATR–FTIR and DART–MS to obtain the chemical composition of organic additives present in toner samples in a comprehensive and simple methodology. Discrimination of 86.3% and 81.8% was achieved by analysing 76 samples individually with ATR–FTIR and DART–MS, respectively. The combined approach provided 96.8% discrimination with ATR–FTIR studying the resin profile of toners depending on their origin, and DART–MS helping to characterise toners on the basis of PPG and PEG components in the toner.[80]

Subedi *et al.* (2015) utilised tandem LIBS/LA–ICP–MS with the aim of screening the samples rapidly in order to save time and attain characterisation for elemental profiling of the samples. It was concluded that the use of the tandem mode reduces the problems otherwise related to the examination of printed documents. Moreover, the author developed a methodology to analyse toner samples with a large degree of similarity in a comprehensive manner with the involvement of a small number of samples as compared to other methods and produced the analytical basis for the combined method to be used in real-world cases to help forensic document examiners (FDE).[81]

Lennard *et al.* (2015) performed chemical characterisation studies for elemental detection of toner samples using LIBS and LA–ICP–MS, with the help of element emission lines and their respective peak ratios. In comparison with LA–ICP–MS, chemometrics accounted for 99.5% discrimination, with 97.4% and 98.4% variation for the 3-sigma criterion and ANOVA with Tukey's *post hoc* test, respectively. The author highlighted the enormous ability of laser ablation methods to be adopted for forensic document

Figure 8.7 Visualization of PLSDA classification outputs for FTIR data of a total of 76 toner samples using a comparison sample as a duplicate toner ink (class 57). The bars represent the PLSDA output magnitude for each replicate (largest bars) for each of 76 different classes. The large similarity scores for sample (class) 57 indicate that the comparison sample is correctly associated with the same sample in the database Reproduced from ref. 78 with permission from John Wiley & Sons, Copyright 2016.

examination and insists that they will be helpful to examiners to reduce the existing case pendency, stating LIBS as a powerful method.[82]

Verma *et al.* utilized diffuse reflectance UV-Vis spectroscopy and chemometrics for the discrimination of 100 toner samples collected from laser as well as photocopier machines. The multivariate methods provide 99.59% and 99.84% pairwise discrimination of laser and photocopier toners respectively.[83] The same group have also utilized Schottky field emission scanning electron microscopy with energy-dispersive X-ray spectroscopy (FE-SEM–EDS) for discrimination and classification of black toners on the basis of their elemental compositions. The obtained SEM images as well as spectra of toner revealed morphological properties and an elemental profile of the toners and helped in the discrimination of photocopier and laser printouts. The multivariate analysis successfully classified the printed documents on the basis of the similarities and differences in their composition.[84]

Kumar *et al.* investigated inkjet, laser and photocopier printed documents by utilising ATR–FTIR and chemometric methods. It was revealed that the printing inks contain polystyrene, methyl acrylate, bisphonol A and ethers. The authors utilized PCA and HCA in the discrimination and LDA for classification of printed documents. The utilized methodology is non-destructive, reproducible and robust as compared to other approaches.[85]

The use of chemometrics in analysing the data (toner, ink, paper) obtained from spectroscopic or other methods of constituent analysis has gained importance and is now being utilised often in questioned document examination. The utilization of data handling techniques for bulk documents has addressed the challenge of managing extensive data in forensic science. Yet, there remains a need to promote and apply these methods routinely in forensic science labs during document examinations.

8.3.3 Paper Analysis

Many times, the fraudulent characteristics of a suspected document are conclusively revealed by the examination of the paper sample alone. Therefore, the paper used to create a document also plays a vital role in questioned document examination. In many legal cases, the paper itself becomes critical evidence in determining the relationship between forged and genuine documents. It is very important in some cases to determine whether two paper sheets are identical or have a common origin or not. The dating of paper is also one of the significant queries in the cases of fraud in historical documents, wills, deeds and financial documents.

Paper is a composite, made up of several distinct components. For printing and writing papers, a fibre content of bleached softwood (75%) and bleached hardwood (25%) is preferred.[86,87] Varieties of paper are developed utilising diverse coating substances, additives, as well as other elements. The surface characteristics of papers diverge from one another. The texture (roughness or smoothness) is primarily crucial for absorption, ink dispersion over the paper, and coating volume estimation.[88] Physical and chemical examination of paper is employed to find the correlation between counterfeit and authentic documents. The work of a document examiner is not only concerned with the comparison of handwriting or typescript, but they may also be asked to decipher or restore erased handwriting or typescript, to estimate the age of a document, and to authenticate the various questioned documents on the basis of their paper quality or characteristics.

Modern paper is generally made up of cellulose, hemicellulose, lignin, and a filler material such as calcium sulphate, calcium carbonate, titanium dioxide, kaolin, and ZnO (zinc oxide), *etc.*[39–41] Some kinds of paper, including currency notes. contain almost 100% cellulose. However, the manufacturers of paper do not reveal its constituents and also the chemical composition largely varies. For this particular reason, paper has a complex quality both physically (*i.e.*, fibre network) and chemically (*i.e.*, choice of additives, *etc.*).

The texture, colour, thickness, appearance, watermarks, cutting, size, chemical composition, and physical properties of the fibres used in its manufacture all contribute to the paper's uniqueness and dissimilarity. Oblique radiation and UV light provide an easy approach to observing the differences in the paper sheet, but this does not provide satisfactory results where one can link the source with great accuracy.

Microscopic examination can be used to identify the physical parameters of papers effectively. Dissimilar papers have different fibre compositions that can lead to positive results. The fibres in paper samples are ascertained by their shape, length, thickness, the presence of characteristic cells, and colour characteristics when fibres are brought into contact with staining agents. However, such methods are destructive in nature and hence can't be used in real forensic casework as the integrity of the document has to be maintained in such cases. Similarly, chemical tests also provide significant results but are destructive in nature. One can use the destructive technique if there is no other possible way to answer certain queries asked in a court of law.

Xia *et al.* explored the non-destructive application of the ATR–FTIR technique in the discrimination of 15 types of paper relics. The authors have utilised five statistical algorithms, *i.e.*, PCA–LDA, PLS-DA, soft independent modelling of class analogy (SIMCA), least square support vector machine (LS–SVM) and PLS–LDA. Among the utilised chemometric methods, PLS–LDA and LS–SVM provide 100% accurate classification, while PCA–LDA, PLS-DA, and SIMCA give accuracy of 98.67%, 97.33%, and 95.56%, respectively.[89]

Ortiz-Herrero *et al.* presented a micro-destructive approach for the estimation of the age of documents by applying Py-GC/MS combined with statistical methods. The prerequisite of the developed model is that the documents should be kept under similar conservation conditions. In this approach, 5 h of ageing in the chamber was equivalent to one natural year of aging. The presence and absence of some compounds in paper samples may be related to the age of those particular documents.[93]

Sharma *et al.* utilised ATR–FTIR spectroscopy coupled with chemometrics for the discrimination of paper samples. The spectra of three types of paper samples depict peaks of cellulose, hemicellulose, kaolin, calcium carbonates, calcium sulphate, *etc.* having different peak intensities that might be caused by the presence of different chemical constituents during their manufacturing. The authors have also predicted the age of office writing paper (A4 size) by utilising statistical models like curve estimation (CE), multiple linear regression (MLR), and partial least squares regression (PLSR). The CE model resulted in a maximum dating error (see Figure 8.8), while the MLR and PLSR models predicted the age of the document significantly well. However, some prerequisites should be followed while predicting the accurate age of unknown paper samples, as stated by the authors.[94]

Paper degrades with ageing because of the polymerisation process of cellulose. Zieba-Palus *et al.* studied degradation and, hence, ageing of paper samples by using spectroscopic methods. The authors artificially aged five paper samples under different environmental conditions. It was revealed

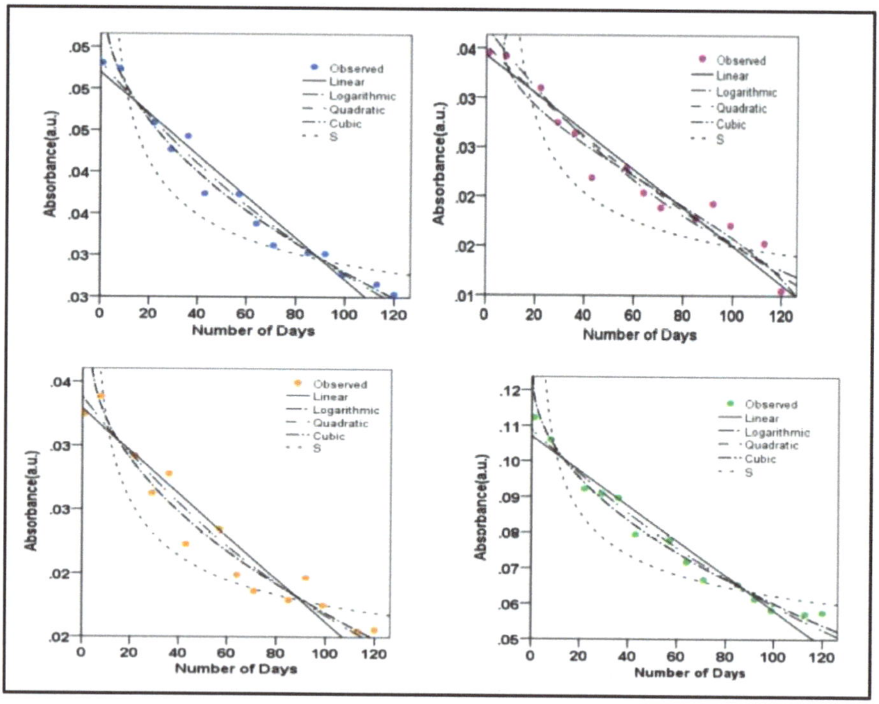

Figure 8.8 Curve estimation models giving information about the "goodness of fit" regression line, used for artificial ageing of office/writing paper samples. This cubic model provides the highest correlation and *R*-square values among all curve estimation models of paper samples. Reproduced from ref. 94 with permission from Taylor & Francis on behalf of Australian Academy of Forensic Sciences, Copyright 2019.

Table 8.2 Paper analysis using analytical methods combined with multivariate analysis.

Ref.	Sample type(s)	Analytical techniques	Chemometric methods	Objective of study
89	Paper	ATR–FTIR spectroscopy	PCA, PLS-DA, SVM, *etc.*	Characterisation and discrimination of papers
90	Writing/printing papers	TGA	PCA	
91	Writing/printing papers	ATR–FTIR spectroscopy	PCA	
92	Books	FTIR spectroscopy	PCA, SVM & PLS	Paper dating
93	Paper (white, recycled and notebook)	Py-GC/MS	PCA	
78	Reports	FTIR spectroscopy	PLS & PCA	
94	Papers	ATR–FTIR Spectroscopy	CE, MLR & PLSR	
95	Historic paper	THz time-domain spectroscopy	PLS	
96	Papers	ATR–FTIR & UV-Vis spectroscopy	PCA	

through UV-Vis spectroscopy that the paper samples differed in peak intensity at 280 nm and 350 nm respectively. The results were further verified by using 2D correlation analysis and PCA methods.[96]

In this chapter, we have discussed destructive and non-destructive approaches such as diffuse reflectance UV-Vis-NIR, FTIR, thermogravimetric analysis (TGA), and GC/MS (see Table 8.2). In general, the chemical characterisation of the paper samples is done by using the ATR–FTIR technique. The results are absolutely quantitative, which eliminates the difficulty that conventional light/radiation scanning processes present. These procedures are straightforward, reliable, and require little to no sample preparation. These procedures might also be implemented for heritage and cultural artefacts in circumstances wherein heritage documents need to be swapped for the benefit of the beneficiaries. Again, in cases where no other method can be used except a destructive technique, chromatography and thermogravimetric analysis have been used by the researchers to predict the date of a document as well as to characterise and differentiate various paper samples.

8.3.4 Analysis of Miscellaneous Exhibits

In this chapter, research based on examination of pen inks, toners, and papers has been highlighted and discussed. However, some more exhibits come under investigation in criminal cases, such as bank notes, cheques, stamp ink, polymer-based documents, *etc.* Among these, only banknotes and

Table 8.3 Banknotes analysis and chemometric approaches.

Ref.	Sample type(s)	Analytical techniques	Chemometric methods	Objective of study
97	Banknotes	Mobile phones	PCA	Discrimination of authentic and fake banknotes
98		XRF and Raman spectroscopy	PCA & PLS	
99		XRF	PCA	
100		Raman spectroscopy	PLS-DA	
101	Driving licence	ATR–FTIR spectroscopy	PCA, SVM & LDA	Examination of polymer-based documents

driver's licences (polymer based) have been studied by using multivariate analysis (see Table 8.3), For their authenticity and discrimination between genuine and fake currency notes, spectrometric techniques, especially FTIR and Raman, as well as X-ray fluorescence, have been employed.

Besides the elemental techniques mentioned in Table 8.3, Vittorazzi *et al.* classified banknotes by using smartphone image analysis. The authors procured four authentic banknotes and 12 falsified banknotes (printed by laser jet, wax-Color Qube and inkjet printer). The authors selected different regions on the bank notes (*i.e.* holographic, region of animal figure, numbers, and effigy micro impressions), different focal lengths of camera (*i.e.* 8, 12, 15 and 20), region of interest (ROI) (*i.e.* 16×16; 32×32 and 64×64), and different smartphones (*i.e.* Motorola, Samsung, LG) for the discrimination. Promising results were obtained from ROI 16×16 pixels with focal length fixed at 10 cm.[97]

McGann *et al.* (2020) were the first to use ATR–FTIR spectroscopy in conjunction with chemometric PCA, SVM, and LDA models to investigate polymer-based Western Australian driver's licence cards for chemical variety. They gathered 38 samples from three series: Swiss (2001), Victorian (2011), and New South Wales (2014). All three series were collected from Western Australia and individual cards from Switzerland, Victoria, and New South Wales were then used as test samples to see if they could be distinguished from any of the Western Australian cards. All three revealed a significant amount of intra-class dispersion during PCA analysis. For this investigation, the multiclass SVM model with 98.7% accuracy surpasses the LDA model with 79.6% accuracy. Within a particular issuance period, it was indicated that different card compositions were employed by different licencing agencies.[101]

8.4 Discussion

The studies related to ink examination emphasise both destructive as well as non-destructive techniques in conjunction with chemometrics for discrimination of ballpoint/gel/tip pen ink samples. Under ideal circumstances, the spectral features of inks from identical and distinct manufacturers were investigated to discover the differences between them. With the implementation of multivariate examination, the FTIR technique has been shown to be an excellent tool for the characterisation of complicated ink formulations as well as systematic discrimination of all ink samples. The most frequent chemometric approaches performed on the data were PCA and HCA analysis to determine whether or not the pen inks could be separated by analytical techniques. Similarly, the dating of a questioned document is a challenging assignment faced by forensic professionals. In this chapter, a number of regression models have been discussed for the evaluation of writing inks by quantifying the fading of ink over time by implementing the analytical approach. The regression model thus created facilitates the estimation of ink arrival chronology. Analysing the existing chronological information should be used to cross-check the hypothesised model. The limitations of such types of studies are that the model is applicable to one type of ink sample only. The prior knowledge of the storage of the document in different environmental conditions is also necessary and therefore these variables are important to formulate the universal model for ink dating.

The classification and differentiation of paper samples, which might be effective in giving forensic assistance in the resolution of legal issues, has also been addressed. For classification and differentiation, the subjective attributes and chemometrics of collected spectral information are exploited. Classification is accomplished by comparing the peaks with standards of cellulose and inorganic fillers such as kaolin, titanium dioxide, calcium carbonate, calcium sulphate, *etc.*, and the general composition of paper. It was determined that cellulose is present in all the paper samples. Paper manufacturers use different inorganic fillers in different brands of paper. It is the quantity of organic and inorganic materials which shows dissimilarity and hence facilitates the paper source differentiation. The presented results confirm that these findings would be extremely beneficial in questioned document inspection activity in criminal circumstances when the document's genuineness is disputed. Use of chemometrics makes the analysis of large data simple and systematic and hence solves the difficulty that comes in light/radiation scanning procedures. The technique is cost-effective, repeatable, requires no sample preparation, and is easy to operate. The diffuse reflectance UV-Vis-NIR, ATR–FTIR, and Raman spectroscopy are the instrumentations of preference for questioned document analysis because of their minimal cost of examination, ease of operation, and non-destructive, trustworthy, and high sensitivity. Moreover, destructive approaches also provide conclusive results and, hence, can be used in cases where non-destructive analytical methods fail to provide significant results or are not available.

8.5 Conclusion

Questioned document examination (QDE) is a forensic science specialty that primarily deals with documents that might have been challenged in a court of law.

The rapid increase in the manufacture of writing inks and paper led to the increase in their use in legal documents such as cheques, wills, financial agreements, stamp papers, passports, tickets, first information reports (FIR), *etc.*, and it was followed by different levels of forgery by offenders or any individual. As a result, there will always be the potential for deliberately counterfeit documents in the community.

The key objective of QDE is to adopt scientific procedures and methodologies to supply proof concerning a suspicious or doubtful document. The investigation and comparison of papers, inks, disputed penmanship, hand prints, typewriting, professional printing, photocopying, and other documentary proof with particular compounds is done to determine the veracity of the disputed content and also identify manipulations. Also, there may be cases where one or more papers from a document in a deed or will are replaced in order to provide an advantage to someone without seeking the prior consent of all parties. Here the differentiation of counterfeit paper out of the whole document can be pointed out. Similarly, in ballpoint pens, all the ingredients vary from product to product in their composition and combinations. If more than one ink is identified to be involved, a questioned document examiner can employ these variances to figure out how many inks have been used to create the document. The two ink samples are compared, as well as differences and similarities between them highlighted.

In forensic science laboratories (FSLs), document examination is carried out by using oblique visible light or UV radiations for preliminary examination to find any writing/paper differences. However, the method fails in the cases where the physical properties of materials are the same, and the chemical properties are different. In this direction, this chapter explains developing methods based upon analytical measurement with statistical confidence so that forensic experts can utilize these instead of qualitative light scanning methods. This chapter highlights the growing trend of chemometric methods for the evaluation of disputed/questioned documents. Thus, questioned document examiners can augment their analytical results with chemometric methods in order to get better interpretation of the results. Additionally chemometric methods will provide useful complementary information, especially in those situations where there are challenges with conventional examinations alone. It should be further emphasised that a library for inks, paper, toners, *etc.* should be developed by using various analytical methods so that it can be used as a standard reference for comparison purposes.

References

1. E. Katz, *Forensic Science: A Multidisciplinary Approach*, Wiley-VCH, Cop, Weinheim, 2016.
2. A. S. Osborn, *Questioned Documents*, The Lawyers' Co-operative Publishing Co., Rochester, New York, 1910.
3. V. Sharma, R. Kumar, K. Devgan, P. K. Mishra, A. Ekielski, V. Kumar and V. Kumar, *Spectrosc. Lett.*, 2018, **51**, 205–215.

4. C. Weyermann and B. Spengler, *Forensic Sci. Int.*, 2008, **180**, 23–31.
5. R. Kumar and V. Sharma, *TrAC, Trends Anal. Chem.*, 2018, **105**, 191–201.
6. National Academy of Sciences, Committee on Identifying the Needs of the Forensic Sciences Community, in *Strengthening Forensic Science in the United States: A Path Forward*, National Research Council, Washington DC, 2009.
7. Indian Evidence Act, 1872, https://legislative.gov.in/sites/default/files/A1872-01.pdf.
8. C. Weyermann, Mass Spectrometric Investigation of the aging processes of ballpoint ink for the examination of questioned documents, http://geb.uni-giessen.de/geb/volltexte/2006/3044/.
9. D. Ellen, *The Scientific Examination of Documents: Methods and Techniques Methods and Techniques*, Taylor & Francis, 2014.
10. M. Calcerrada and C. García-Ruiz, *Anal. Chim. Acta*, 2015, **853**, 143–166.
11. V. Causin, R. Casamassima, C. Marega, P. Maida, S. Schiavone, A. Marigo and A. Villari, *J. Forensic Sci.*, 2008, **53**(6), 1468–1473.
12. H. S. Chen, *J. Forensic Sci.*, 2005, **4**, 15.
13. J. Zimmerman, D. Mooney and M. J. Kimmett, *J. Forensic Sci.*, 1986, **31**(2), 489.
14. W. J. Egan, R. C. Galipo, B. K. Kochanowski, S. L. Morgan, E. G. Bartick, M. L. Miller, D. C. Ward and R. F. Mothershead, *Anal. Bioanal. Chem.*, 2003, **376**(8), 1286.
15. S. Materazzi, R. Risoluti, S. Pinci and F. S. Romolo, *Talanta*, 2017, **174**, 673–678.
16. K. Saini and J. S. Saroa, *J. Forensic Identif.*, 2011, **560**(61), 6.
17. A. Assis, M. Barbosa, J. V. Nabais, A. F. Custodio and P. Tropecelo, *Forensic Sci. Int.*, 2012, **214**, 59.
18. R. Corzo, K. Subedi, T. Trejos and J. Almirall, *J. Forensic Sci.*, 2016, **61**, 725.
19. A. Metzinger, R. Rajko and G. Galbacs, *Spectrochim. Acta, Part B*, 2014, **94–95**, 48–57.
20. M. I. Szynkowska, K. Czerski, T. Paryjczak and A. Parczewski, *Surf. Interface Anal.*, 2010, **42**, 429–437.
21. C. G. G. Aitken and F. Taroni, *Statistics and the Evaluation of Evidence for Forensic Scientists*, UK John Wiley & Sons, Ltd, Chichester, 2004.
22. G. Zadora, A. Martyna, D. Ramos and C. Aitken, *Statistical Analysis in Forensic Science: Evidential Value of Multivariate Physicochemical Data*, John Wiley & Sons Inc., Chichester, West Sussex, 2014.
23. Forensic Science Regulator, Annual Report November 2017–November 2018, 2019.
24. President's Council of Advisors on Science and Technology, *Forensic Science in Criminal Courts: Ensuring Scientific Validity of Feature-Comparison Methods*, Washington DC, 2016.
25. National Institute of Standards and Technology, New NIST Center of Excellence to Improve Statistical Analysis of Forensic Evidence, 2015, http://www.nist.gov/forensics/center-excellence-forensic052615.cfm.

26. J. Morgan, Deputy Director National Institute of Justice, Office of Justice Programs, U.S. Department of Justice, Presentation to the committee, 2007.
27. H. F. Kaiser, *Educ. Psychol. Meas.*, 1960, **20**, 141–151.
28. D. L. Massart, *Chemometrics: A Textbook*, Elsevier, Amsterdam, 1988.
29. P. Gemperline, *Practical Guide to Chemometrics*, CRC/Taylor & Francis, Boca Raton, 2006.
30. B. K. Lavine, *Encyclopedia of Analytical Chemistry*, Wiley, Chichester, 2000.
31. L. A. Berrueta, R. M. Alonso-Salces and K. Héberger, *J. Chromatogr. A*, 2007, **1158**, 196–214.
32. F. Santana, A. Souza, M. Almeida, M. Breitkreitz, P. Filgueiras, M. Sena and R. Poppi, *Quim. Nova*, 2020, **43**, 371.
33. C. Cortes and V. Vapnik, *Mach. Learn.*, 1995, **20**, 273–297.
34. C. Silva, A. Braz and M. F. Pimentel, *J. Braz. Chem. Soc.*, 2019, **30**, 2259.
35. S. B. Gadžurić, S. O. Podunavac, M. B. Kuzmanović, M. Vraneš, T. Petrin, S. Z. Bugarski and S. Z. Kovačević, *Iran. J. Pharm. Res.*, 2016, **15**, 725.
36. P. Geladi, *Spectrochim. Acta, Part B*, 2003, **58**, 767.
37. M. Otto, *Chemometrics : Statistics and Computer Application in Analytical Chemistry*, Wiley-VCH, Weinheim, 2007.
38. S. F. Møller, J. von Frese and R. Bro, *J. Chemom.*, 2005, **19**, 549–563.
39. L. M. Larsen, *Industrial Printing Inks*, Reinhold, New York, N.Y., 1962.
40. R. H. Leach, R. J. Pierce, E. P. Hickman, M. J. Mackenzie and H. G. Smith, *The Printing Ink Manual*, Springer Netherlands, Dordrecht, 1999.
41. J. A. G. Drake, *Chemical Technology in Printing and Imaging Systems*, Royal Society Of Chemistry, Cambridge England, 1993.
42. C. Weyermann, D. Kirsch, T. Andremann and B. Spengler, *Forensic Sci. Int.*, 2003, **136**, 73–74.
43. R. L. Brunelle and K. R. Crawford, *Advances in the forensic analysis and dating of writing ink*, Charles C Thomas, Springfield, Ill., 2003.
44. R. P. Gorziza, C. M. B. Carvalho, M. González, R. S. Ortiz, G. A. Helfer, M. F. Ferrão and R. P. Limberger, *Braz. J. Forensic Sci. Med. Law Bioethics*, 2020, **9**, 331–355.
45. G. Sauzier, J. McGinn, T. Trubshoe and S. W. Lewis, *Forensic Sci. Int.: Rep.*, 2019, **1**, 100021.
46. C. A. Teixeira and R. J. Poppi, *Microchem. J.*, 2019, **144**, 411.
47. G. Sauzier, P. Giles, S. W. Lewis and W. van Bronswijk, *Anal. Methods*, 2015, **7**, 4892–4900.
48. M. Kamil, M. Asri, W. Desa and D. Ismail, *Malays. J. Forensic Sci.*, 2015, **6**, 48.
49. V. Sharma and R. Kumar, *Vib. Spectrosc.*, 2017, **92**, 96–104.
50. M. N. Mohamad Asri, W. N. S. Mat Desa and D. Ismail, *Aust. J. Forensic Sci.*, 2016, **49**, 175–185.
51. L. C. Lee, M. R. Othman, H. Pua and A. A. Ishak, *Malays. J. Anal. Sci.*, 2012, **3**, 5.

52. R. Kumar and V. Sharma, *Spectrochim. Acta, Part A*, 2017, **175**, 67.
53. M. I. A. Halim, N. S. A. Aziz, N. Saim, R. Osman and H. Jasmani, *Malays. J. Fundam. Appl. Sci.*, 2014, **8**, 159.
54. F. S. L. Borba, T. Jawhari, R. S. Honorato and A. D. Juan, *Forensic Sci. Int.*, 2015, **19**, 73.
55. F. Alamilla, M. Calcerrada, C. García-Ruiz and M. Torre, *Forensic Sci. Int.*, 2013, **228**, 1–7.
56. V. A. G. da Silva, M. Talhavini, J. J. Zacca, B. R. Trindade and J. W. Braga, *J. Braz. Chem. Soc.*, 2014, **25**, 1552.
57. P. K. Yadav and R. M. Sharma, *Vib. Spectrosc.*, 2020, **108**, 103054.
58. L. Rodrigues e Brito, A. Braz, R. Saldanha Honorato, M. F. Pimentel and C. Pasquini, *Forensic Sci. Int.*, 2019, **298**, 169–176.
59. R. E. Goacher, L. G. DiFonzo and K. C. Lesko, *Anal. Chem.*, 2016, **89**, 759–766.
60. A. R. Martins, C. S. Dourado, M. Talhavini, A. Braz and J. W. B. Braga, *Forensic Sci. Int.*, 2019, **296**, 91–100.
61. L. Rodrigues e Brito, A. B. Chaves, A. Braz and M. F. Pimentel, *Spectrochim. Acta, Part A*, 2019, **223**, 117287.
62. G. Sauzier, J. McGann, S. W. Lewis and W. van Bronswijk, *Anal. Methods*, 2018, **10**, 5613–5621.
63. L. Ortiz-Herrero, A. C. de Almeida Assis, L. Bartolomé, M. L. Alonso, M. I. Maguregui, R. M. Alonso and J. S. Seixas de Melo, *Chemom. Intell. Lab. Syst.*, 2020, **207**, 104187.
64. C. M. B. Carvalho, R. S. Ortiz, M. dos Reis, M. F. Ferrão and R. P. J. Limberger, *J. Am. Soc. Questioned Doc. Exam.*, 2019, **22**(1), 19–35.
65. L. Ortiz-Herrero, L. Bartolomé, I. Durán, I. Velasco, M. L. Alonso, M. I. Maguregui and M. Ezcurra, *Microchem. J.*, 2018, **140**, 158–166.
66. V. Sharma and R. Kumar, *Microchem. J.*, 2017, **134**, 104–113.
67. S. Senior, E. Hamed, M. Masoud and E. Shehata, *J. Forensic Sci.*, 2012, **7**, 1087.
68. A. S. Diamond and D. S. Weiss, *Handbook of imaging materials*, CRC Press, Boca Raton, 2019.
69. J. S. Harris, *Int. J. Forensic Doc. Exam.*, 1998, **4**, 182.
70. M. Kriss, *Handbook of Digital Imaging*, United Kingdom John Wiley & Sons, Inc., Chichester, West Sussex, 2015.
71. J. Yang, T. J. Wang, H. He, F. Wei and Y. Jin, *Ind. Eng. Chem. Res.*, 2003, **42**, 5568–5575.
72. A. Braz, M. López-López, G. Montalvo and C. G. Ruiz, *Forensic Sci. Int.*, 2013, **232**, 206.
73. L. Gál, M. Oravec, P. Gemeiner and M. Čeppan, *Forensic Sci. Int.*, 2015, **257**, 285–292.
74. R. Kumar and V. Sharma, *TrAC, Trends Anal. Chem.*, 2018, **105**, 191.
75. C. F. Poole, *Reference Module in Chemistry, Molecular Sciences and Chemical Engineering*, 2013.
76. C. S. Silva, M. F. Pimentel, J. M. Amigo, C. García-Ruiz and F. Ortega-Ojeda, *Anal. Chim. Acta*, 2018, **1031**, 28–37.

77. C. E. Johnson, P. Martin, K. A. Roberts, T. Trejos, R. Corzo, J. R. Almirall and A. M. Safer, *J. Forensic Sci.*, 2017, **63**, 66–79.
78. T. Trejos, P. Torrione, R. Corzo, A. Raeva, K. Subedi, R. Williamson, J. Yoo and J. Almirall, *J. Forensic Sci.*, 2016, **61**, 715–724.
79. J. R. Almirall and T. Trejos, *Elements*, 2016, **12**, 335–340.
80. R. Williamson, A. Raeva and J. R. Almirall, *J. Forensic Sci.*, 2016, **61**, 706–714.
81. K. Subedi, T. Trejos and J. Almirall, *Spectrochim. Acta, Part B*, 2015, **103–104**, 76–83.
82. C. Lennard, M. M. El-Deftar and J. Robertson, *Forensic Sci. Int.*, 2015, **254**, 68–79.
83. N. Verma, R. Kumar and V. Sharma, *Spectrochim. Acta, Part A*, 2018, **196**, 40–48.
84. N. Verma, V. Sharma, R. Kumar, R. Sharma, M. C. Joshi, G. R. Umapathy, S. Ohja and S. Chopra, *Anal. Bioanal. Chem.*, 2019, **411**, 3477–3495.
85. R. Kumar, A. Samkaria and V. Sharma, *Sci. Justice*, 2020, **60**, 347–357.
86. F. M. Udristioiu, I. G. Tanase, A. A. Bunaciu and H. Y. Aboul–Eneinods, *Appl. Spectrosc. Rev.*, 2012, **47**(7), 550–570.
87. J. C. Roberts, *Paper Chemistry*, New York, Glasgow, Blackie, 1991.
88. P. Vernhes, S. Rolland Du Roscoat, A. Blayo, B. Pineaux, X. Thibault and J.-F. Bloch, *J. Imaging Sci. Technol.*, 2008, **52**(6), 010502.
89. J. Xia, J. Zhang, Y. Zhao, Y. Huang, Y. Xiong and S. Min, *Spectrochim. Acta, Part A*, 2019, **219**, 8.
90. R. Kumar, V. Sharma, N. Verma, P. K. Diwan, V. Kumar and V. Kumar, *Aust. J. Forensic Sci.*, 2019, **51**, 22.
91. R. Kumar, V. Kumar and V. Sharma, *Spectrochim. Acta, Part A*, 2017, **170**, 19–28.
92. J. Xia, Y. Huang, J. Zhang, X. Du, H. Yan, Q. Li, Y. Li, Y. Xiong and S. Min, *Cellulose*, 2020, **27**, 5323–5335.
93. L. Ortiz-Herrero, M. E. Blanco, C. García-Ruiz and L. Bartolomé, *J. Anal. Appl. Pyrolysis*, 2018, **131**, 9–16.
94. V. Sharma, J. Kaur and R. Kumar, *Aust. J. Forensic Sci.*, 2020, **53**(6), 703–726.
95. T. Trafela, M. Mizuno, K. Fukunaga and M. Strlic, *35th International Conference on Infrared, Millimeter, and Terahertz Waves*, 2010.
96. J. Zięba-Palus, B. Trzcińska, A. Wesełucha-Birczyńska, P. Moskal and J. Sacharz, *J. Mol. Struct.*, 2020, **1205**, 127651.
97. B. V. Vittorazzi, R. A. Costa, L. M. Coelho, M. M. Isidoro, K. M. G. Lima, P. R. Filgueiras and W. Romão, *Quim. Nova*, 2020, **43**, 447.
98. A. R. N. Rodrigues, F. L. Melquiades, C. R. Appoloni and E. N. Marques, *Forensic Sci. Int.*, 2019, **302**, 109872.
99. C. R. Appoloni and F. L. Melquiades, *Appl. Radiat. Isot.*, 2014, **85**, 92–95.
100. M. R. de Almeida, D. N. Correa, W. F. C. Rocha, F. J. O. Scafi and R. J. Poppi, *Microchem. J.*, 2013, **109**, 170–177.
101. J. McGann, M. Willans, G. Sauzier, M. J. Hackett, S. W. Lewis, J. McGinn, T. Trubshoe and W. van Bronswijk, *Forensic Sci. Int.: Rep.*, 2020, **2**, 100149.

CHAPTER 9

Biological Materials

H. LIN,[a] P. HUANG[b] AND Z. WANG*[c]

[a] Department of Forensic Medicine, School of Basic Medical Sciences, Fudan University, Shanghai 200032, PR China; [b] Shanghai Key Laboratory of Forensic Medicine, Shanghai Forensic Service Platform, Academy of Forensic Science, Shanghai, 200063, PR China; [c] Department of Forensic Pathology, College of Forensic Medicine, Xi'an Jiaotong University, Xi'an, 710061, PR China
*Email: wzy218@xjtu.edu.cn

9.1 Introduction

According to a report based on the FBI's Crime in the US reports, the holistic violent crime rate in the US dropped by nearly a third from 1999 to 2018, but the rate, which includes murder, manslaughter, rape, and aggravated assault, is still high.[1] Searching for the key evidence to identify the suspects in such cases is the burden of forensic investigators. Biological evidence (*e.g.*, bodily fluids, tissues, hair, and fingermarks) is an important type of key evidence encountered at such crime scenes. It can not only assist forensic investigators in adjusting the investigation direction but also help narrow the pool of suspects by effective characterization of the donor's basic information (*e.g.*, age, sex, race and DNA profile).

Bodily fluid traces recovered at crime scenes are among the most important types of biological evidence to forensic investigators. In routine forensic investigations, presumptive and confirmatory tests are two conventional approaches to analysing the bodily fluid evidence. For example, presumptive assays, such as the Kastle–Meyer test, leucomalachite green, and luminol, and confirmatory assays, such as the Takayama and Teichmann crystal assays, are commonly used to identify blood but are often inefficient

Chemometric Methods in Forensic Science
Edited by Vishal Sharma, Georgina Sauzier and Simon W. Lewis

Published by the Royal Society of Chemistry, www.rsc.org

and non-specific.[2–4] Commercial immunochromatographic assays, such as HemaTrace® and ABAcard®, are highly specific for the identification of human blood.[5,6] However, the disadvantage of this approach is that it is not suitable for wildlife forensic investigation and animal wounding cases where the identification of the species of the animal is very important. Currently, although forensic DNA analysis, as a mature and conventional technique, has proven accurate and reliable for blood identification,[5,6] the tests need to be performed in a laboratory by professional technicians, which is time-consuming and cannot satisfy the need for immediate results at the crime scene. The current state of semen analysis is similar to blood analysis. Serology methods, such as seminal acid phosphatase and prostate-specific antigen, are not specific.[3] Moreover, in some special cases involving azoospermia samples, sperm cells cannot be detected, making it impossible to perform DNA genotyping and mitochondrial sequencing of the cytochrome *b* gene. Apart from blood and semen, the two most commonly found bodily fluids at crime scenes, other kinds of bodily fluids, such as urine, saliva, sweat and vaginal secretion, also play an important role in crime scene investigation. However, the detection approaches for these bodily fluids are very limited. For instance, the Phadebas® test is always used to ascertain the presence of saliva, but it cannot confirm the presence of human saliva, which should be further confirmed by other methods, such as the ELISA method for the detection of statherin.[7] Lugol's staining method has been considered effective for vaginal fluid detection because abundant stratified squamous epithelial cells found in vaginal fluid can be stained. However, cells of the oral cavity can also be stained, and careful evaluation of Lugol's staining result is required.[8] Except for DNA profiling, the information provided by all the existing mature methods that are applied in forensic practice for bodily fluid investigation is very limited. It is impossible to obtain the donor's basic information, such as sex, race, age, and deposition time, by these methods.

Apart from bodily fluids, other biological materials such as hair, fingermarks, and human remains found at crime scenes are also important evidence for forensic investigation. Hair is among the most common types of biological evidence found at crime scenes. Microscopic examination is commonly used in forensic hair investigation to determine specific properties like race, somatic region, and hair dyes.[9] However, such an approach is subjective. Skeletal remains can also disclose information about the individual's age, gender, and race, as well as injuries.[10] Nevertheless, conventional forensic anthropological techniques are qualitative, and they are also subject to investigator bias. Fingermarks are another important piece of biological evidence. Fingermark composition can provide a wealth of information about the individual, like sex, age, ethnicity, and health status. In order to get a better understanding of the fingermark composition, a variety of analytical approaches (*e.g.* gas chromatography mass spectrometry, liquid chromatography coupled with mass spectrometry, and laser desorption ionization mass spectrometry) have been applied.[11] Despite their utility, knowledge of fingermark composition is still incomplete. Given the important role biological evidence plays in crime scene investigation, methods that are non-destructive to the samples, capable of directly measuring samples without other chemical labels, and that can also obtain as much information as possible, are especially welcomed by forensic investigators. Vibrational spectroscopy, as a fast, simple, non-destructive and label-free chemical detection technology that can sensitively detect the spectral chemical information of all molecular components (including proteins, lipids, nucleic acids and carbohydrates) in the sample and reflect the basic state of the biological sample being tested,[12,13] is becoming increasingly popular in the fields of biological medicine[14] and forensic science.[15] Infrared (IR) and Raman spectroscopies are two main vibrational spectroscopy technologies used in the field of forensic science. Infrared spectroscopy is defined as the analysis of infrared light as it interacts with the covalently bonded atoms and groups of molecules within a sample,[16] while Raman spectroscopy is narrowly defined as spontaneous inelastic light scattering originating from the chemical groups within the sample when irradiated with light.[13] The IR/Raman spectrum, usually plotted in the format of intensity against wavenumber, is composed of a large number of variables. The chemical information of all components within the sample is directly reflected in these variables and their intensities (see Figure 9.1). Professionals can pick out useful variables manually by observing the peak positions, intensities, and widths of the spectra according to their goals, perform statistical analysis on them, and finally obtain the results. This analytical process is suitable for conventional and easy qualitative studies, such as the identification of pure inorganic/organic compounds with a known reference spectrum, but it is not applicable to complex analysis tasks, such as multicategory classification scenarios in which the spectral profiles between the categories are very similar. For such situations, conventional

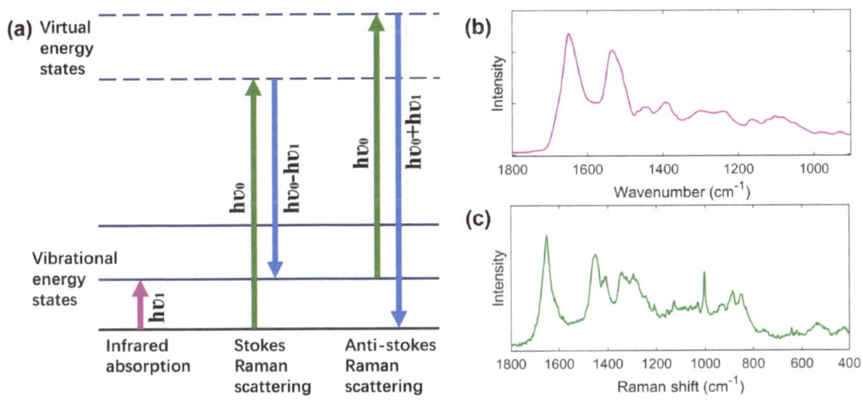

Figure 9.1 (a) Energy diagram showing transitions involved during infrared absorption, Raman Stokes and anti-Stokes scattering, (b) FTIR biological spectrum, and (c) Raman biological spectrum.

statistical approaches are not enough to identify important information, recognize patterns within sets of data and obtain new insights into the data sets. Chemometrics, a popular analysis tool consisting of various mathematical methods that shows great power in the extraction and visualization of information from data and recognition of complex relationships between data sets, has been widely used in spectral data analysis tasks, such as signal preprocessing, curving fitting, variable selection, pattern recognition and machine learning.[17] Because the subject of this chapter is to review the use of chemometrics for the spectral analysis of forensic bodily fluids, the involved algorithms only account for a small fraction of the available chemometric methods, which are artificially divided into three main categories according to research purpose: exploratory, regression and classification. Details about these algorithms will be introduced in the next section.

Several reviews have been written in the past that concentrate on using vibrational spectroscopy for forensic investigations.[15,18–20] However, the majority of these reviews have been broad in their scope, which involves not only bodily fluids but also additional forensic evidence. Moreover, the chemometric analysis of spectral data is not their focus. Our review focuses on the application of chemometrics for spectral analysis of forensic bodily fluids. Studies involving common forensic bodily fluids, such as blood, semen, urine, saliva, and sweat, are discussed in detail in this review, while the other biological materials-related studies are covered briefly. It should be emphasized that studies that do not utilize any chemometrics or chemometric algorithms involving spectral data preprocessing were not considered in this review. Furthermore, an in-depth discussion on the challenges of vibrational spectroscopy in combination with chemometrics from the aspect of practical application for forensic bodily fluid investigation is also included. The goal of this review is to help forensics/chemical experts work in their fields to better understand the applications of vibrational spectroscopy in combination with chemometrics in forensic bodily fluid investigation and promote the integration of such methods into the forensic science environment in the near future.

9.2 Chemometrics for Spectral Investigations of Forensic Bodily Fluids

The primary aim of chemometrics is to extract useful information from data consisting of a large number of measured variables and find the crucial patterns. Chemometrics includes many algorithms that can be used for various types of chemical data, such as spectral data, mass data, and chromatographic data. In this chapter, our focus is on chemometric analysis of vibrational spectral data originating from forensic bodily fluids. The involved chemometric algorithms make up a small part of the analysis and can be simply divided into three categories according to data analysis purposes: unsupervised exploratory data analysis, supervised regression modelling and classification modelling.

9.2.1 Exploratory Data Analysis

Exploratory data analysis is often the first step in chemometric processing, whose purpose is to summarize the main characteristics of data in an easy-to-understand form, often with visual graphs. Principal component analysis (PCA) is the most popular chemometric method for exploratory spectral data, and it allows us to explore possible outliers and patterns in the data.[21]

Multivariate curve resolution (MCR) is a group of chemometric methods that are capable of extracting underlying pure component spectra from spectroscopic mixture data through bilinear model decomposition.[22] MCR is ideal for the spectral identification of forensic bodily fluids because the bodily fluid stains found at crime scenes are not always pure and can be contaminated with other substances and/or bodily fluid stains. MCR can greatly help interpret the results. MCR is usually combined with other algorithms to improve the decomposition ability. Multivariate curve resolution-alternating least squares (MCR-ALS) is the most classic combination.

9.2.2 Regression Modelling

In forensic practice, bodily fluid investigation very often relies on some form of quantitative prediction. For example, estimation of the time since deposition (TSD) of bodily fluid stains is an important task for forensic investigators. This goal can be achieved by combining spectral measurements with chemometric regression analysis like PLS regression and principal component regression (PCR).[23,24]

9.2.3 Classification Modelling

The goal of classification analysis is to build a classifier by using a dataset with predefined categories and then use this classifier to predict unknown samples. The most commonly used algorithm for spectral classification tasks is a modified PLS algorithm, also known as partial least squares discriminant analysis (PLS-DA), which recodes the categorical output variables into continuous variables to turn the classification problem to a special PLSR problem.[25] Besides PLS-DA, there are a number of additional noteworthy classification algorithms used for spectral classification tasks, like linear discriminant analysis (LDA), support vector machine discriminant analysis (SVM-DA), and random forest (RF).[26–28]

Although the multiple variables of spectral data can offer more information about the testing objects and help improve the prediction performance of the spectral classification/regression models, the noise contained in the variables may weaken the outcome performance of the chemometric models in some cases. Therefore, in these cases, discarding some variables is necessary. However, sometimes, the choice of what variables to discard is not clear. The genetic algorithm (GA) is an optimization algorithm inspired by the process of natural selection that can iteratively discard and locate the subset of variables that results in the best performance of the chemometric models.[29]

Once the classification/regression model is built, the quality of the model should be properly evaluated. External tests and cross-validation (CV) are two main ways to accomplish this.[30] An external test is performed by using a completely independent data set for evaluating the performances of the model, and it is the preferred validation approach because it best mimics how the model will be routinely used. There are several important statistical metrics to evaluate the model performance. Sensitivity, specificity and accuracy, defined from the number of true positives (TP), true negatives (TN), false positives (FP) and false negatives (FN), are often used to evaluate the classification, while root mean square error (RMSE), which is a metric that tells us the average distance between the predicted values (PV) from the model and the actual values (AV) in the dataset, is often used to monitor the prediction performance of the regression methods.

9.3 Key Research Findings

9.3.1 Blood

Blood is one of the most common bodily fluids at violent crime scenes and is widely investigated to find more valuable clues, such as the species, TSD and the donors' basic information. Vibrational spectroscopy is a fast, non-destructive and user-friendly way to investigate blood traces. The effect of contamination on the spectra of blood is the first problem that should be solved. In a study conducted by Sikirzhytskaya *et al.*,[31] blood samples contaminated with three types of contaminants (dust, sand, and soil) that are commonly found in outdoor crime scenes were analysed to extract blood signatures using Raman spectroscopy. The results demonstrated that a pure blood signature can be extracted from the mixture spectra of high heterogeneity by MCR-ALS. Subsequently, the classification model for blood identification was built by PLS-DA, resulting in 100% accuracy. Even heavily contaminated bloodstain samples can be identified with no prior knowledge of the fluid or contaminant. Vibrational spectroscopy is not only capable of generating pure blood spectra of blood extracted from non-absorbent surfaces but has also proven feasible to recognize signals of bodily fluid that are embedded in strong substrate interferences. In a study conducted by Takamura *et al.*,[32] postmortem and antemortem blood samples deposited on three common absorbent surfaces (polyester, cotton, and denim) were analysed using attenuated total reflection Fourier transform infrared (ATR–FTIR) spectroscopy combined with innovative multivariate processing algorithms. The results demonstrated that regardless of the severe interference of the absorbent substrates, the pure postmortem and antemortem blood signals can still be extracted successfully by the weighted linear regression algorithm that was originally developed by the authors (see Figure 9.2). Then, the postmortem and antemortem blood spectra

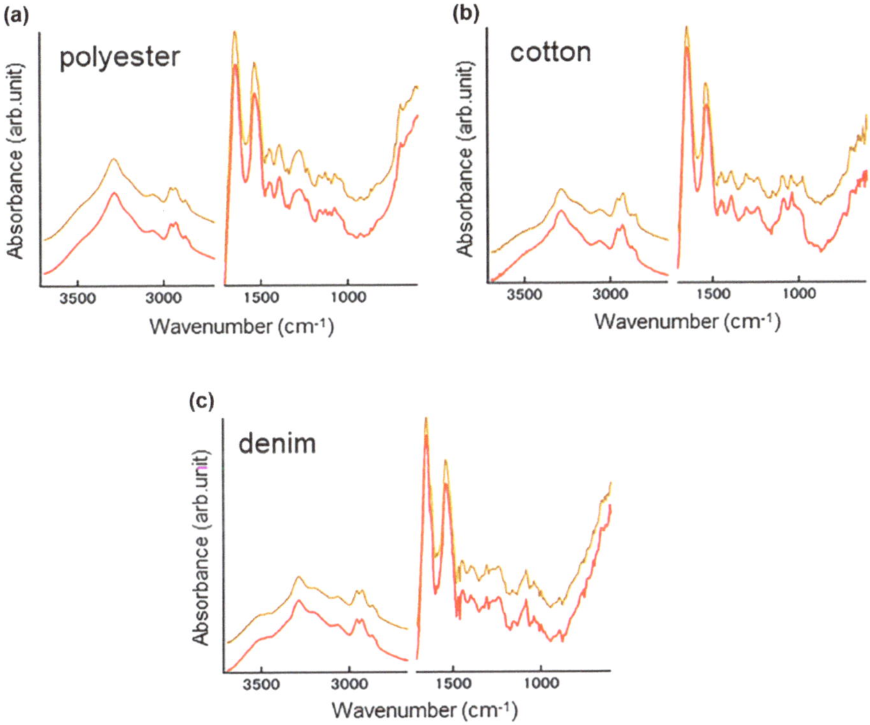

Figure 9.2 Extraction of blood-originated signals from blood stain spectra. Average blood-originated signals extracted from postmortem bloodstain (camel) and antemortem bloodstain (red) stains on (a) polyester, (b) cotton, and (c) denim. Reproduced from ref. 32 with permission from American Chemical Society, Copyright 2017.

reconstituted using weighted least-squares regression were classified by PLS-DA, demonstrating high discriminant accuracies of 95%, 86%, and 91%, respectively, for blood stains on polyester, cotton, and denim. ATR–FTIR combined with innovative chemometrics is capable of identifying bodily fluid evidence independent of the substrate type. The detection limit of vibrational spectroscopy for bloodstain investigation was evaluated by Muro *et al.*,[33] who demonstrated that the spectra of single red blood cells can be easily collected and correctly assigned to blood by using Raman spectroscopy combined with chemometrics. In their experiment, 50 Raman spectra of individual red blood cells were loaded into an SVM-DA model, which was trained on the spectra of five bodily fluids (blood, saliva, semen, sweat, and vaginal fluid), for forensic bodily fluid identification. The results showed that all the spectra were correctly predicted to be blood, demonstrating that Raman spectroscopy is a good tool to identify trace blood, even if the amount of blood is as small as a single cell, which is 5000× smaller than the amount of blood required for forensic DNA profile analysis. Surface-enhanced Raman spectroscopy (SERS) is a well-established technique that is also capable of detecting trace amounts of blood. It has been reported that the established PLS-DA classification model based on SERS data of blood, vaginal fluid, urine, semen, and saliva could identify blood diluted by 10^3, 10^4 and 10^5 with 100% accuracy (see Figure 9.3).[34]

Vibrational spectroscopy in combination with chemometrics can not only identify blood but also distinguish between different types of blood. Sikirzhytskaya *et al.*[35] employed Raman spectroscopy to distinguish menstrual and peripheral blood samples collected from 40 donors. The results

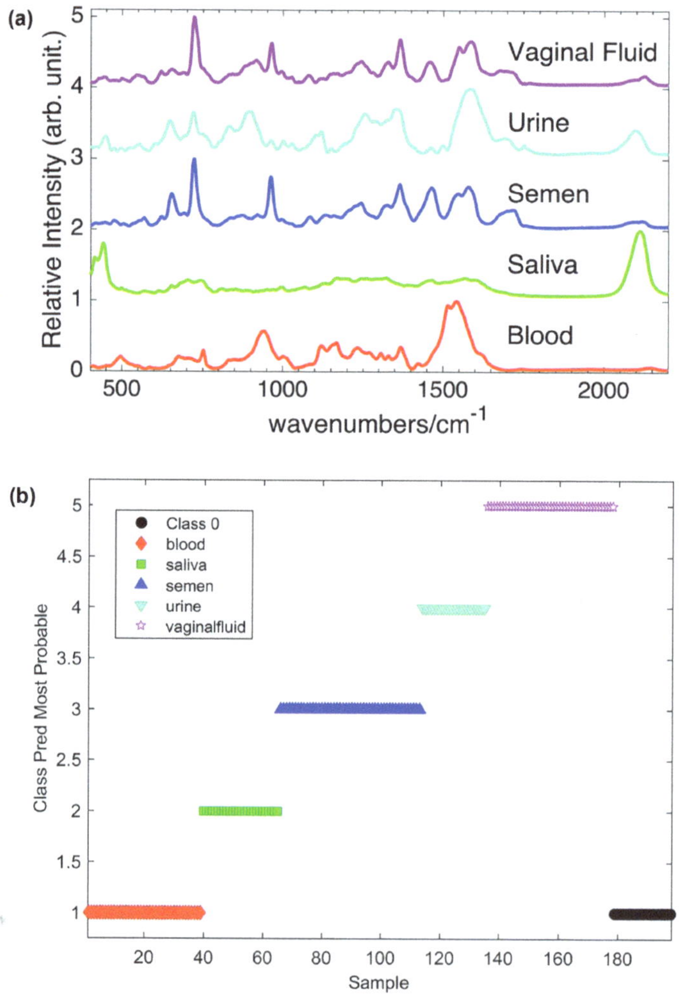

Figure 9.3 Prediction for blood class using PLS-DA model. (a) SERS spectra of bloodstain, vaginal fluid, urine, semen, and saliva. (b) PLS-DA body fluid classification comparison of blood, saliva, semen, urine, vaginal fluid, and unknown body fluid class. The unknown samples are spectra of neat blood and blood diluted by 10^3, 10^4 and 10^5. Reproduced from ref. 34 with permission from the Royal Society of Chemistry.

showed that PCA was capable of distinguishing the differences between the menstrual and peripheral blood spectra. Then, the advanced classification algorithms PLS-DA and SVM-DA were used to build the classification model to discriminate menstrual and peripheral blood, resulting in 100% discrimination. Infrared spectroscopy is also capable of discriminating samples of menstrual and peripheral blood. In a study conducted by Sharma *et al.*,[36] ATR–FTIR in combination with the chemometric tools PCA, PLSR, and LDA was employed to distinguish menstrual and peripheral blood. PCA resulted in 93.3% accuracy, whereas PLSR and LDA resulted in 100% accuracy for the discrimination of peripheral blood from menstrual blood. In a recent study,[37] the ability of Raman spectroscopy to identify bloodstains obtained from donors with blood-affecting diseases was investigated by Lednev and Nichols. The GA-SVM model constructed using samples from healthy donors was validated using Raman data of blood samples acquired from donors with celiac disease, sickle cell anaemia, and type 2 diabetes. The results showed that the model was 100% accurate in classifying all spectra.

Vibrational spectroscopy has been widely used in research for species identification of blood. A list of studies shows that Raman spectroscopy, a non-destructive technique, shows promise in the forensic investigation of blood species. In a study conducted by McLaughlin *et al.*,[38] blood samples were collected from humans and 12 species of animals that were forensically relevant (cat, dog, horse, cow, chicken, cow, pig, rabbit, mouse, rat, opossum, and raccoon), spectra of the samples were obtained, and then the spectra were analysed by chemometrics. The PLS-DA-based species identification model was built in a binary manner by classifying each spectrum as either human or animal. The CV results, with regard to human identification ability, had 98% specificity and 100% sensitivity, and the external test results showed 100% accuracy in human blood identification. In a further study also conducted by McLaughlin *et al.*,[39] the species discrimination ability of PLS-DA was re-evaluated by building a new classification model by classifying each spectrum as the corresponding species. The authors also report a satisfactory result: the blood samples were correctly classified according to their species of origin. Because it is not realistic to collect blood spectra from all species of wild animals and then use them to train the classification model, evaluation of the robustness of PLS-DA models based on known and limited species for the identification of unknown species is necessary. In their experiment, Doty *et al.*[40] tested the performance of the PLS-DA model built in previous work using Raman spectra of blood from six species of animals that were previously unaccounted for, including chimpanzees, deer, elk, ferrets, fish, and macaques. A sensitivity of 1.00 and specificity of 0.93 for human class predictions and the accuracy of 99% for both human and non-human blood identifications were achieved from the PLSDA model. This work is a significant advancement over previous work because it demonstrates that the PLS-DA algorithm is capable of extracting the spectral pattern of animal blood that is different from that of human blood and identifying the blood of unknown animal species. Apart from the high-performance desktop Raman spectrometers mentioned in the above studies, a portable Raman spectrometer has also been used for blood identification and discrimination between human and non-human blood. In a preliminary study conducted by Fujihara *et al.*,[41] Raman spectra of bloodstains from 11 species (human, rat, mouse, cow, horse, sheep, pig, rabbit, cat, dog, and chicken) were collected using a portable Raman spectrometer and analysed by chemometrics. The results showed that PCA can successfully distinguish human bloodstains from non-human bloodstains even if the TSDs of bloodstain samples were as long as 3 months. ATR–FTIR spectroscopy is an alternative tool for species blood differentiation. In contrast to Raman spectroscopy, infrared measurements are not subjected to fluorescence interference, and data preprocessing is much easier. In a study conducted by Mistek *et al.*,[42] IR spectra of bloodstains from three species (human, cat, and dog) were used to build PLS-DA models. The binary human–non-human PLS-DA model demonstrated 100% accuracy, and the species-specific PLS-DA model also showed 100% accuracy in distinguishing between all three individual classes. In the external validation, IR spectra from a new animal species were loaded into the model, and none of the spectra were assigned as humans. In a study conducted by Lin *et al.*,[43] bloodstains originating from six species (human, dog, rabbit, rat, chicken, and duck) were created and stored in mimetic outdoor and indoor environments for up to 107 days. Then, IR spectra of the bloodstains were collected and analysed using the PLS-DA algorithm. The human–mammal–domestic fowl trilateral PLS-DA model and species-specific PLS-DA model demonstrated complete separation among the three classes (human, mammal, and domestic fowl) and distinguished six species (human, rat, rabbit, dog, chicken, and duck). Then, these two PLS-DA models were subjected to external validation, resulting in 100% and 94.2% accuracy, respectively; even human bloodstains placed in an outdoor environment for up to 107 days were successfully identified. This study demonstrated that the ATR–FTIR approach in combination with PLS-DA is capable of identifying bloodstains involving different deposition times and environments; the bloodstains used in this study are actually more similar to the bloodstains found at crime scenes. In addition, bloodstains and non-blood substances that had a similar appearance or easily produced false positives were easily distinguished by the squared Euclidean cosine method. PLS-DA is the most commonly used chemometric algorithm for blood species classification regardless of whether Raman or IR data are used. However, the disadvantage of PLS-DA is that it is difficult to explain the difference in chemical compositions between different classes. In their paper,[44] Bian *et al.* proposed a self-reference algorithm to discriminate between human and non-human blood samples. Their results illustrated that both the self-reference algorithm and PLS can achieve similar satisfactory results. Moreover, the model construction by the self-reference algorithm suggests that the primary difference between the Raman spectra of human and non-human blood was at 1341 cm^{-1} (corresponding to the chemical composition of tryptophan), indicative of amino acid differences between human and non-human blood.

Vibrational spectroscopy has also proven capable of mining a donor's basic information from bloodstains. In a proof-of-concept study conducted by Sikirzhytskaya *et al.*,[45] Raman spectra were collected from female and male blood samples and then subjected to SVM and artificial neural network (ANN) analysis for sex identification. The classification models built on these two algorithms demonstrate high accuracy in sex differentiation. More specifically, ANN combined with GA could promote the sex differentiation accuracy for individual bloodstains to 98%. Race identification based on Raman investigation of peripheral blood is also possible. In a study conducted by Mistek *et al.*,[46] Raman spectroscopy was applied to measure peripheral blood samples from 20 Caucasian and African American donors varying in sex and age, and SVM-DA was used to analyse the spectra for differentiation of the two races. The performance of SVM-DA showed an 83% probability of correct classification for both races. This preliminary study shows promise in distinguishing between different races using human blood. In addition, vibrational spectroscopy can also be used to distinguish age groups based on the blood from donors. In a study,[47] IR spectra of blood from newborn (<1), adolescent (11–13), and adult (43–68) donors were collected. Although differences in the levels of lipids, glucose, and proteins in whole blood between these three age groups were minor according to the spectrochemical analysis, the discrepancies were enhanced by GA-PLS-DA, which was used to classify the ATR–FTIR spectra of blood from newborn, adolescent, and adult donors, resulting in a 95% correct classification of donors in the cross-validation. In another study,[48] Raman spectroscopy was reported to be good at identifying a blood donor's age. The SVM-DA model constructed for age determination showed 96% sensitivity and 97% specificity in the cross validation.

The accurate aging of bloodstains could help forensic investigators establish the relevant association of multiple bloodstains, reconstruct the crime scene, estimate time since death/injury, and finally, reduce the numbers of suspects. In a preliminary study conducted by Doty *et al.*,[49] a PLSR model for bloodstain age estimation was built on Raman spectral data collected from bloodstain samples ranging from 1 hour to 7 days, demonstrating high cross-validation performance with RMSE and R^2 values of 0.13 and 0.97, respectively. In a study conducted by Lin *et al.*,[50] ATR–FTIR combined with PLS-based regression and classification methods was utilized to determine the age of bloodstains that were stored in natural indoor and outdoor environments. Two PLSR models based on indoor and outdoor bloodstains stored for 7 to 85 days showed good performance for external validation, with low values of predictive RMSE (5.83 and 4.77), respectively. Two PLS-DA classification models based on indoor and outdoor bloodstain samples were also built, demonstrating good separation of fresh (age $\leq$1 day) bloodstains from older (age >1 day) bloodstains. Bloodstain age can be distinguished between not only hours, days, and weeks but also months or years. In their study, Doty *et al.*[51] confirmed that Raman spectroscopy in combination with chemometric regression algorithms was effective in monitoring specific kinetic changes in aged bloodstains and further estimating the TSD of aged bloodstains for up to 2 years. PLSR- and PCR-based models showed low error (the RMSE values were 0.29 and 0.31 log(hours), respectively) and high linear fits (the R^2 values were both 0.97) with an overall accuracy of 70% to correctly predict the TSD at each time point in the external test. Kumar *et al.*[52] compared three chemometric algorithms (curve estimation (CE), multiple linear regression (MLR), and PLSR) in regression modelling for bloodstain TSD estimation based on ATR–FTIR spectral data of bloodstains up to 175 days. They found that the curve estimation models represented the maximum dating errors of $\sim 6 \pm 5$ days, whereas MLR and PLS estimation models showed lower errors of $\sim 3 \pm 1$ days and $\sim 4 \pm 1$ days, respectively.

9.3.2 Semen

Identification of semen stains is a critical step in evidence analysis in cases of sexual assaults (rape, child molestation, sexual harassment, *etc.*). In a study conducted by Sharma *et al.*,[53] ATR FTIR spectroscopy combined with chemometric tools was used to discriminate semen stains from vaginal fluid, other bodily fluids (menstrual blood, peripheral blood, and saliva) and non-biological substances that look like semen. The results showed that 98% accuracy was achieved in distinguishing semen from vaginal fluid by PCA, and 100% accuracy was achieved by LDA. Semen samples were discriminated from other bodily fluids with 100% accuracy by PLSR and LDA. PCA–LDA resulted in 100% accuracy for the discrimination of seminal fluid from non-biological substances. A significant challenge of using Raman spectroscopy to detect semen stains is substrate interference. The severe heterogeneity of both substrates and semen fluid makes it difficult to eliminate the Raman signals of substrates by an automatic subtraction algorithm. McLaughlin *et al.*[54] proposed a new strategy consisting of PLS-DA and an automatic subtraction algorithm that can reduce and eliminate spectroscopic interferences encountered by substrates. In addition to the complications with substrates, the identification of mixed traces of semen and other bodily fluids and the detection of minor contributions are also problems that need to be solved. In the study conducted by Sikirzhytski *et al.*,[55] Raman spectra of mixtures with different blood/semen ratios were collected and analysed by a classification methodology based on the combination of SVM regression and SVM-DA. The results demonstrated that the detection limit of the minor contributor is as low as a few percent (5% or less). In addition, using Raman spectroscopy, healthy semen can be distinguished from abnormal semen with a sensitivity of 73% and specificity of 82% by PCA–LDA,[56] different levels of sperm nDNA damage can be separated from each other by PCA,[57] and azoospermatic semen can be correctly assigned to its semen class by SVM-DA.[58] The application of IR spectroscopy for semen stain species identification has also been explored. In a study conducted by Wei *et al.*,[59] ATR–FTIR spectra of semen stains varying in substrates and ages were collected from six species (human, rabbit, dog, boar, bull, and ram). PCA demonstrated that various substrates and different TSD values (1–30 days) did not affect the identification of humans and other species (see Figure 9.4). The PLS-DA model demonstrated exceptional classification ability to distinguish between humans and other species. In the external validation, excellent prediction ability with 100% sensitivity and specificity for classifying humans and other species was achieved. Race identification based on Raman investigation of semen was also possible. In a study conducted by Muro *et al.*,[60] semen samples from 18 Caucasian and Black donors were collected by Raman spectroscopy, and the data set was then subjected to model calibration by SVMDA combined with GA. The final classification results revealed that all calibration donors were classified correctly. Based on the spectral regions selected by GA, it appears that the source of the differentiation may be the relative concentrations of choline, tyrosine, and spermine phosphate hexahydrate. Age estimation of human semen stains plays an important role in forensic work. However, it is rarely studied due to a lack of well-established methods. In a study conducted by Zha *et al.*,[61] ATR–FTIR in combination with the PLSR algorithm was employed to determine the age of semen stains deposited on three different substrates (glass slides, tissues and fabric) for up to 6 days. The regression results showed good linear fit ($R2$ values of CV and external test close to 0.81 and 0.74) and low error (RMSE of cross-validation and external test close to 0.77 and 1.02).

9.3.3 Sweat, Saliva, Vaginal Fluid and Urine

In addition to blood and semen, two of the most commonly encountered bodily fluids on crime scenes, other bodily fluid evidence, such as sweat, saliva, vaginal fluid and urine, is also highly suggestive of how crimes are committed. The ability to identify sweat stains quickly and non-destructively at crime scenes remains a largely unsolved issue. Sikirzhytski *et al.*[62] developed a non-destructive technique to identify sweat stains using Raman spectroscopy. The results of significant factor analysis and PCA of Raman spectra indicated intrinsic heterogeneities of the sweat stain and donor-dependent variations of its biochemical composition. MCR-ALS analysis was used successfully to resolve undesired fluorescent spectral signals from Raman spectra that can be subtracted to obtain clearer spectral signals of sweat. Vaginal fluid traces are usually found in rape and other sexual assault cases, and correct identification of them can help support allegations of sexual assault. Sharma *et al.*[63] employed ATR–FTIR spectroscopy to detect vaginal fluid stains. They found that spectra of vaginal fluid samples and non-vaginal fluid substances that look like vaginal fluids could be separated with 100% accuracy, specificity, and sensitivity rate using the PCA–LDA modelling algorithm. In another study conducted by Zapata *et al.*,[64] three types of sexual crime-related stains, semen, vaginal fluid, and urine stains, were placed on different coloured cotton fabrics and analysed using external reflection FTIR spectroscopy to evaluate the capability of external reflection

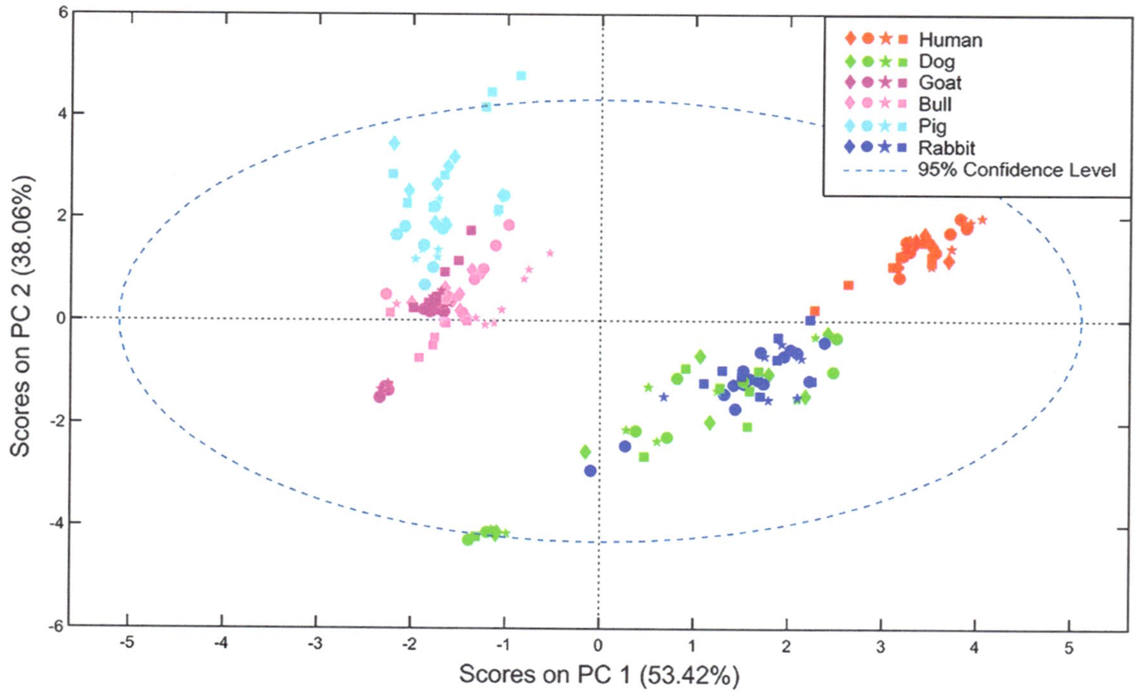

Figure 9.4 PCA score plots of infrared spectra from six species semen samples involving four different TSD points, including 8 days (diamond), 16 days (dot), 21 days (star), and 30 days (square). Reproduced from ref. 59 with permission from Springer Nature, Copyright 2021.

FTIR spectroscopy for detecting human bodily fluids on superabsorbent substances. The results demonstrated that, by means of a soft independent modelling of class analogy (SIMCA) classification algorithm, the IR spectral signatures of vaginal fluid, urine and semen stains were classified with 100% accuracy. However, vaginal fluid semen mixture stains were only classified as vaginal fluid. In forensic practice, determination of the donor sex from urine evidence is challenging because of the unavailability of DNA-based profiling. Takamura *et al.* proposed a novel method for determining donor sex based on ATR–FTIR spectroscopy of urine traces.[65] The classification model based on ATR–FTIR spectra collected from male and female urine samples was built using the GA-PLS-DA algorithm and it showed the discrimination rate for both male and female donors of 0.97. It has been difficult to discriminate sex from urine traces because male and female donors share higher compositional similarity and DNA-based urine testing is insufficient. Sharma *et al.*[66] developed a chemometrics-based ATR–FTIR approach for sex discrimination from dry urine traces. They demonstrated that the PLS-DA model achieved 95.3% accuracy in discriminating male and female urine spectra and outperformed the PCA–LDA (85.2% of accurate discrimination) and PCA models. Additionally, the validation study was carried out by using look-alike non-urine substances, which resulted in 0% rates of false-positive and negative assignments. Saliva-based Raman analysis provides another approach for donor sex determination. Muro *et al.*[67] conducted a state-of-the-art study that employed the SVM-DA algorithm to model the Raman spectra of saliva collected from 60 donors. The SVM-DA model results demonstrated the sex of 45 (94%) of the 48 donors for model calibration and 11 (92%) of 12 donors for external validation to be correctly identified. In a study conducted by Takamura *et al.*,[68] five types of bodily fluids (blood, semen, sweet, saliva, and urine) with different deposition times ranging from fresh to eight months were measured by ATR–FITR, and two novel chemometric discrimination strategies were applied to classify these types of bodily fluids. The results showed that unlike the hard modelling methods PLS-DA and LDA, which may mistakenly classify non-bodily fluid samples as one of the modelled bodily fluids, the soft modelling method PLS-LDA followed by a Q-statistics test (PLS-LDA-Q) could not only provide strong discrimination performance for the target bodily fluids but also allow the exclusion of unexpected non-bodily fluid samples as outliers. Furthermore, the other novel strategy (see Figure 9.5), PLS-LDA-Q equipped with hierarchical clustering analysis, was proven capable of reducing the complexity of the model structure and improving the discrimination ability of the model, and it allowed the researchers to classify the aged bodily fluid spectra into the correct bodily fluid group.

9.3.4 Other Biological Materials

The application of vibrational spectroscopy to bodily fluid samples has received the most attention, but spectroscopic analysis of other biological

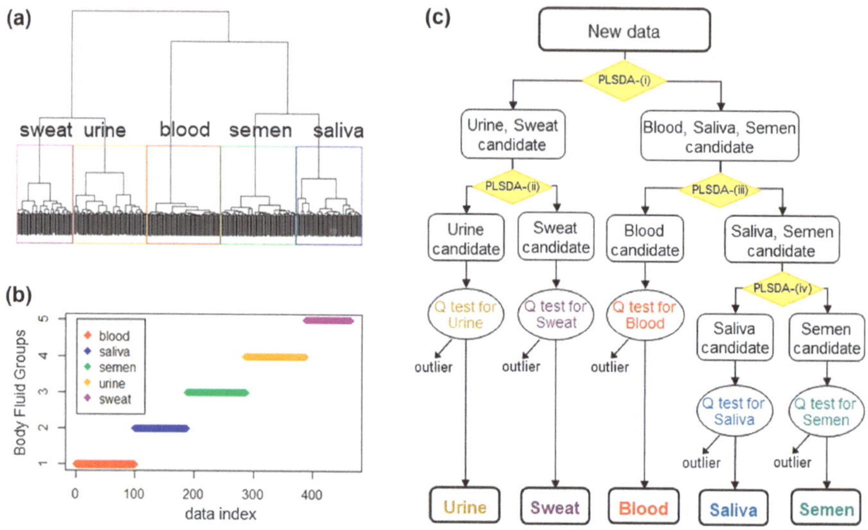

Figure 9.5 A novel strategy of PLS-LDA-Q in combination with hierarchical clustering analysis for classifying ATR–FTIR spectra of five body fluid types. (a) A dendrogram built on the PLS-DA scores of bodily fluid spectra predicted by the PLS-LDA-Q model. (b) Assignment of each spectrum into the respective bodily fluid group. (c) Classification tree built according to the dendrogram. Reproduced from ref. 68 with permission from Springer Nature, Copyright 2018.

materials has also made significant progress. Several studies stand out, despite the fact that they are not the focus of this review. Wang *et al.* researched the use of FTIR spectroscopy and chemometrics for forensic anthropology analysis.[69–72] The group demonstrated that bone species could be discriminated using PCA and PLS-DA.[69] Additionally, they demonstrated that the post-mortem interval of buried and unburied bones could be estimated using GA-PLS regression.[70] More recently, the group reported the use of FTIR spectroscopy in combination with PLS-DA to determine the species of a tooth.[71] The use of vibrational spectroscopy in forensic hair analysis has also been investigated. Panayiotou *et al.*[73] developed a technique to elucidate hair discrimination using FTIR microspectroscopy. They showed that hair samples with the same type of hair treatment, hair colour, and hair length from two individuals could be discriminated using PCA and SIMCA. Manheim *et al.*[74] employed the PLS-DA algorithm to model hair spectra for discriminating human hair from animal hair. The resulting classification model demonstrated a satisfactory accuracy. There was no error in predicting the human samples. Boll *et al.*[75] demonstrated the successful use of ATR–FTIR spectroscopy in combination with PLS-DA to identify if hairs were dyed or undyed. Vibrational spectroscopy can also be used in forensic wildlife investigations. Sharma and co-workers developed a technique to identify Shahtoosh and Royal Bengal Tiger claw using ATR FTIR spectroscopy and the PLS-DA algorithm.[76,77] Vibrational spectroscopy has the ability to analyse the fingermark composition. The Girod research group employed the PLS regression algorithm to model infrared spectral data for estimating the age of fingermarks.[78] The resulting regression models demonstrated an overall precision of ±3 days. Mammalian soft tissue decomposition occurs after death, resulting in the chemical breakdown of the body's main constituents like lipids, proteins, and carbohydrates. Wang *et al.* used IR spectroscopy to reveal postmortem biochemical changes in a variety of tissues and established ideal chemometric models for estimating the time since death.[79,80]

9.4 Future Prospects

9.4.1 Important Considerations

In the above section, we reviewed several studies that employed vibrational spectroscopic methods combined with chemometrics to investigate forensic bodily fluids. The workflow in Figure 9.6 illustrates different steps for IR and Raman investigation of forensic body fluids. Raman spectrometers equipped with 785 nm near infrared lasers are extensively applied in bodily fluid studies, as 785 nm lasers have relatively low photon energy and generally do not cause substantial photodamage. An infrared spectrometer equipped with an ATR accessory is the preferred tool for bodily fluid investigation because the measurement is relatively simple. As long as the sample is directly attached to the surface of the ATR crystal, the spectrum can be obtained. To obtain spectra of optimum quality and a high signal-to-noise ratio, the parameters of spectral measurements need to be optimized according to the specific experiment. For instance, the laser power should be adjusted according to the types of samples when starting Raman measurements. If the Raman scattering is intense, the laser power should be attenuated to prevent detector saturation; otherwise, it should be raised. Increased accumulation parameters may give satisfactory spectral results but extend the experimental period, thus resulting in a potential trade-off between spectral quality and time availability. Prior to chemometric modelling, the preprocessing procedure is always applied to uncover valuable biological information concealed by instrumental and sample-specific variances. There are many chemometric algorithms available for spectral data preprocessing. For example, Savitzky–Golay is the denoising algorithm that is most commonly used to smooth spectra. Asymmetric least squares (AsLS) and rubber band algorithms are often applied to correct the baselines of Raman spectra. Multiplicative scatter correction (MSC) and the standard normal variate transform (SNV) are used to remove scattering effects from infrared spectra. Although the chemometrics involved in spectral data preprocessing are not discussed in this review, we still want to emphasize that any algorithm should be performed at a minimum level because a higher level of data preprocessing could lead to the loss of minute vital

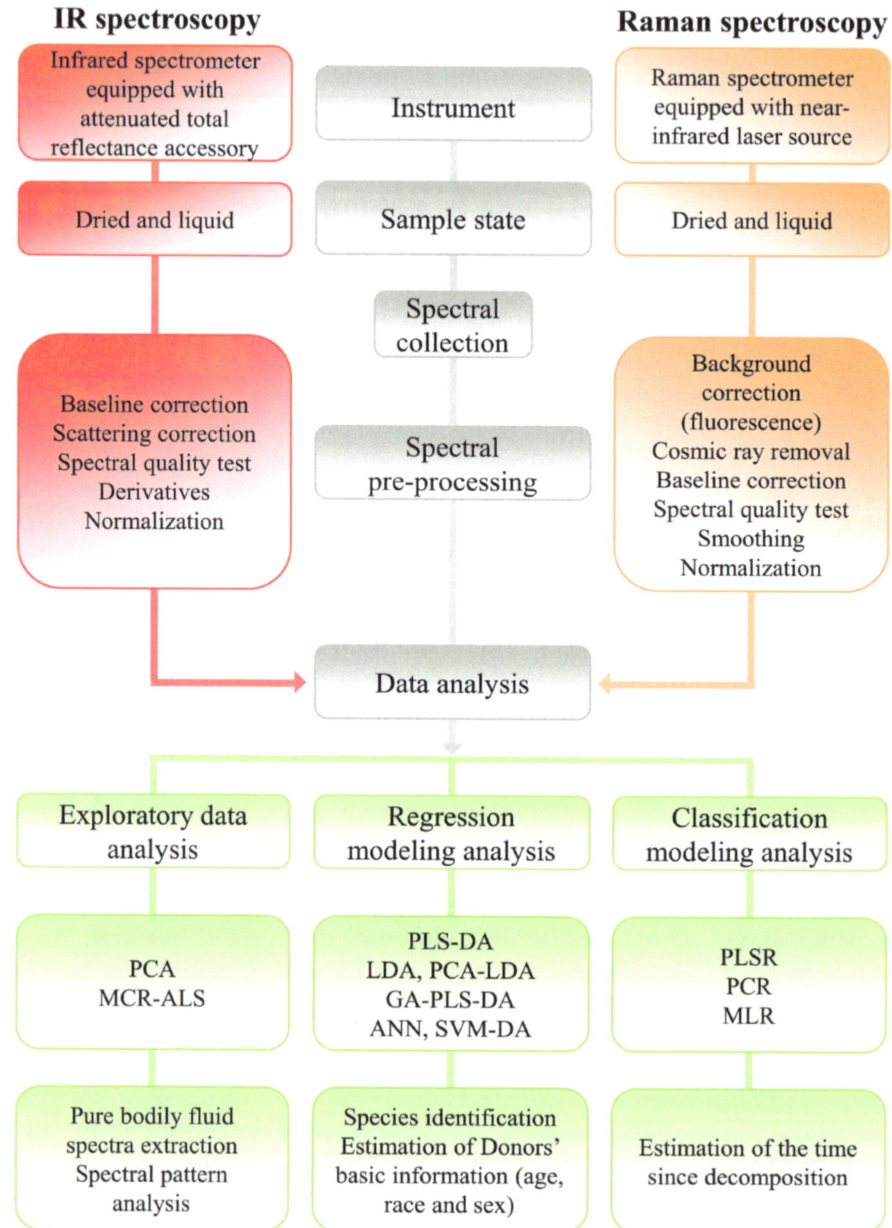

Figure 9.6 Workflow of vibrational spectroscopy for bodily fluid investigation.

information. Nevertheless, sample preparation and measurement conditions are still the key to high-quality spectra. Chemometric modelling is the core of bodily fluid spectral analysis. The chemometric algorithms that should be used to explore the patterns within a spectral data set depends not only on the quality and type of the data but also on the data analysis objective. Among the methods for chemometric modelling used in the studies cited in this paper, PLS-DA is the most commonly used classification algorithm, and it has demonstrated high performance for discriminating between bodily fluid samples. Similarly, for the dating of bodily fluid stains, PLSR provides much better results than other regression algorithms. However, the sample size, data type and data preprocessing pipeline considerably affect the results of chemometric modelling. One should be aware of all these parameters/prerequisites to obtain significant and accurate results with statistical confidence so that the results will be well accepted in a court of law.

9.4.2 Future Challenges

Bodily fluid investigation plays an important role in the investigation of violent crimes, wildlife trafficking, and mass disaster cases. Determination of the nature, species, and deposition time of bodily fluids is the key task of bodily fluid investigation. Utilization of spectroscopic techniques and chemometrics has shown promise. However, a few future challenges for bodily fluid investigations are as follows:

(i) Most of the studies used samples deposited on smooth glass. However, bodily fluids can be found on various substrates in real forensic cases. It is suggested to include more substrates, such as fabric or paper, to determine the effects of bodily fluid-collecting materials on the spectra.
(ii) Spectral studies on mixed bodily fluid stains should be undertaken in the future. The ability to identify all kinds of bodily fluids should be evaluated.
(iii) The minimum detection limit of the vibrational spectral method in bodily fluids should be evaluated in further studies, which is very significant because not all the stains in the cases are adequate.
(iv) Environmental factors, such as temperature, humidity, illumination and contaminants, should be included in the developed models for bodily fluid investigations.
(v) Except for the PLS-DA algorithm, which is most commonly used in the studies cited in this review, more advanced machine algorithms should be included in further studies. The ideal algorithms should be good at modelling colossal, imbalanced and multiclass data.

9.5 Conclusion

Forensic bodily fluid evidence is a very important part of the evidence chain in criminal cases. Obtaining as much information from bodily fluids as possible would greatly benefit criminal investigations. Conventional serological and forensic DNA profiling methods are time-consuming, expensive, and destructive in nature. Thus, forensic investigators and researchers turn their attention to time-saving and non-destructive vibrational spectroscopy techniques, which could provide repeatable and reliable results, especially in combination with chemometrics. The most commonly used chemometric algorithms for spectral discrimination and dating studies of forensic bodily fluids are PLS-based algorithms, such as PLS-DA and PLSR. Other advanced algorithms, such as SVM and ANN, are also proven to have good performance in modelling the spectra of bodily fluids. However, the most pressing issue at present is that the effect of these algorithms on complicated samples, large datasets, and multiple classes has not yet been validated. To obtain more precise and subjective results, more research on these issues is required. Nevertheless, information mining using vibrational spectroscopy with chemometrics for forensic bodily fluid investigation is still open, and cooperation from the interdisciplinary fields of biophotonics, analytical chemistry, and forensic science is worth investigating.

References

1. SafeHome.org Team, https://www.safehome.org/resources/crime-statistics-by-state/.
2. H. C. Lee, R. E. Gaensslen, E. M. Pagliaro, M. B. Guman, K. M. Berka and T. P. Keith, *J. Forensic Identif.*, 1989, **39**, 339.
3. R. Li, *Forensic biology*, CRC Press, Florida, 2015.
4. L. Kobilinsky, *Forensic chemistry handbook*, John Wiley & Sons, New Jersey, 2011.
5. S. H. James and J. J. Nordby, *J. Forensic Sci.*, 2006, **51**, 205.
6. N. N. Daeid and M. Houck, *Interpol's forensic science review*, CRC Press, Florida, 2012.
7. T. Akutsu, K. Watanabe, Y. Fujinami and K. Sakurada, *Int. J. Legal Med.*, 2010, **124**, 493.
8. K. Sakurada, K. Watanabe and T. Akutsu, *Diagnostics*, 2020, **10**, 693.
9. C. T. Oien, *Forensic Sci. Commun.*, 2009, **11**.
10. S. N. Byers, *Introduction to forensic anthropology*, Taylor & Francis, 2016.
11. S. M. Bleay, M. J. Bailey, R. S. Croxton and S. Francese, *Wiley Interdiscip. Rev.: Forensic Sci.*, 2021, **3**, e1403.
12. M. Pilling and P. Gardner, *Chem. Soc. Rev.*, 2016, **45**, 1935.
13. H. J. Butler, L. Ashton, B. Bird, G. Cinque, K. Curtis, J. Dorney, K. Esmonde-White, N. J. Fullwood, B. Gardner, P. L. Martin-Hirsch, M. J. Walsh, M. R. McAinsh, N. Stone and F. L. Martin, *Nat. Protoc.*, 2016, **11**, 664.
14. M. J. Baker, H. J. Byrne, J. Chalmers, P. Gardner, R. Goodacre, A. Henderson, S. G. Kazarian, F. L. Martin, J. Moger and N. Stone, *Analyst*, 2018, **143**, 1735.
15. C. K. Muro, K. C. Doty, J. Bueno, L. Halámková and I. K. Lednev, *Anal. Chem.*, 2015, **87**, 306.
16. M. J. Baker, J. Trssevisan, P. Bassan, R. Bhargava, H. J. Butler, K. M. Dorling, P. R. Fielden, S. W. Fogarty, N. J. Fullwood, K. A. Heys,

C. Hughes, P. Lasch, P. L. Martin-Hirsch, B. Obinaju, G. D. Sockalingum, J. Sule-Suso, R. J. Strong, M. J. Walsh, B. R. Wood, P. Gardner and F. L. Martin, *Nat. Protoc.*, 2014, **9**, 1771.

17. E. Szymańska, *Anal. Chim. Acta*, 2018, **1028**, 1.
18. M. A. Fikiet, S. R. Khandasammy, E. Mistek, Y. Ahmed, L. Halámková, J. Bueno and I. K. Lednev, *Spectrochim. Acta A Mol. Biomol. Spectrosc.*, 2018, **197**, 255.
19. A. V. Ewing and S. G. Kazarian, *Analyst*, 2017, **142**, 257.
20. C. Muehlethaler, M. Leona and J. R. Lombardi, *Anal. Chem.*, 2016, **88**, 152.
21. R. Bro and A. K. Smilde, *Anal. Methods*, 2014, **6**, 2812.
22. J. Felten, H. Hall, J. Jaumot, R. Tauler, A. de Juan and A. Gorzsás, *Nat. Protoc.*, 2015, **10**, 217.
23. R. B. Keithley, R. M. Wightman and M. L. Heien, *TrAC, Trends Anal. Chem.*, 2009, **28**, 1127.
24. P. Geladi and B. R. Kowalski, *Anal. Chim. Acta*, 1986, **185**, 1.
25. L. C. Lee, C.-Y. Liong and A. A. Jemain, *Analyst*, 2018, **143**(15), 3526–3539.
26. R. Kumar and V. Sharma, *TrAC, Trends Anal. Chem.*, 2018, **105**, 191.
27. H. J. Byrne, P. Knief, M. E. Keating and F. Bonnier, *Chem. Soc. Rev.*, 2016, **45**, 1865.
28. R. Tauler and H. Parastar, *Angew. Chem., Int. Ed.*, 2022, **61**(44), e201801134.
29. M. Yong-Jie and Y. Wen-Xia, *Appl. Res. Comput.*, 2012, **29**, 1201.
30. A. Biancolillo and F. Marini, *Front. Chem.*, 2018, **6**, 576.
31. A. Sikirzhytskaya, V. Sikirzhytski, G. McLaughlin and I. K. Lednev, *J. Forensic Sci.*, 2013, **58**, 1141.
32. A. Takamura, K. Watanabe, T. Akutsu, H. Ikegaya and T. Ozawa, *Anal. Chem.*, 2017, **89**, 9797.
33. C. K. Muro and I. K. Lednev, *Anal. Bioanal. Chem.*, 2017, **409**, 287.
34. M. L. Shaine, W. R. Premasiri, H. M. Ingraham, R. Andino, P. Lemler, A. N. Brodeur and L. D. Ziegler, *Analyst*, 2020, **145**, 6097.
35. A. Sikirzhytskaya, V. Sikirzhytski and I. K. Lednev, *J. Biophotonics*, 2014, **7**, 59.
36. S. Sharma, R. Chophi and R. Singh, *Int. J. Legal Med.*, 2020, **134**, 63.
37. N. A. Nichols and I. K. Lednev, *Am. J. Anal. Chem.*, 2022, **13**, 1.
38. G. McLaughlin, K. C. Doty and I. K. Lednev, *Forensic Sci. Int.*, 2014, **238**, 91.
39. G. McLaughlin, K. C. Doty and I. K. Lednev, *Anal. Chem.*, 2014, **86**, 11628.
40. K. C. Doty and I. K. Lednev, *Forensic Sci. Int.*, 2018, **282**, 204.
41. J. Fujihara, Y. Fujita, T. Yamamoto, N. Nishimoto, K. Kimura-Kataoka, S. Kurata, Y. Takinami, T. Yasuda and H. Takeshita, *Int. J. Legal Med.*, 2017, **131**, 319.
42. E. Mistek and I. K. Lednev, *Anal. Bioanal. Chem.*, 2015, **407**, 7435.
43. H. Lin, Y. Zhang, W. Qi, L. Bing, S. Fan and Z. Wang, *Int. J. Legal Med.*, 2018, **132**, 667.
44. H. Bian, P. Wang, J. Wang, H. Yin, Y. Tian, P. Bai, X. Wu, N. Wang, Y. Tang and J. Gao, *J. Biomed. Opt.*, 2017, **22**, 095006.

45. A. Sikirzhytskaya, V. Sikirzhytski and I. K. Lednev, *Anal. Chem.*, 2017, **89**, 1486.
46. E. Mistek, L. Halámková, K. C. Doty, C. K. Muro and I. K. Lednev, *Anal. Chem.*, 2016, **88**, 7453.
47. S. Giuliano, E. Mistek-Morabito and I. K. Lednev, *ACS Omega*, 2020, **5**, 27026.
48. K. C. Doty and I. K. Lednev, *ACS Cent. Sci.*, 2018, **4**, 862.
49. K. C. Doty, G. McLaughlin and I. K. Lednev, *Anal. Bioanal. Chem.*, 2016, **408**, 3993.
50. H. Lin, Y. Zhang, Q. Wang, B. Li, P. Huang and Z. Wang, *Sci. Rep.*, 2017, **7**, 1.
51. K. C. Doty, C. K. Muro and I. K. Lednev, *Forensic Chem.*, 2017, **5**, 1–7.
52. R. Kumar, K. Sharma and V. Sharma, *Sci. Justice*, 2020, **60**, 538.
53. S. Sharma and R. Singh, *Int. J. Legal Med.*, 2020, **134**, 411.
54. G. McLaughlin and I. K. Lednev, *J. Forensic Sci.*, 2015, **60**, 595.
55. V. Sikirzhytski, A. Sikirzhytskaya and I. K. Lednev, *Forensic Sci. Int.*, 2012, **222**, 259.
56. Z. Huang, X. Chen, Y. Chen, S. Feng, R. Chen, J. Chen, M. Dou and H. Zeng, *J. Biomed. Opt.*, 2011, **16**, 110501.
57. V. Sánchez, K. Redmann, J. Wistuba, F. Wübbeling, M. Burger, H. Oldenhof, W. F. Wolkers, S. Kliesch, S. Schlatt and C. Mallidis, *Fertil. Steril.*, 2012, **98**, 1124–1129.e33.
58. M. A. Fikiet and I. K. Lednev, *Talanta*, 2019, **194**, 385.
59. X. Wei, K. Yu, D. Wu, P. Huang, Q. Sun and Z. Wang, *Int. J. Legal Med.*, 2021, **135**, 73.
60. C. K. Muro and I. K. Lednev, *Anal. Chem.*, 2017, **89**, 4344.
61. S. Zha, X. Wei, R. Fang, Q. Wang, H. Lin, K. Zhang, H. Zhang, R. Liu, Z. Li and P. Huang, *Forensic Sci. Res.*, 2020, **5**, 119.
62. V. Sikirzhytski, A. Sikirzhytskaya and I. K. Lednev, *Anal. Chim. Acta*, 2012, **718**, 78.
63. S. Sharma and R. Singh, *Int. J. Legal Med.*, 2020, **134**, 1591.
64. F. Zapata, M. Á. F. de la Ossa and C. García-Ruiz, *Appl. Spectrosc.*, 2016, **70**, 654.
65. A. Takamura, L. Halamkova, T. Ozawa and I. K. Lednev, *Anal. Chem.*, 2019, **91**, 6288.
66. S. Sharma, H. Kaur and R. Singh, *Int. J. Legal Med.*, 2022, 1.
67. C. K. Muro, L. de Souza Fernandes and I. K. Lednev, *Anal. Chem.*, 2016, **88**, 12489.
68. A. Takamura, K. Watanabe, T. Akutsu and T. Ozawa, *Sci. Rep.*, 2018, **8**, 1.
69. Q. Wang, W. Li, R. Liu, K. Zhang, H. Zhang, S. Fan and Z. Wang, *Int. J. Legal Med.*, 2019, **133**, 269.
70. Q. Wang, Y. Zhang, H. Lin, S. Zha, R. Fang, X. Wei, S. Fan and Z. Wang, *Forensic Sci. Int.*, 2017, **281**, 113.
71. X. Wei, K. Yu, H. Wu, C. Shen, H. Li, R. Liu, Q. Sun and Z. Wang, *Forensic Sci. Int.*, 2022, 111205.

72. K. Yu, H. Wu, C. Shen, H. Li, X. Wei, R. Liu, W. Cai, G. Wang, Q. Sun and Z. Wang, *Int. J. Legal Med.*, 2021, **135**, 2385.
73. H. Panayiotou and S. Kokot, *Anal. Chim. Acta*, 1999, **392**, 223.
74. J. Manheim, K. C. Doty, G. McLaughlin and I. K. Lednev, *Appl. Spectrosc.*, 2016, **70**, 1109.
75. M. S. Boll, K. C. Doty, R. Wickenheiser and I. K. Lednev, *Forensic Chem.*, 2017, **6**, 1–9.
76. C. P. Sharma, S. Sharma, V. Sharma and R. Singh, *Sci. Justice*, 2019, **59**, 622.
77. C. P. Sharma, S. Sharma, G. S. Rawat and R. Singh, *Sci. Justice*, 2022, **62**, 349.
78. A. Girod, L. Xiao, B. Reedy, C. Roux and C. Weyermann, *Forensic Sci. Int.*, 2015, **254**, 185.
79. K. Zhang, Q. Wang, R. Liu, X. Wei, Z. Li, S. Fan and Z. Wang, *Int. J. Legal Med.*, 2020, **134**, 565.
80. K. Yu, H. Zhang, Y. Liu, H. Wu, W. Cai, X. Wei, R. Liu, G. Wang, Q. Sun and Z. Wang, *Microchem. J.*, 2021, **164**, 105977.

CHAPTER 10

Environmental Forensics

A. KOLOBARIC,[a] D. COZZOLINO[b] AND J. CHAPMAN*[a]

[a] School of Science, RMIT University, Melbourne, 3001, Australia;
[b] Centre for Nutrition and Food Sciences, Queensland Alliance for Agriculture and Food Innovation, The University of Queensland, St. Lucia, Brisbane, QLD 4072, Queensland, Australia
*Email: drjameschapman@gmail.com

10.1 Introduction

Identification of contaminant sources represents the crux of all environmental forensic investigations. All environmental analyses and environmental tasks are multifactorial issues, including sampling plan design, sample collection, and field methodologies. Furthermore, analytical chemistry and analytical methods, including standard methods, provide in-depth knowledge of the problem and the surrounding environmental inputs and processes. The environment can be considered an overly complex matrix, and multiple sources provide input into a receiving environment. This multifactor input often makes it challenging to discriminate between the source being investigated and existing contamination, highlighting that even meticulous planning and project design may not be adequate to remove all contamination issues. If the source of contamination cannot be pinpointed to a defined chemical, then individual contaminants and their concentrations can be mapped to identify the point source. However, many sources exhibit unique characteristics defined as "chemical fingerprints" that assist the environmental officer or environmental scientist in their quest to determine the point source, understand the contamination event, and finally deduce the impact on the receiving environment. Owing to the complexity of the environment as a multimatrix and multi-analyte system, multivariate methods are the best approach to chemical fingerprinting. These multivariate

Chemometric Methods in Forensic Science
Edited by Vishal Sharma, Georgina Sauzier and Simon W. Lewis

Published by the Royal Society of Chemistry, www.rsc.org

approaches help us to understand: (1) the chemistry and the fingerprints of chemicals that exist in the environment under investigation; (2) chemical fingerprints can be multivariate in composition, be it biological or chemical; and (3) the collected samples' relative contribution in the environmental system for the chemical fingerprints. However, the use of chemometrics has advanced the development of mathematical methods used to determine these types of variables.[1] Most early applications of chemometrics targeting environmental issues focused on solving relatively simple problems.

In contrast, over the last decade, procedures and methods that have been developed for environmental analyses tend to elucidate broader environmental problems that consider "complications" such as insufficient or bad data, mixing effects such as commingled plumes or unknown sources, and any chemical alteration, such as weathering. These methods and procedures are primarily due to the advancement of instrumentation and the ability to produce large environmental datasets in a classical thinking approach. Therefore, the two objectives of this chapter are:

(1) To discuss multivariate procedures instead of thinking and approaching an environmental forensic problem classically, to discuss any limitations or strengths to be able to guide the environmental scientist through appropriate procedures focusing on classical instrumentation which is now deemed field-deployable, and rapid techniques (such as spectroscopy).
(2) To discuss how these approaches are being used in the literature to analyse some of the latest emerging contaminants in the context of the field-deployable advances in environmental forensics and chemometrics.

The assignment of the contamination source to the forensics of the environment falls into two categories: comprehensive knowledge of the potential source (*i.e.*, location, chemical composition, history, activity, among others) or nothing is known with any degree of certainty. Receptor contamination (*e.g.*, the groundwater and estuary sediments) is often the result of activities that may have been historical or well removed from the site. The benefit of *a priori* knowledge makes determining the contamination event much more straightforward; however, this is seldom the case with environmental analysis as we deal with multifactorial issues. For example, environmental forensics can be more complex than initially thought, even when all knowledge is known about the contributing factors. For example, environmental chemicals typically never retain their original compositions; they may biodegrade, volatilise, solubilise, or migrate, which is also true of recalcitrant pollutants such as perfluoroalkyl and polyfluoroalkyl substances (PFAS).

While there are multiple approaches to help solve environmental forensics problems of "who or what was responsible", we require methods that cover the temporal and spatial aspects of the pollution event, *i.e.*, "where to start" and "when did it happen." Often, these questions are hampered by the situation in which the scientist or site finds itself. For example, conventional experimental analysis relies on the hypothesis and appraisal of the data in terms of distinct assays and solid arguments pertaining to the probability. If this is not possible, exploratory data analysis (EDA) can be used. EDA can be used to determine relationships or correlations in each data set, with very few assumptions or hypotheses proposed. Potentially EDA can be used as a tool to define new hypotheses.

The same is true for the data itself; it is well known that if the input data is wrong, then the entire environmental forensic result will fall. Common issues associated with the so-called "bad data" include:

- Multi-laboratory analyses could introduce systematic biases (*i.e.*, chain of custody breaches, certified analytical method breaches).
- Errors and issues with the instrumentation used to collect the data (*i.e.*, false positives, contamination, incorrect quantitation).
- Reactive selection of certified analytical methods that do not meet the limits of detection required for the investigation.
- Non-target analytes (unexpected matrix effects, presence of other analytes that pose more risk to human and environmental health).
- Human error (*e.g.*, data entry or peak integrations).

When these errors are present, they contribute to systematic variability. If this systematic variability goes unrecognised, chemical fingerprints that have no association with the system manifest; outliers need to be identified to remedy this.

10.2 Classical Approaches to Environmental Forensic Analyses

Environmental forensics and the process of the forensic investigation rely on gaining results to assist in interpreting changes to the environment and different issues from the scene using lab-based methodologies. It is prudent to do this as many instruments are advanced and can generate very low-limit of detection data, and with high accuracy and precision. However, the main reason for laboratory-based methods for the identification of environmental pollutants, and therefore the ability to perform the required forensic operations, was because field-deployable equipment was unavailable, until recently. The standard basic measurements involving pH, conductivity, and dissolved oxygen measurements, have been routinely made in the field, and although these measurements do offer some information and fingerprints on environmental contamination events or changes to the environment, they seldom provide any robust information on chemical entities. However, they are also used classically to generate these types of inferences.

Therefore, the standard tactic in environmental analysis is sending samples to a suitable laboratory who is accredited for the range of analytes, limits of detection, threshold limits, *etc.*, providing a more detailed analysis and report of the data.

Subsequently, chemical fingerprinting of the samples will provide the ability to identify the contamination and event.[2,3] Chemical fingerprinting compares the chemical profiles in a given sample to identify if the samples are chemically identical. This process determines where the chemicals are exogenous, which have been introduced into the environment because of a contamination event. Chemical fingerprinting affords the comparison of contaminated samples with source samples, thus ascertaining the contaminant's origin. If confirmation shows that the environmental contaminant is the same chemical profile as the source sample, then the pollutant identified will be profiled along with the potential links to the site or point source. When such events happen, the longevity of the contaminant is subjected to chemical degradation or transformation. Therefore, breakdown products must be considered for contamination issues, making environmental forensics difficult. Some common environmental degradation factors include chemical evaporation, chemical biodegradation (degradation due to bacterial activity) and chemical photooxidation.[4] For example, if we examine oil fingerprinting methods, where changes influence the chemical profile, these factors can also rapidly change in the natural environment (*e.g.* oils being influenced by microbial degradation);[5] thus, forensically establishing the culprit and the point source of the actual pollution event can be extremely difficult.

Many analytical approaches are commonplace for the analysis of broad classes of pollutants. Some of these methods are well established contamination with heavy metals such as mercury or arsenic, sewage discharges, asbestos illegal dumping, organic agricultural products such as pesticides and herbicides, polycyclic aromatic hydrocarbons (PAHs) from anthropogenic energy creation, crude oil refineries, radionuclides from energy creation or warfare, and personal care products (PCPs) as part of daily healthcare and wellbeing.[6] Recently, the fate of pharmaceutical and PCPs in environmental matrices have been extensively tracked, and these compounds have shown to be persistent in environmental and human health impact.[7] Additionally, and perhaps of broader interest, is the ability to track illicit drug consumption.[8] Many of these studies have merged environmental forensics with general forensic sciences and thus provide a comprehensive outlook using combinatory disciplinary approaches.

The United States Environmental Protection Agency, the EU, and the UK have specific regulations and detailed analytical methodologies for multi-element or multicompound libraries and are deemed the "gold-standard" for operational laboratories as it provides continuity across the world. Many of these methods are standardised, which can then be replicated as "standard methods of analyses". These regulating countries have vast detailed knowledge of common contaminants, and a well-established environmental laboratory can therefore provide the ability to ascertain unknown pollutants which may not be in the library databases. This approach requires understanding of analytical chemistry and instrumentation, which can detect, characterise, and then, using the analyst's knowledge, identify these unknown pollutants/chemicals. However, typically the response rates are slower and turnaround times much longer. The trade-off is low limits of detection and sensitivity; however, if time is of critical importance, more rapid techniques would cater for improving the damage caused to the environment and potentially health. To this extent, we will concentrate most of this chapter on the development and emerging techniques of field-portable equipment.

10.3 Site-specific Requirements and Analyses

Environmental forensics typically requires a broad knowledge of the scenario, the matrix, and the analyte of interest, whether human fluids or pollution, *etc.*[9] Often, some sites have historical environmental events that have occurred and may not have been reported, therefore the environmental impact has not been truly appreciated. For example, legacy mines that have been disused and closed for some time have a whole raft of pollution sources that will be present. These types of cases are often denoted as a "historical-cases". For example, a mine that has been "left to rot" is deemed a legacy contamination event, and has the potential to generate serious environmental issues, especially if not remediated properly. To illustrate, a tailings dam eroding can leach acidic waters into waterways and creeks.[10] These legacy issues will always require monitoring and significant datasets for the site to be correctly managed and confirm potential sources of contaminants. The matrix and the sample type require a range of different storage conditions to ensure that changes to the analytes contained in the matrix don't alter from collection to analysis. For example, if a site has been deemed to have historical influences, many pollutants may have already broken down (through microbial, photochemical, *etc.*). So, the use of rapid and field portable techniques can surmount this issue. For example, those based on vibrational spectroscopy, the near-, the mid-infrared, and also Raman, or chemical noses or tongues, the benefits of which are no transportation is required and also no sample storage needed before the analysis.[11]

Environmental forensics will move towards implementing new mobile platforms providing near-instant results in the field and to a range of matrices.[2] This chapter focuses on using field-portable equipment combined with chemometrics to gain vital information for environmental forensics. One obvious application is in the field of emergency response environmental scenarios where data acquisition is required rapidly. To achieve this, rapid field analyses could provide prediction and insight into pollution distribution and movement at a site. Additionally, the rapid sampling method would enable accurate sample for detailed accredited laboratory analyses, thus saving time and money.[12]

10.4 Field Portable Equipment/Instrumentation and Its Applications

Several commercially available field portable instruments have been released and allows contaminants to be measured in broad samples matrices from the environment. Much field portable instrumentation can reproduce any necessary laboratory competences currently offered and will be extremely useful for environmental forensic investigations.

This portable equipment should provide the ability to identify the contamination event/contaminant or analyte of interest and provide a pollutant chemical fingerprint which in turn permits the point-source of the analyte to be tracked. We will discuss some of this equipment and the requirements needed to highlight this environmental forensic capability. Here we have reviewed relevant studies showing the potential for these portable instruments' ecological applications. Methodologically, some techniques cannot analyse the pollution event directly. Instead, mathematical models and algorithms in chemometrics must be applied to bring out significant trends and information in the data. The application of this powerful tool has been demonstrated in some of our previous work examining microbial activity, growth, and treatment in complex matrices.[13]

While there are several chemometric models available, environment forensics prescribes the methods needed to incorporate the following actions:

(1) Provide a model of mixtures from multiple point-sources, which could include the physico-chemical information of the matrix and the analyte.
(2) Allow for the resolution of contributing chemical fingerprints outside of any assumed knowledge aligned to the system's geographic, temporal, or chemical composition information.

Since the 1970s, chemometric methods have readily produced data reduction which has allowed for the interpretation of multicomponent systems involving chemical and physical measurements, see Figure 10.1.

Given this, an environmental scientist is often presented with large tasks: assessing ecosystems, pollution sources, human and health impacts, and new and emerging contamination events, which constantly evolve as new contaminants of concern emerge from the ever-growing list of anthropogenic chemicals. They also must have the ability to sense and determine lower concentration ranges. In the essential stages of data collection, interpretation of the data and its structure, combined with the predictive model development, will allow the environmental scientists, and aligned forensic teams, to gain meaningful insights into the sources and the migration of contamination, the events ensuing, and provide better insight into historical cases. Without addressing these variables, they are unlikely to sufficiently understand the risks or consequences to both environmental and human health. The final stages of remediation and control strategies are

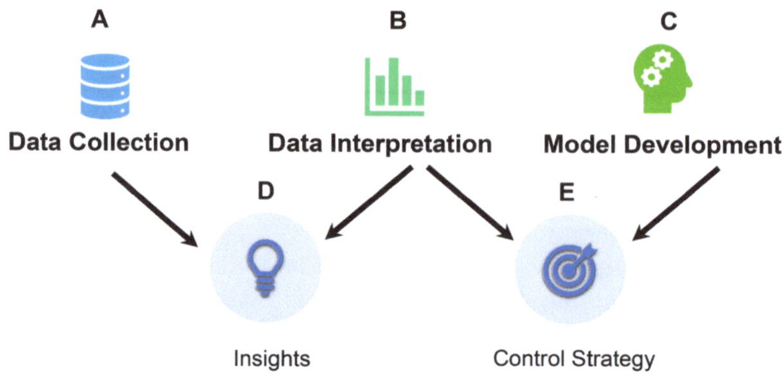

Figure 10.1 Environmental assessment workflow and example of chemometric methods that are pertinent to each. (A) Signal transforms, curve resolution, calibration; (B) pre-processing, cluster analysis, factor analysis, data visualisation; (C) proportional modelling, classification modelling, non-linear approaches; (D) health issues, migration pathways, contamination source; (E) source apportionment, waste recovery, engineering processes, mitigation.

vitally important in this job role, as the initial information will inform the end-decision implementation. Chemometrics is the vital step for the ability to perform data reduction and data interpretation for multicomponent physical and chemical problems that are typical in environmental forensics, which will be discussed below.

10.5 Chemometrics in Environmental Science

Chemometrics is the umbrella term applied to multivariate statistics and using mathematical tools and models to study chemical-based problems or answer hypotheses. Chemometrics therefore provides multivariate mathematical methods and algorithms to aid environmental scientists in basic environmental steps and analyses.[1]

Applying chemometric methods to chemical "fingerprint recognition" is a valuable tool to identify point-source and chemical interaction in the environment, where chemometrics can provide fundamental differences and similarities between air, water, and soil compartments to identify or explain variation.

The use of chemometrics can be applied to a range of instrumental output data, for example, spectroscopic or chromatographic data which are multichannel in nature, or combinations of multivariate and univariate instrumentation whose data provide vast arrays of hidden complexities in these complex environments.[14] The chemical fingerprint pattern variations when discussing spectral or chromatographic chemicals from environmental samples offer the ability to reveal relationships resulting from natural environmental phenomena or contamination impacting these chemical profiles.

10.5.1 Chemometric Analysis for Contamination

Chemometrics is readily applied to UV-Visible (UV-vis) spectroscopy, high-performance-liquid chromatography (HPLC), gas chromatography mass spectrometry (GC–MS), liquid chromatography–mass spectrometry (LC–MS), among several other techniques used for environmental monitoring. The aim of using chemometrics can be to reduce noise in the spectra, and improve the precision and accuracy to determine the measurement of the specific analyte (contaminant) in its given sample matrix.[15] Chemometrics can provide vital data like the concentration of single analytes in complex matrices that would usually require standard concentration graphs. Estimating concentration, kinetics and intermediate production are essential for environmental remediation applications. Chemometrics enables the remediation process to have less of a requirement placed on expensive instrumentation and their associated techniques. Pre-processing the data is a critical stage in chemometrics, especially when the data has been obtained using analytical instrumentation. UV-Visible spectrophotometric analysis of a sample, with repeated time intervals, when pre-processing techniques are applied, can reduce the noise in the system. Table 10.1 provides some of the available pre-processing methods generating low signal-to-noise (STN) ratios.

Data pre-processing will increase the STN ratio preventing intense signal influence in the data. Prior to any chemometric modelling, the step of data pre-processing, *i.e.*, identifying spectral subintervals, generates improved model capability. An example by Li *et al.* is where the authors used partial least squares (PLS) to determine and optimise the chemical oxygen demand (COD) using UV-Visible spectroscopy.[16] Synergy interval PLS (siPLS) identified any optimal

Table 10.1 Chemometric pre-processing mathematical functions. Reproduced from ref. 15 with permission from Elsevier, Copyright 2003.[a]

Pre-processing method	Equation used
Transformation	
Logarithmic	$\log 10xij$
Box–Cox	$xij\lambda - 1/\lambda$, where λ is the exponent value
Power	$xijp$, where $p = 0.5$ gives the square root
Row pre-processing	
MSC	$x = \mathrm{bo} + \mathrm{brefxref} + \mathrm{e}$
	$\mathrm{xcorr} = (x - \mathrm{bo})/\mathrm{bref}$
SNV	$(xij - \bar{x}i)/\sigma i$, where σ is the standard deviation
Normalisation	$xij/\bar{x}$, where $\bar{x}$ is the estimated value
Column pre-processing	
Mean centring	$(xij - \bar{x}j)$
Autoscaling	$(xij - \bar{x}j)/\sigma j$
Pareto scaling	$(xij - \bar{x}j)/\sigma j$

[a] MSC: multiplicative scatter correction, SNV: standard normal variate.

spectral regions known as PLS to develop a model to identify mixtures of chemicals using spectroscopic methods such as UV-vis, fluorescence, MS, or FTIR. The output spectra of pure and individual chemical compounds could then be applied to generate the chemometric method to predict unknowns.[17–19]

Measurement and monitoring of contaminants like antibiotics, pharmaceuticals, pesticides, heavy metals, and other emerging contaminants can be delivered using chemometrics in laboratory-simulated and real environments. Instrumentation like UV-vis, FTIR, HPLC, and GCMS for analysing both organic and inorganic pollutants are all candidate instruments for measuring these contamination events. Chemometrics can be used to analyse individual pollutant molecules that may also have interference from other compounds; therefore, singly identifying compounds in complex matrices.

When conducting quantitative analysis using chemometric methods, and when contaminants are measured using UV-vis spectroscopy, chemometrics can provide vital process data in the environment. Mohamed *et al.*, used binary mixtures for lactam antibiotics and their concentrations were resolved using a classical least squares (CLS) method and principal component regression (PCR) methods.[20] The following antibiotics were used: ampicillin, amoxicillin, flucloxacillin, and dicloxacillin as binary mixtures where the pure components and mixture component spectra were found to show spectral overlaps of 71% and 94%, respectively. The model determined antibiotic concentration values and was corroborated using both the CLS and PCR methods. With a growing need to monitor pollution from pharmaceutical agents, PCR methods have been used to determine the contamination of common antibiotics ampicillin, flucloxacillin, dicloxacillin, and amoxicillin.[21] This work showcases that eyedrops are a contamination problem in some waters. Sajjadi and co-workers used chemometrics to determine the degradation of phenolic agents *ortho*- and *para*-nitro phenols.[22] The authors employed a matrix augmentation multivariate curve resolution-alternating least squares (MCR-ALS) method to simulate the ability to overcome spectral overlapping to generate individual components in the matrix or dataset.

We will now discuss the current experiences in this critical area, based on the specific examples of field-portable and rapid instrumental analysis with a specific focus on emerging contaminants based on search terms performed in 2022 using Google Scholar "emerging contaminants". We have selected "broad-church" contaminants for interest in this chapter; these are: organics (oils and PFAS, for example), plastics, microorganisms, and inorganics (heavy metals, *etc.*).

10.6 Field Deployable and Rapid Instrumentation – New Approaches for Environmental Forensics Using Chemometrics

Field-portable instrumentation and equipment has rapidly become commonplace in environmental forensic investigations, and an excellent review of field-deployable instrumentation was written by Galuszka *et al.*[23] However, field deployability is only one facet in gaining meaningful data and related "hidden" trends about the environmental problem. The classical approach in presenting samples to an instrument include: (1) sampling, (2) sample (pre)treatment, (3) analyte separation, (4) presentation to the instrument or analyte detection. Chemometrics can reduce the need for some of these stages, just in the way the data is analysed/pre-processed. The use of field-portable equipment as a suitable measurement technique is reliant on the environmental situation, where the advancing technologies will provide explosive applications soon. The ability to rapidly sense with minimal sample preparation will provide rapid field information, providing the ID of pollution as the priority, and then performing in-depth laboratory-based methods to generate a more comprehensive environmental forensic outcome. Without field-portable equipment, responding to the emergency will rely on scientist's training and the sample methods – these are all limiting steps to the response. Field portable instrumentation can therefore offer viable solutions for emergency responses.[24] The field portable instrumentation and the generated data is not there to generate the prosecuting court proceeding data, but to gather active data to inform and then manage the environment and protect human health through the development of policy and risk assessment in a timely fashion. The field portable instrumentation results will not always be confirmatory but also exploratory, especially when quality control and quality assurance obligations are not as strict. Again, reiterating that lab-based measurement will still take its place, until advanced field portable devices come to fruition.

For example, if an environmental incident necessitates a more thorough investigation (if rapid and field deployable devices have been used), and the hazard is under control, the forensic environmental scientist will still take samples and then send them to the laboratory to obtain evidence suitable for court and legislation.[25]

Field-deployable equipment can also be applied to other situations, such as historical or non-emergency cases.[2] For these investigations, a visual interpretation of the scene will provide the initial "screening" and subsequent reasons for improved sample knowledge and why specific samples were collected when presenting evidence in court. FPE and their results provide the point-source identity or distil the source or site. Nowinski *et al.*, pioneered the use of field deployable instrumentation to provide elemental screening to guide the sampling regimen, which could then be used as guiding samples for subsequent laboratory analysis. However, little reporting has been provided on the use of chemometrics in these environmental analyses.

The complexity of the environment spills into the way many scientists analyse data. The use of complex statistics to prove methods, trends, or validate environmental systems is fraught with labour intensive and often unfruitful data, never reported to the public. Chemometric data analysis methods provide a powerful tool to overcome this. For example, when datasets contain concentration datapoint information for the multiple chemical pollutants collected from the various sampling sites, they are often arranged in large tables, or more complex data structures according to the may dimensions and experimental measurements taken. For example, one of the most frequently used exploratory tools for environmental data is principal component analysis (PCA). PCA allows for the transformation of the data to visualise complex datasets that make the science more obvious. For example, if we use PCA for a contamination problem, sources may be identified based upon their geographical or spatial region, and then the temporal distributions can also be estimated.[26] In work by Cozzolino *et al.*, recent and historical chemical data was analysed using multivariate statistics for a mine-site. The authors concluded that rainfall played a major role with dissolved oxygen (DO) interacting as part of the "churning process" of the weather. There were significant correlations related to calcium and copper and DO. In this work, optimum principal components were determined using leave-one-out cross validation to avoid overfitting the models.[27] Additionally, partial least squares discriminant analysis (PLS-DA) was performed, dummy variables against the parameters measured were assigned and then standardised using a pre-processing method divided by the standard deviation (SD).

We are observing that future trends are moving towards using chemometrics in the "day to day analysis" of chemicals or polluting species. For example, many spectral datasets require a lot of effort to process to uncover the interesting features. Pre-processing strategies should highlight features in the data for further analysis, removing the unwanted variation, data artefacts, noise, baseline shifts, and standardising signals. Automation of chemometric pre-processing will have a place in laboratory methods, especially for discrimination purposes. For example, Martyna *et al.*, developed a discrimination analysis algorithm for the assessment of forensic comparison studies. The authors define the quality parameter ratio between source and within source for the pre-processed data which then selected the highest variation between the sources (*i.e.* the groups). This action reduces the data dimensionality that improves the discrimination of the data. Regularisation (defined as the sum of absolute values of the parameters) of the method makes it feasible for handling singularity problems of variance–covariance matrices for highly multidimensional data. The grid search process as well as the genetic algorithm are used to find the optimal strategy.

Microbial community profiling has been used in the soil sciences to measure soil evidence since the early 1990s. Microbial diversity can play an important role in generating forensic analysis as this examines the biological dimension of soils, predominantly based on the analysis of the DNA, whereby nothing has to be cultured. MCP can be performed in normal laboratory settings; however, the ability to measure DNA outside of the laboratory is attractive. New systems such as the handheld MinION sequencer from Oxford Nanopore Technologies (ONT), which measures ionic current changes as DNA passes through an array of protein nanopores that are embedded in a special polymer membrane, is one such system.[28] For example, Johnson *et al.*, performed real-time DNA sequencing measurements in Antarctic samples for offline DNA measurement. The system required libraries in order to ascertain the sequencing data. If cluster analysis (CA) was applied to this system, exploratory pattern recognition would group MCP samples together. One such type of CA used in environmental studies is hierarchical clustering analysis (HCA), where objects are clustered based on either an agglomerative or divisive method. The agglomerative method would start with single objects such as microbes and begin to group them based on their similarity until a cluster forms. Divisive methods would start with the entire cluster of microbial profiles and then in each step, the least similar sample would be divided into clusters until we reach single objects, providing a dendrogram.[29]

10.7 Advantages and Disadvantages of Field Deployable Equipment (FPE)

FPE has numerous advantages and is changing environmental forensic analysis into an initiative-taking method, rather than being reactive to a situation. FPE is therefore more suitable and advantageous for pollution events that are hazardous and are classified as emergency events.[30] Laboratory sample analysis will always be conducted after an emergency and pollution event. Therefore, the optimum information acquisition window is better placed during the event rather than afterwards. By taking the initiative, emergency events can be better protected. The proactive approach will ensure the initial responders will benefit from full chemical protection and risk awareness prior to visiting the site, with data generated from the field equipment.[2,31] Clear benefits of this information include the on-site intelligence to provide safety for the first responders, enabling the selection of appropriate measures to provide personal protection (PPE) or methods and techniques to again reduce the risk of exposure before arriving at the situation. In addition, field deployable instrumentation would allow the responders to assess hazards presented to the public or ongoing hazards which may present later, all of which would be validated by the environmental forensic scientist. Using this initial information gathering or "triage" of the event allows for a more informed reaction to the initial environmental forensic response.

Besides all the advantages, some aspects must be considered in employing FPE. The capability requires personnel who can manage the equipment at the scene and be an experienced analytical scientist to be hands on in the data interpretation.

FPE and the generated results should not be considered as an exact confirmatory tool, but as an initial screening tool for a presumed contamination or pollution event as it is occurring. The reason being is that the environment is dynamic and presents a raft of changing parameters which could affect the FPE datasets. Furthermore, field-based instrumental data and analysis will correspond to the equivalent laboratory-based techniques, and therefore performance levels may vary due to the size and power constraints when comparing lab based to field deployable methods. For example, long-term deployed sensors in the marine environment, while excellent for confirming environmental changes, may undergo biofouling, which causes data drift in the instrument.[32] Therefore, we feel that laboratory analysis will always be required if data is intended for a legal capacity.

FPE is not necessarily the cheapest or cost-efficient method. By virtue of miniaturisation, portable versions of a laboratory technique could be more costly to make. Additionally, with any equipment that needs to be deployed in a timely fashion, the instruments need to be upkept, calibrated well, and have the appropriate standards in the field – all of which may be drivers of increasing costs. A field-portable instrument could be more expensive in some analysis types when compared to classical laboratory analysis. One would assume this is not the case as the FPE may have non-destructive capabilities like spectroscopy. However, these are all key considerations in making decisions as environmental forensic departments. Certain portable instruments, when coupled with newer techniques such as artificial intelligence and machine learning, may provide new avenues for a significantly cheaper laboratory method comparisons.[33] We will now discuss chemometrics using instrumentation that has sampled the environment.

10.7.1 Chemometrics Added to FPE

When discussing field deployable systems, it will be important to state what chemometric methods can value add to each type of instrumental analysis. In this section, we briefly discuss the benefits of certain chemometric methods for environmental analysis and forensics.

10.7.2 Unsupervised Methods

10.7.2.1 Principal Component Analysis (PCA)

PCA reduces dimensionality in the dataset, and therefore is a method used to transform a dataset with high-dimensionality containing inter-correlated variables into smaller independent variables. For example, PCA is particularly useful for the analysis of the pollution of rivers, coastal regions, creeks, and lagoons. The dataset from water quality is often highly intercorrelated owing to the chemical and physical nature of the pollution source, the vector, chemistry, mixing, or microbial breakdown to provide numerous examples of the variables we are referring to. PCA can reduce the data into principal components (PCs) which can then be used to identify types of pollution and their identity.[34]

10.7.2.2 Cluster Analysis (CA)

Much the same as PCA, CA is used as an exploratory method to primarily determine general relationships between the data. CA can answer more complex environmental science problems such as, do the samples fall into groups? CA is well established to compare and group data. For examples, Shields *et al.*, discuss hierarchical cluster analysis (HCA) to reduce and simplify a complex set of variables such as those found in air pollution events, for example, using congeners (individual compound within a family of compounds). Individual profile comparisons of chemicals are useful when the chemical profiles are different. However, subtle patterns among histogram plots are hard to determine using the naked eye. For example, congeners or, as an extreme example, the peaks of semi-volatile compounds on a gas chromatogram, would be difficult to distinguish, but CA and PCA could be used to condense and simplify a complex set of variables.[35]

HCA clusters the data within a dendrogram by using a similarity criterion and a clustering rule within the algorithm. The similarity criterion is a measure of the distance between samples. In a study by MrGregor *et al.*, the Euclidean distance was used to highlight the similarity between samples for coal tar. Coal tar manifests as a by-product of the manufactured gas industry, and produces a number of hazardous compounds such as coal tar or ammoniacal liquor rich in PAHs or dense non-aqueous-phase liquids. Environmental forensic based fingerprinting of different types of coal tars is not well reported. This under-reporting is due to the limit in the analytical techniques used to analyse the by-products. Enhanced separation techniques such as two-dimensional gas chromatography (GC × GC) provides a wealth of information, Several different clustering rules exist, describing the way in which samples are linked in a dendrogram (Figure 10.2). McGregor used a single, average, and complete linkage method to provide optimal

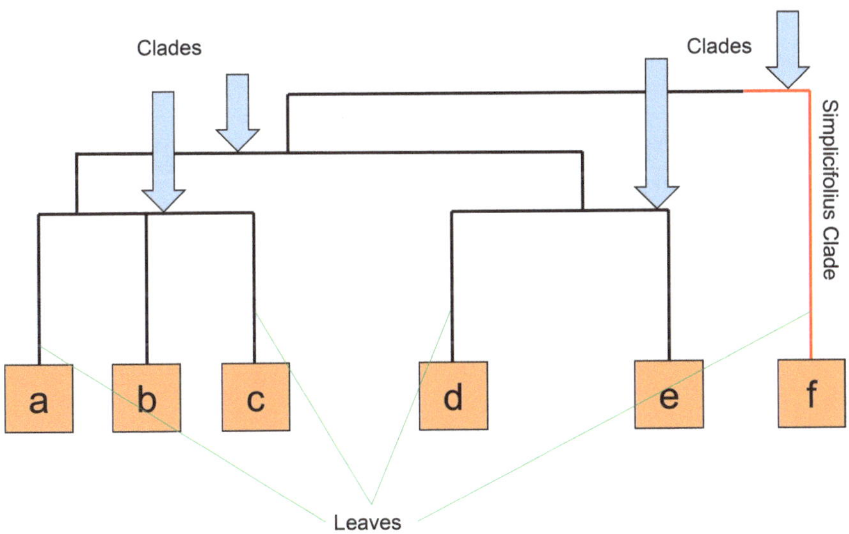

Figure 10.2 Dendrogram showing the grouping associated with the HCA algorithm. Samples are classed based on their similarities in leaves. (a–f= variables), clades, and leaves represented to show the separation and similarity, respectively.

clustering, using a repeated HCA, on the observations (samples) within the data set as well as on the variables (chromatographic peak areas).

Liu *et al.*, performed a spatial distribution-PCA model that integrates the spatial characteristics of soil pollution with linear data transformation by eigenvector-based PCA. This technique will serve well in the identification of heavy metal contamination in soils, for example. Here the authors used the global Moran index to analyse similar attributes in the samples to cluster the samples together.[36]

10.7.2.3 Factor Analysis (FA)

FA is another dimensionality reduction method, assuming that the variance in measured samples is caused by some other smaller unobservable underlying factors, better known as latent factors. FA is similar to PCA; however, FA assumes that the latent factor is caused by the variance in the *Y*-dimension and that the total variance can be partitioned into common variance.[37] As an example of the use of FA, Baragaño *et al.*, examined industrial sites affected by anthropogenic contamination using MVS and environmental forensic techniques. The authors applied FA to identify sub-areas with distinct geochemical behaviours. Random sampling was applied to 66 samples, where total concentrations for Ag, Al, As, Ba, Bi, Ca, Cd, Co, Cr, Cu, Fe, Ga, Hg, K, Li, Mg, Mo, Ni, Pb, Rb, light rare earths (LREE; as the sum of Ce, Eu, La, Nd, Pr, and Sm), heavy rare earths (HREE; as the sum of Dy, Er, Gd, Ho, Lu, Tb, Tm and Yb), Sb, Sc, Sn, Sr, Th, Ti, Tl, U, V, Y and Zn were analysed. This database was then subjected to FA employing PCA extraction by the Kaiser–Gutmann criterion and Varimax rotation to minimise any variables with high loadings. FA and subsequent HCA led to the establishment of clearly grouped samples, providing a powerful combinational tool for understanding pollution source identification.[38]

10.8 Equipment Required for Field Deployment Forensic Analysis

Commercial FPE allows multiple contaminants to be analysed in a broad range of environmental sample matrices. Some FPE/instrumentation can now readily replicate the exact capabilities of the laboratory representing their suitability for environmental forensic investigations. The FPE should generate chemical fingerprints to chemically profile the contamination event and therefore the point source to be identified. Some of the instrumentation and equipment discussed herein is still undergoing improvement and will have early demonstration for forensic applications; we have included methods for chemometrics for each of these FPE.

10.8.1 Field Portable X-ray Fluorescence Instrumentation

X-Ray fluorescence (XRF) and portable versions of XRF are already well established for environmental applications and XRF is already considered a standard forensic technique for this reason. XRF is popular owing to its low(er) cost and rapid result generating ability. The technique has been used to identify copper, zinc, lead, arsenic, mercury, chromium, and tin. XRF produces chemical profiles that are useful for point source contamination, for example, metals in soils.[39] XRF is also a valuable non-destructive technique, which is clearly attractive for use in environmental forensics. Over the last decade XRF has been refined in determining point source contamination tracking of metals; one example is for coal contamination investigations, where the portable nature of XRF demonstrated a tracking and point source of the mobile metals back to the coal station.[40]

10.8.1.1 Positive Matrix Factorisation (PMF)

The PMF algorithm determines the source and quantity of pollution, making PMF useful for source tracking in environmental forensics. The PMF model analyses a range of environmental data such as sediments, wet deposition, surface water, and air (all as complex matrices). The PMF model reduces variables in complex analytical datasets to combinations of species called "source types" and "source contributions". The identified source types are compared to the measured profiles. Then, the source contributions apportion the involvement of each source contributed to a sample. The PMF model calculates fingerprints, source contributions, and source profile uncertainties. However, the PMF model output is limited to providing positive source contributions, and therefore the uncertainty of the weighted difference between the observed and predicted species concentration is minimised. As an example, Gunchin *et al.*, reported source apportionment results (184 samples, each of fine (PM2.5) and coarse (PM2.5–10) fraction atmospheric particulate matter) over a 3 year period in Ulaanbaatar, Mongolia. PMF was applied using the concentrations of 16 elements measured by an EDS-XRF spectrometer along with the black carbon content measured by a reflectometer as the input data. The PMF results showed mixed sources dominate the coarse fraction, soil, and traffic sources as the "contributors" to the fine fraction. The source profiles and the seasonal variations of their contributions indicate that fly ash deemed to be emanating from coal combustion, mixes with traffic emissions and resuspended soil, resulting in variable chemical source profiles.[41]

10.8.1.2 XRF in Environmental Analysis

Soil and sediments have been routinely analysed for their elemental mapping using portable XRF systems.[42] Some of these uses were first showcased by Bosco in 2013.[43] The instrument's versatility is also shown in using XRF to detect the level of water contamination, but also the contamination on extraction disks and water filters.[44,45] However, XRF in its portable form is not suitable for trace elemental analysis; accredited laboratory methods will need to be employed.

XRF portability gives rise to several other applications including analysing elements in air concerning occupational health and safety, where it would be suitable to determine metal speciation in air samples. Food contamination, such as in oyster shells, also determines heavy metal accumulation.[46] Elemental analysis in paint waste using XRF can confirm heavy metals. Heavy metals are usually incorporated in paints and reduce the effects of rusting or oxidation. As an environmental forensic example, antifouling paints containing tri-butyl tin has been banned since 2007, highlighting the possibility that these paints could still be used by illegal entities and would therefore require forensic style analyses.[47]

Spent nuclear fuel (SNF) is a challenging task to be able to quantify in environmental analyses simply because of the safety issues surrounding the samples. Therefore, there is an urgent need for methods that would provide remote and on-line quantification of elements in SNF and its reprocessing technological solutions. X-ray fluorescence could be one way to generate this need, X-ray excitation conditions will obviously vary from sample to sample, where the resulting spectra will be a complex superposition of numerous signals from soft gamma emitters and X-ray fluorescence signatures. These complex spectra can be treated with chemometric data processing techniques for quantification of elements. For example, Panchuk *et al.*, normalized the sum of the intensities of registered spectra for 20 mixtures; X-ray fluorescence is excited by the intrinsic radioactivity of a sample and therefore sample variability may occur. The target analytes were identified and then a PLS algorithm applied. The aim of the PLS was to provide a predictive model based on latent variable decomposition of the data. The authors proved the validity of the regression models and in this situation we applied the following three-step procedure: first we make variable selection and optimization of PLS models with respect to the number of LVs based on the minimum of RMSECV (root mean squared error of cross-validation) in the course of leave-one-out cross-validation, then we employ double loop cross-validation to evaluate model performance. Finally, we check the model for chance correlation through permutation testing, using the R environment and using the mdatools package for PLS modelling.[48]

These results confirm the statistical validity of the models and prove the potential of the proposed analytical technique based on the employment of intrinsic radioactivity of SNF samples for XRF excitation. This approach does not compete in accuracy with ICP-AES measurements; however, X-ray fluorescence combined with chemometric modelling can be implemented quickly and allows for fast evaluation of the analyte of interest and content. The accuracy and overall simplicity of the proposed protocol was extremely encouraging.

10.8.1.2.1 Applications of Kohonen Neural Networks or Self-organising Maps (SOMs) in XRF Measurements. Many XRF methods and techniques are time consuming and can be subjective.[49] A Kononen neural network can be used to explore metallic fingerprints. As we have stated, XRF measurements are routine for non-invasive elemental analyses. However, many of these instruments only collect one spectrum at a time at a single isolated spot and are subjective. While some efficient data processing methods exist, they only deal with a small number of elements; for more complexity in the data, such as spatial and spectral data acquisition, other methods are required, such as Kohonen neural networks or SOMs. For example, Kogou *et al.*, used the unsupervised SOM algorithm based on neural networks, applied to XRF spectral imaging data. The SOM method reduced hundreds of thousands of XRF spectra in a spectral image dataset to a handful of distinct clusters that share similar spectra, demonstrating how clustering and the combination of spatial and spectral information can be used to aid material analysis and deduce the layers in a material sequence.[50]

Portable XRF provides elemental information such as the concentration of a given element in a sample, making this technique highly relevant for a forensic environmental scientist. However, matrix interferences, spectral interferences and sample geometries often interfere in this process of analysis. If the user requires quantitation, care must be taken to ensure the instrument has been calibrated and the detection limits are clearly defined. To combat this, several chemometric approaches to XRF data have been conducted to explore the matrix effects and any matrix corrections. XRF is one such technique that succumbs to the issue of matrix effects when performing quantitative analyses. XRF data pre-processing using chemometric algorithms leads to intensity correction of the data, highlighting the ability to overcome these shortcomings. PLS can manage simple spectral overlaps, additionally the application of a non-linear artificial neural networks (ANNs) reduces any matrix effects. However, for appropriate network training, calibration samples should be included in the dataset. As in many of our previous papers, we always state that care must be taken to understand the performance of the model.[51]

10.8.1.3 Applications of XRF in Microplastic Analysis

XRF spectrometry can provide the elemental composition of synthetic polymers, in this case, we focus on microplastics. In this work, XRF can provide critical chlorination termination and brominated flame retardants or identify inorganic pigments that may have been sourced from restricted metals.[52]

Some problems will eventuate when analysing microplastics with a portable XRF. By their very nature, microplastics can have smaller diameters than the source-detector window of the instrument (<10 mm); therefore, samples will need to be presented on a suitable level surface, otherwise damage will occur on the instrument presentation window. Radiation can be poorly absorbed when using low-density plastics, generating poor detector values of the emitted fluorescent X-rays.[53] Another primary concern is the generation of secondary X-rays; Compton scatters in low-density materials are brought about by photoelectric absorption in these material types.[54] To overcome some constraints, some manufacturers have introduced a fixed point stand. For example, the Thermo Scientific (Niton) which uses a tungsten-PVC casing to irradiate the sample.

10.8.1.3.1 Heavy Metals and Micro/Nano Plastics. The very nature of plastic as a material means that other chemicals can readily absorb onto the material. We have seen in the literature that heavy metals, persistent organic pollutants, and other hydrophobic organics can achieve this interaction. In doing so, this presents new unknown hazards to organisms that accidentally ingest them or, worse still, can be a risk to the analyst tasked with measuring them. XRF has been used to determine heavy metal concentrations in a range of polymers, using a portable field device.

For example, PCA is a versatile method providing an overview of complex multivariate data, and it is widely adopted to treat μXRF data. XRF data and the variables come from the whole spectrum whereby the intensities of several spectral lines, or the elemental concentration, can be obtained. When PCA is not used as a sample classification technique distance metrics are introduced. The most widely known classification tool based on PCA is soft independent modelling of class analogies (SIMCA). SIMCA projects new samples with unknown class assignment onto the PCA models for known classes. The advantage of SIMCA classification is that the algorithm provides a multiclass classification by simply projecting unknown samples onto several PCA models of established classes. The SIMCA can be used as a one-class classifier separating the samples with features from all other samples regardless of the nature of the differences.

An example of the application of the SIMCA algorithm to XRF data is given in the work of Farquharson *et al.*[55] where 77 normal and cancer breast tissue samples were studied using XRF and EDXRD (energy-dispersive X-ray diffraction). PCA models were constructed for both normal and diseased tissue classes and were further analysed using the SIMCA classification. Neither XRF nor EDXRD allowed for the reliable classification between normal and cancer tissues, while the combined data from the two methods yielded better results. However, other authors have implemented a multi-analytical method, including XRF, to understand the biodistribution of metals in tissues. This could prove to be beneficial when understanding the fate of metal toxicants. Gillois *et al.*, was able to combine NIR and XRF to predict the uptake of meso silver and colloidal silver, Mesosilver™ or AgC. When the NIR spectra were processed using chemometrics the authors could monitor changes in the intestine at the content and tissue levels, and also demonstrate that the liver had high-dose Mesosilver™ or AgC. The NIR spectra were mathematically pre-processed to remove the effects of light scattering and to compensate for baseline offsets and bias. Several classical pre-processing methods were tested (derivatives and multiplicative scatter correction, MSC) The mathematical pre-processing that gave the best performances of CART models here were the second derivative.

Sediments with microplastic contamination using an energy dispersive field portable XRF (Thermo model: Niton XL2 GOLDD), shown in Figure 10.3, has been used in the field.

The Thermo model: Niton XL2 GOLDD uses a low-power X-ray tube and possesses a silver (Ag) transmission anode as a sample excitation source at

Figure 10.3 Left, Hitachi X-MET8000 pXRF, and right, Thermo Fisher Scientific NITON XL5 pXRF.

100 μA current and 45 kV voltage. The instrument has a large area silicon drift detector, providing X-ray detection of the sample matrix. Robin *et al.* utilised XRF to corroborate FTIR–ATR microplastic contamination for marine applications (fish, sand, and water). The authors showed that FTIR–ATR characterised polyethylene (P.E.) and polypropylene (P.P.) were the most abundant plastic types in the marine samples. The fish and their digestive tract showed that of 15 out of 70 samples, 22 microplastic particles were present. Polyethylene (38%) followed by cellulose (23%), rayon (15%), polyester (15%), and polypropylene (7%) were the major ingested microplastics. XRF determined a range of heavy metals present in the polymer samples (among other metalloids and elements). The highest concentrations recorded for PE were found to be Cd, Cr, Hg, and Pb. For PS, Fe, Sb, Ti, and Zn concentrations were found to be most abundant. Robin *et al.* highlighted that those dual analyses are required to determine plastic contamination in multi-matrices.[56]

10.8.1.3.2 Overlapping Peaks Resolved Using Algorithms. Peak overlapping effects can be observed in some XRF data. To overcome the overlapping some authors have used a crow searching algorithm or CSA. The CSA is a meta-heuristic algorithm proposed by Askarzadeh in 2016, and it is inspired by the behaviours of storing and stealing food.[57] For example, in soil, the overlapping peaks of Pb and As are relatively serious, and there are three overlapping peak positions; therefore, the information processing method is more reasonable. As and Pb can be resolved using the CSA by several pre-treatment methods using the characteristic net peak area or peak height from the decomposing overlapping peak to establish the model. The spectral pre-treatment is therefore imperative to be correct. Much of the work we have performed with spectral pre-processing has been centred on noise filtration such as the Savitsky–Golay method, then you can filter the outlying datapoints using an isolation-forest algorithm,[58] and for XRF data, you can remove the spectral base line by the linear background method – resulting in the element being determined by the basic law of Moseley.[59]

The classification of microplastics using cathodoluminescence (CL) and machine learning was developed by Hoppener *et al.* (2021). While XRF was not used as the direct technique for this characterisation, XRF is a corroborative tool for further characterisation of microplastics. The authors trained machine learning classifiers that learned to predict the type of plastics given the CL spectra of those plastics. In short, 111 plastic samples from reference and consumer plastics with different sizes (0.001–1 mm), colours (*e.g.* black, blue, green, red, white/transparent and yellow) and shapes (*e.g.* irregular, fibre, spheres) were added to a database. They then then trained multiple classification models using an ANN approach using three methods to the spectral analysis and R-packages (pracma):[60]

- Peak finding on the aggregated spectra
- Peak finding on the derivative plots
- Finding anomalies in the aggregated spectra using the "local outlier factor score" method.[61]

Machine learning (ML) algorithms are now frequently used in combination with numerous analytical techniques for identifying microplastics, including FTIR, laser-induced breakdown spectroscopy and, in this section, XRF.[62]

10.8.1.3.3 Portable X-ray Fluorescence and Contamination Detection. Tailings dams can have prolific effects on the environment if leachates manage to migrate through dam walls and into the surrounding environments. For example, in 2015, in southeast Brazil, waste from an iron mine was found to be contaminating the Doce River system. Using a portable-XRF (pXRF) a primary screening method was established, and subsequently the data was processed and analysed using a principal component analysis (PCA). The XRF data was corroborated with laboratory based methods using more sensitive techniques (ICPMS) where As, Ba, Cr, Cu, Mn, Ni, Pb, V, and Zn were found in the samples.[63] The PCA was used to classify samples using the iron-signal based on the PXRF spectra. It was found that mine samples could then be classified based on their location of discharge in the river system. The As content (up to 164 $mg\,kg^{-1}$) and the Mn (as high as 2410 $mg\,kg^{-1}$) were also shown to be dominating in the Doce River system. The authors used the Tracer III-SD, Bruker portable hand-held spectrometer.[63] The interface facilitates interaction with raw and processed data and provides support for many standard instruments and data exchange file formats, making importing data painless, but the software can also handle thousands of subsets created from a single data file.

Another material that has caused some serious health concerns is asbestos. Asbestos was used globally in abundance during 1900–1990s. Many materials contain asbestos (asbestos containing materials, ACM) and were produced in significant amounts and used to produce insulation, improve the flame-retardant properties in materials, advance the mechanical and chemical characteristics when used for construction materials, for example, roofs, walls, and flooring. Due to its overuse, fibres are released in abundance especially during demolition and construction of buildings. Asbestos removal is expensive and requires a great deal of expertise to remove the material. Therefore, an illegal dumping market has occurred.

To counter this, ACMs require rapid analysis and characterisation. Some authors have developed a micro-XRF method to generate fingerprint spectra, and then have coupled this with multivariate statistical approaches. For example, Serranti *et al.*, examined six ACMs containing three diverse types of asbestos fibres (chrysotile, crocidolite, and amosite). The XRF analyses provided micro-XRF maps of the materials, which were also processed and then analysed on a benchtop spectrometer (M4 Ronado, Bruker) system. The generated spectral data (*i.e.*, hyper-maps) were processed using a range of chemometric methods in MATLAB using the PLS_Toolbox (version 8.6, Eigenvector Research, Inc). The authors then used a calibration dataset acquired by the micro-XRF to build a classification model to classify the different asbestos fibres, automatically. A set of three asbestos fibres clearly identified in the ACM samples formed the training dataset, Figure 10.4, left, which could then be used to build the classification model, Figure 10.4, right.[64]

Analysis of the spectra showed that the data could be used to quantify the spectral information and then pushed into a classification model. A PCA and a further method using the SIMCA, which proceeds to classify new samples based on the classifications provided in each PCA. PCA is a fundamental multivariate data analysis technique used for data handling, outlier detection, exploratory and rank (dimensionality) reduction, graphical clustering, classification, and regression. The results from this work highlight the potential of environmental forensics, where an automatic classification model can be created. The results showed that asbestos fibres could be classified according to their chemical composition and therefore identified in the model. When XRF is used in combination with a classification model that is automatic elemental maps can be produced and provide effective methods as non-destructive techniques for the analysis of asbestos detection, alleviating the need for a trained operator.

10.8.1.4 Field Deployable Gas-chromatography–Mass Spectrometry (GC–MS) Methods

Analysing organic and volatile chemicals is typically performed using gas-chromatography–mass spectrometry (GC–MS). About a decade ago, portable GCMS instrumentation became a reality, and suddenly, by enabling laboratory-based capabilities to be miniaturised and translated into portable systems, GC–MS in the field could be realised. Some authors have also used MS alone, without GC, where excellent results have been obtained on different sample matrices, such as air contamination. However, the hyphenation

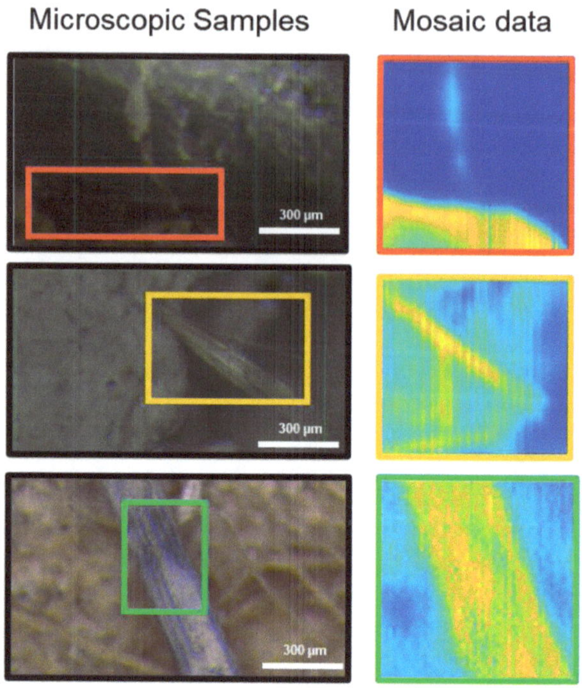

Figure 10.4 Left, microscopic samples of the asbestos-containing material (ACM) microscopic images: chrysotile (top), amosite (middle) and crocidolite (bottom). Right, mosaic data image of ACM samples used for calibration and validation. Adapted from ref. 64, 10.31025/2611-4135/2020.14007, under the terms of the CC BY-NC-ND 4.0 license, https://creativecommons.org/licenses/by-nc-nd/4.0/legalcode, with permission from the author.

of GC to MS includes additional information often desirable for complex mixtures.[65] Portable GC–MS can be good for organics, to generate elevated levels of certainty and chemical fingerprints for point source chemical contamination determination. Many authors have used portable GC–MS as a method for confirmatory and near-real time data analysis.[66] However, even though these systems have dramatically improved, they are still not as capable at quantitation as laboratory methods. For example, laboratory-based GC–MS can still separate compounds more efficiently based on capabilities such as triple-quad, better column technology, and SIMS methods.

Notwithstanding, portable GCMS undoubtedly provides a good screening method for investigative forensic environmental work, which is what we shall focus our discussion on.

10.8.1.4.1 GC–MS and Chemometrics. In forensic analysis, GC–MS methods are not effective in dealing with isomers. For example, isomeric compounds frequently co-elute because they have identical masses, therefore possessing similar electron ionization (EI)-MS fragmentation spectra and patterns, therefore selectivity for forensic applications is low.[67] For example, PCA on raw spectral data was also applied to real-time MS spectra for direct analysis. The PCA can be employed to reduce the number of mass spectral peaks to more manageable datasets. Additionally, background-subtracted mass spectra can be acquired from the raw data. Recently, Forbes and Verkouteren, extracted a mass spectrum consisting of 64 592 data points from *m*/*z* 30 to 600. They designed a custom Matlab (Mathworks Inc., Natick, MA, USA) code to read in a series of exported mass spectra files, scale the raw data, and perform PCA, generating scoring and loading values. The PCA scoring and loading plots using five replicate samples were used as training data for principal component determinations.

Explosives provide environmental concerns in soil and groundwater, manifesting as contaminants during or after training and manufacturing explosives exercises for example. Field portable GCMS can give immediate responses with approximately one hour of sampling and provides absolute confirmation of munitions constituents. By using field portable equipment explosives can be confirmed in groundwater, even providing rapid quantitation. The use of MS-detection, such as the Griffin 400 GC–MS and the Mini-10 mass spectrometer, samples can be collected, prepared, and analysed directly in the field. Furthermore, compounds in the sample matrix can be identified using MS data and library searches/confirmation, reducing the likelihood of false positive detections.[68]

Oil spills require assessing the environmental risks and preventing further damage. Furthermore, dissolution, weathering, oxidation, and microbial events will tend to affect the profiles of oils in waters.[69] Some authors have used 2D-GC as a tool to separate oils which are usually highly complex samples, and are more than often unresolvable. 2D-GC typically has two columns with orthogonal separation properties, The advantage of 2D-GC when examining crude oils is that chemical fingerprints can be produced for each of the chemical fingerprints in the sample, this is especially useful for geochemistry applications where oil spills need to be sourced. For example, petroleum distillate characterisation for oil spill investigations. Coupling 2D-GC with high-resolution mass spectrometry (GCGC/HRMS) provides accurate target investigation of critical compounds and less ambiguous chromatographic peak identification. It has been shown that data from spill cases can be pre-processed using retention time shift correction methods, which can then be normalised and then translated to pixel-based analysis using a weighted principal component analysis (WPCA), assigning weighted values to the pixels/samples. This process provides point-source probability values.[70] The absence of recalcitrant biomarkers, requires a different modelling method to:

(i) Distinguish from various sources of the same oils.
(ii) Identify weathering effects of spills.

(iii) Propose new modelling methods to highlight weathering effects which are then manifested in the chromatographic and chemical fingerprints.

These methods, shown in Figure 10.5, show the criteria to match the overall chemical similarities with any suspected sources and the oil spills themselves.[69] The 2D-CHEMSIC method used in this work was used as a diagnostic method to obtain oil spills data in a forensic capacity.

Source oils and spill samples can be compared to distinguish oils from the various sources and to identify the point source of the spilt oil, this has been

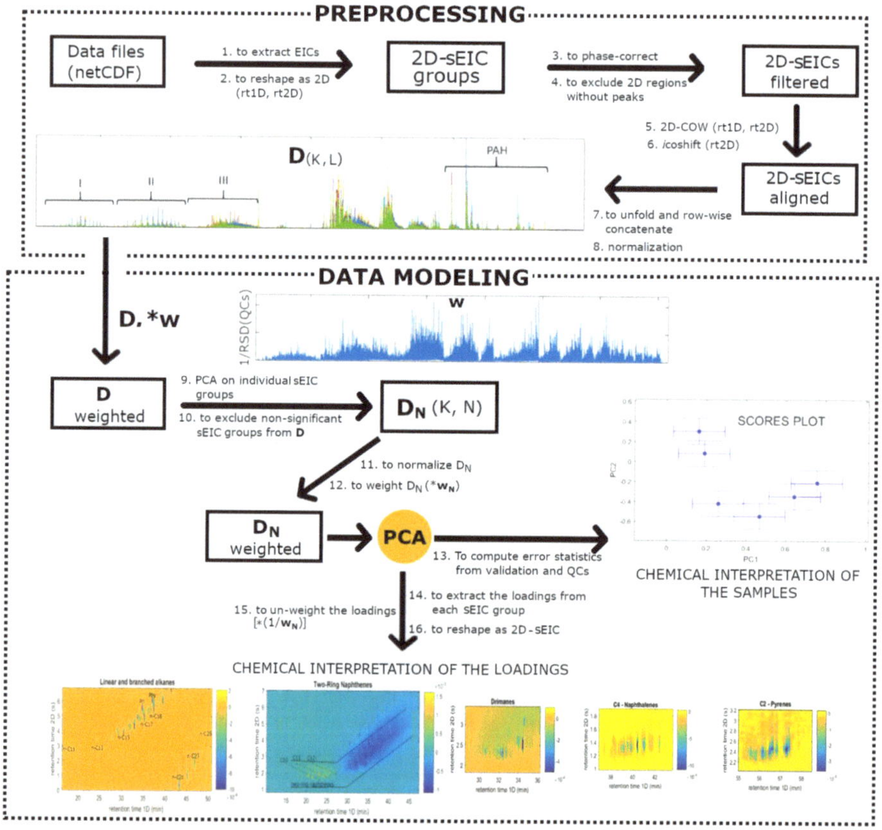

Figure 10.5 Scheme of the data pre-processing and next data modelling by pixel-based analysis using WPCA. The data were normalised to the Euclidian norm = 1 (*i.e.*, the total area of the row-wise chromatograms) before the data analysis. $w(1, L)$ is the weights vector computed as pixel-by-pixel 1/RSD(QCs), RSD = relative standard deviation, and QCs = (unfolded) 2D-sEICs from the QC samples. D_N (K, N) is the data matrix containing only the 2D-sEICs with a significant signal-to-noise ratio, N = number of variables. Reproduced from ref. 69 with permission from American Chemical Society, Copyright 2019.

achieved through pixel based analytical methods. Chromatograms that contain variations in the data which are different to the chemical composition (for example instrumental noise, residual retention time shifts, and chemical noise) all influence the PCA model, negatively. To reduce these negative influences a WPCA can be used, where the weight vector, w, contains the inverses of the pixel-by-pixel relative analytical standard deviations (RSDA). Here the method will only use data from the analytical replicate's quality control samples. If we examine D, all elements in each column will have the same weight. Therefore, the WPCA can use preprocessing (scaling each column by the corresponding element of w). The data can be improved by downscaling the signal D, in this case the pixels that contained little to no chemical information. Furthermore, up-scaling the pixels with a low analytical variation can also be performed. The PCA generates a mean-centred D, excluding samples and replicates ("training set" consisting of samples and QCs). The 10 sample replicates then form the validation dataset which are used to inspect for the absence of model overfitting and check for outliers. The number of PCs containing relevant chemical fingerprints and associated chemical information, which neglect the next PCs, are attributed to the instrumental noise. This method is proof that the modelled oil spill data using the WPCA can be used to explain an oil spill for forensic applications.

10.8.1.4.2 Portable Liquid Chromatography Mass Spectrometry (LC–MS). The growing need for real-time environmental monitoring analysis is becoming more prevalent.[1] Small footprint, lower reagent consumption, and power reduction all point toward the creation and expanding use of field portable analytical systems. Environmental monitoring using these systems will allow an on-site analysis with no need for formal accredited laboratory approaches. The approach will generate cost reductions and fast analysis, and reduces sample degradation or contamination when transporting or storing. Emergency applications will be the best suited for this type of methodology, especially where unknown hazards have not been identified or known. Several optical- and electro-chemical-based sensors are used for field-based and more real-time analyses, particularly for organic chemicals or pollutants. These techniques have already shown promise for herbicides, pesticides, and per- and polyfluoroalkyl substances (PFAS) in different matrices. We discuss the more rapid techniques in the next section of this chapter. In short, while the more rapid optical-based sensors are cheaper, sensitivity and selectivity are sometimes compromised (depending on the instrument). GCMS, as we have seen, is an excellent choice for field-deployable systems; however, these systems can be used to identify organic volatile and semi-volatile compounds. Therefore, LC–MS-based systems have begun to receive more attention, especially with their miniaturisation, whilst retaining the ability to determine structural confirmation.

PFASs are a major source of pollution in coastal sediments. In a recent paper by Oliver *et al.*, concentrations of individual PFASs were analysed using LC–MS coupled to a time of flight (TOF)–MS. Additionally, NMR was used to quantify the chemical nature of organic carbon and thus understand the sorption behaviours of the PFASs. The authors identified the chemical sorption coefficients with the chemical composition using a PLSR analysis. In short, the spectra were normalised to give the total signal intensity of 1000 and then multiplied by their organic carbon concentration. This was achieved using the Unscrambler 10.5 software.[71]

Field-deployable LC–MS systems have not evolved as quickly as other methods, the reason being LC–MS has several services required such as gases and reagents, also requiring elaborate sample pre-treatment protocols incompatible with the field, Portable LC systems coupled with standard nano- or micro-spray ionisation is challenging. LC systems usually require microfluidic connections and emitter-tip optimisation. These steps are sophisticated and are time-consuming. Recently, the invention of a nano-LC-electron ionisation-MS system has given rise to improved potential for on-site (in-field) analysis, with a particular focus on forensics of illicit drug usage. For example, the Viking 573 quadrupole MS and Thermo Easy nLC-1000 instruments are both sensitive, demonstrated by their effective use for illicit drug and mixtures of heroin and synthetic opioids confirmation. However, the system was not capable of measuring a complex real-world environmental sample. For environmental analysis, PFAS are now deemed to be one of the largest anthropogenic contaminants distributed in the environment, this system would prove to be extremely useful in the field.

For example, LC–MS has been routinely used for PFAS analysis. The LC–MS methodology benefits from low thermal degradation and requires less preparation of the samples when compared to a GC–MS counterpart method. A field-deployable LC–MS system has been developed into a miniaturised system by Hemida *et al.*[72] The briefcase-sized capLC system possesses a portable MS (weighing 39.2 kg), capLC columns, and can perform positive and negative modes of nano-electrospray ionisation MS (n-ESIMS), with a built-in split flow interface (SFI) for simple coupling in the field which makes this system very neat.[72] Analysis can be run in short run times of <8 min per sample, and can be placed in a small vehicle for transportation. The system, Figure 10.6, can also perform ultrasound-assisted extraction (UAE) which further develops the method for sample pre-treatment, in this case for spill extraction and analyses.

The authors demonstrate an excellent pioneering attempt at taking LC–MS to the field. The system was able to quantify PFAS with detection limits down to 0.1 $ng\,g^{-1}$, and with high-precision. The complex separation problems that are associated with multi-analyte separation and then detection can be solved with chemometric methods, for example, many authors could use the chromatographic response function (CRF).[73] This function was introduced by Berridge in 1982, which examines the individual chromatograms with respect to resolution and analysis time.[74] The CRF function has also been readily used for separations involving enantiomers.[73]

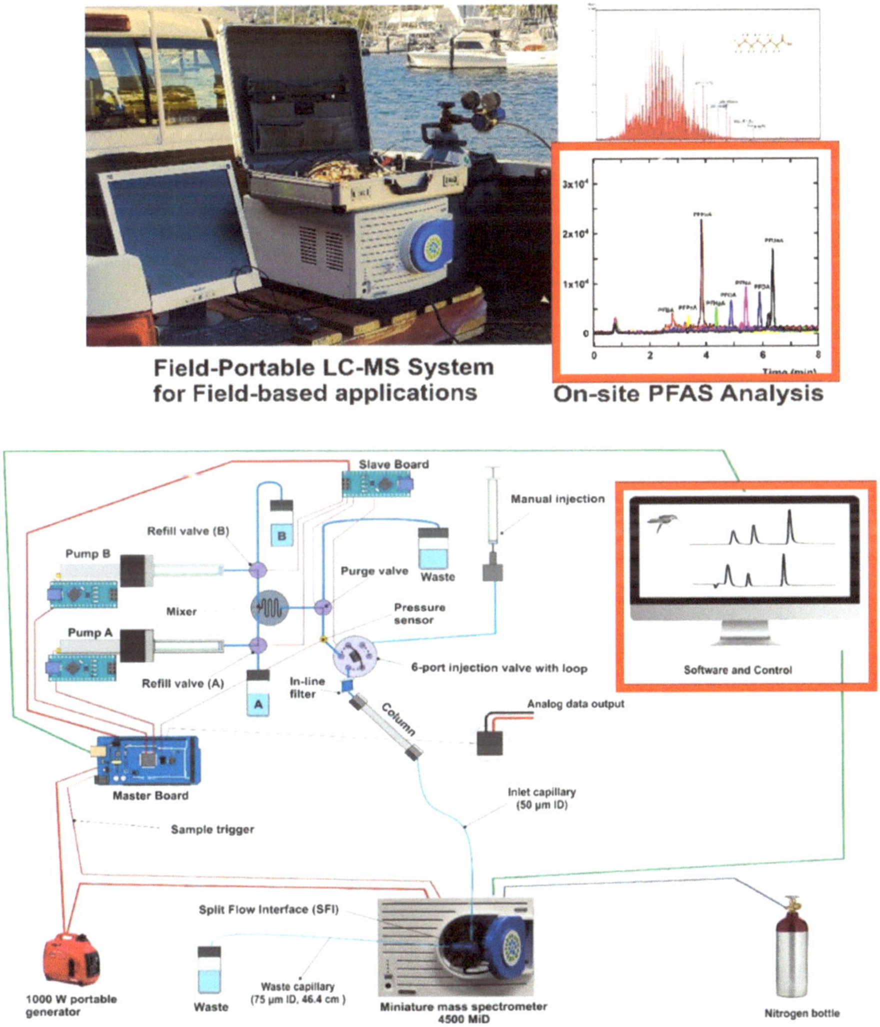

Figure 10.6 Top, field deployable system created by Hemida *et al.*, 2021. Bottom, the schematic architecture of the portable capLC system coupled with a portable mass spectrometer for field-based applications. Microfluidic flow (cyan line), data transfer (green line), electrical power lines (red line), and nitrogen (desert blue) – for the detection of PFAS in the field. Reproduced from ref. 72 with permission from American Chemical Society, Copyright 2021.

10.8.1.4.3 Spectroscopic Methods. The use of mid-infrared (MIR), near-infrared (NIR), hyperspectral imaging (HSI), and Raman spectroscopy are reported abundantly as complimentary techniques. MIR and NIR can scan matrices through the containers or holding vessels such as plastic or glass.[18,75,76] This has the advantage that the first responder does not have to encounter toxic contaminants, providing a safer work environment and limiting the introduction of sampling error associated with the removal of a sample from the site. The non-destructive nature of these approaches ensures that the evidence is not disturbed and produces *in situ* spectral information that can function as triage or a screening step to inform the response or investigation. MIR and NIR, again using chemometrics, can provide an excellent ability to remove the background noise associated with water's spectral band. Raman has a comparatively small water signal.[77] Fourier transform infrared (FTIR) will not suffer from the fluorescence that is exhibited in Raman spectroscopy. Raman also means that heating is an issue and changes the sample matrix as many samples in the environment/food/tissues absorb heat from the laser – removing the truly non-destructive benefits that other spectroscopic methods have. Raman spectroscopy should therefore be used with "an err on the side of caution" as, when certain substances are analysed by the user, burning and explosions can be casued.[78] It also changes the matrix, which is somewhat not practical, especially when trying to determine the true analyte of interest.

IR-based instruments are already typically used for air sample and air monitoring, for example CO_2.[79] Portable FTIR instrumentation is therefore suited for hazardous materials (HAZMAT), and incidents involving HAZMATs where unknown gases and vapours are required to be determined.[80] FTIR is typically coupled with attenuated total reflectance (ATR) attachments. This cell allows liquid and solid samples to be analysed, which makes it an attractive non-destructive technique for environmental forensic investigations. Portable FTIR is already used frequently for police forensic investigations to fingerprint powders such as drugs and explosives or other evidence types such as greases and resins.[78] FTIR is therefore suitable for a range of matrices.[81] Near-infrared can be used to quantify oils in soils.[82]

FTIR are usually light-weight, portable, and operate on batteries.[83] The ability to rapidly accumulate data is also very attractive. Furthermore, small sample aliquots can be analysed using these portable IR systems.[83] Although portable FTIR is not considered in the literature as a method for environmental forensic contexts, it is undoubtedly a competent technique which can provide rich fingerprint data and should be used with chemometrics.

Portable Raman instrumentation is seldom used concerning environmental forensic investigations owing to the severe limitations in sample integrity and fluorescence issues. Raman is suitable for some compounds, including the differentiation of plastics and the analysis of inorganics, these samples are more robust from the effects of heating and burning.[84] Raman spectroscopy has been used for benzene, toluene, ethyl benzene and xylenes (BTEX) by Sharma *et al.* who used the method for environmental contaminants as a proof of concept for safe working and distance from more toxic compounds.[85] In other works, Raman spectra used to analyse BTEX caused band shift appearances. These can be corrected for using MatLab, and therefore background spectra can also be removed to keep the signal intact. This can be achieved in part by pre-processing techniques, such as background corrections. Spectral quantification and then ID can be achieved using the SIMPLISMA algorithm.[86] Compared to other algorithms, such as multivariate curve resolution-partial least square (MCR-PLS), asymmetrical least square (ALS) or classical least square (CLS), the SIMPLISMA algorithm doesn't require the initial estimation of either component spectra or the contribution profiles. This can be advantageous in sensor applications where the concentrations values are unknown and the time and space are dependent.[87]

10.8.1.4.4 Micro- and Nano-plastic Contamination Detection. The most compact NIR system was released in 2012, the MicroNIR1700, at the PittCon® 2012 annual conference.[88] The system is powered by a light source, which makes spectral ranges from 908 to 1676 nm and uses linear variable filters (LVF) as a dispersion element. The Micro-NIR system has been used for several applications, including for explosives, marine plastics, and soil hydrocarbon pollutant determination. For example, Svetlana *et al.* studied plastic litter and created a database for 186 plastic items. When a spectral library was combined with the MicroNIR 1700 the identify of plastic could be easily and rapidly identified.[89] The authors confirmed that 96% of the samples were macro- and meso-plastics, and 73% of the samples analysed were in fact microplastics.[88] This was achieved by applying the standard normal variate (SNV) and second derivative to all spectral data. The authors then used a spectral match value (SMV) for the determination of the unknowns.

10.8.1.4.5 UV-vis for Metal Ion Determination in Plastics. Sales *et al.*, investigated three methods using UV-vis spectroscopy for metal ion determination using piecewise-direct-standardisation (PDS), single wavelength standardisation (SWS) and centring correction (CC); these are applied to the obtained UV-vis spectra.[90] The data showed that Ca(H) and Mg(II) could be readily observed in waters. Conversely, when using NIR the condition changes are established by spectral sensitivity changes. The test models' prediction errors were taken from the better performing standardisation conditions. From the standardisation techniques assayed, the PDS method provided the lowest mean-squared-error (MSE) of prediction and mapped most variability of the spectra. The most influential parameters for this work were the number of samples and the dotted size PDS.

If the sample number and the standardisation subset were effective, the more influential the techniques were. The authors demonstrated that the window size variation does not affect their calculations. They also discovered that when multivariate standardisation was used, they avoided recalibration of the instrument. The multivariate standardisation process can be used, when necessary, if used when there are not significant changes are exemplified, the results will not be satisfactory. Therefore, when spectral data and model development are conducted standardisation must be applied only to the significantly varied model.

10.8.1.4.6 Portable Hyperspectral Image (HSI) Analysis. The emergence of hyperspectral imaging (HSI) systems is fast moving. The HSI system contains the following: an objective lens, a wavelength modulator, the detector, an illumination source, and the acquisition system is seen in Figure 10.7. The HSI system can be adjusted to the requirements of many applications, especially in the environment. HSI systems are abundant in the food industry, and the range of uses can be endless for environmental analysis, ranging from specific sites and pixels of interest to microscopic and landscapes. Similarly, objective lenses for conventional imaging will generate he most suitable spatial resolution required for the desirable application, *e.g.*, macroscopic lenses, zoom lenses, and wide-angle lenses. The HSI system can also be coupled to a microscope which can then in turn analyse environmental samples on a microscopic scale, where rich spatial and temporal data can be obtained.

HSI integrates spectroscopy and imaging of a sample, generating rich spatial and spectral information about a sample. This duality of HSI has attracted the attention of researchers from diverse research fields. HSI provides the chemical composition of trace fingerprint chemicals and simultaneously visualise their spatial distribution in the sample. HSI therefore provides, visualisation, and identification of complex samples, especially when combined with chemometrics.

One such application is the use of explosive detection. For example, Fernandez de la Ossa has used NIR-HSI as a fast, non-invasive, and

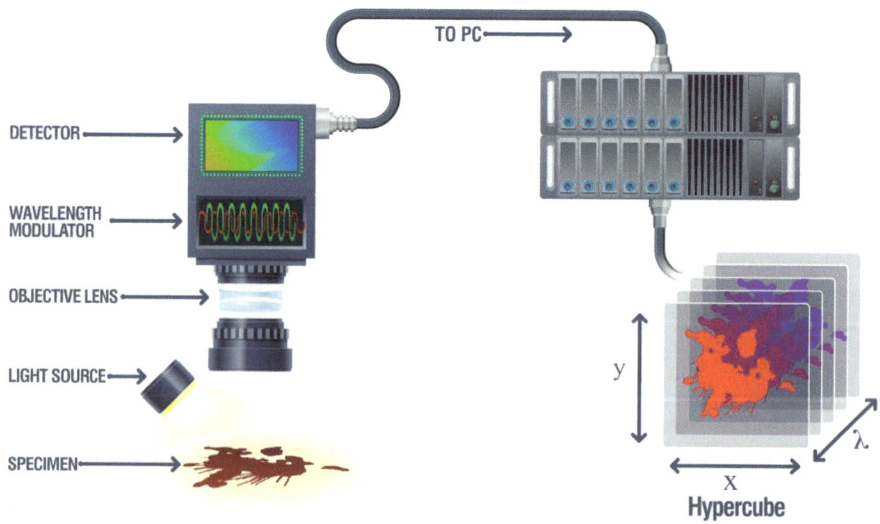

Figure 10.7 Schematic showing components of an HSI system, resulting in a hypercube of the specimen. Reproduced from ref. 91 with permission from Elsevier, Copyright 2021.

non-destructive tool for explosive residue analysis. This method could be used for contamination events also. Additionally, Antonucci *et al.*, used an HSI system coupled with chemometrics for copper analyses in soils capable of determining the concentration. When carefully used, HSI is a complex technology that can provide new insight into a sample and environmental forensic incident. Antonucci *et al.*, processed <2 mm soil fractions, which were air dried and were then contaminated with copper using $CuSO_4$ (0–1000 mg of Cu per kg of soil). Scanned samples showed that the visible–NIR and NIR spectrophotometers could quickly analyse the samples. The authors also predicted Cu concentration using numerous PLS regression models derived from the spectral arrays. In short, the chemometric methods were established to calculate the mean reflectance for each polluted soil sample and was related to the copper concentration by the PLS. The ability to predict using this particular model was dependant on the number of latent vectors used to address the root mean square error (RMSE), standard error of prevision (SEP) and correlation coefficient (r) between the predicted and observed values; calculating the bias.[91]

Our group recently demonstrated that plastic contamination could be found in tissues, waters, and the food chain using simple UV-vis and MIR technology. To this end, a hyperspectral camera system can also be used. For example, Karlsson *et al.* have monitored plastic particles in seawater filtrates.[92] Multivariate data analysis provided the plastic spatial location but also the composition of the plastic materials. Hyperspectral images are collected as hypercubes, containing a large number of datapoints. Therefore data reduction is necessary in order to perform the multivariate analyses. Figure 10.8 shows the decomposition of a hypercube obtaining two score images and two loading vectors for the plastic ID that they were anticipating.

For most datasets, limited components represent 99% of the dataset and the rest of the data can be assumed to be noise, making the data reduction stage large. This vast volume of data means that the plots have to be visualised as density plots. The density plots can therefore easily recognise clustering data/trends. A cluster of pixels representing any background can then be detected and removed from the dataset. The authors use simulated samples spiked with reference plastics representing microplastic filtrate in seawater. The samples have been imaged using a microscope and using multiple wavelength ranges (375–970 nm, 960–1662 nm, and 1000–2500 nm). The HSI data were then processed by HSI analysis. The 1000–2500 nm wavelength was the most applicable as it provided accurate (100%) particle recognition on the reference plastic which had size ranges above 300 μm and 84% pixel recognition on household polyethylene plastic.[92] The real-time applications of the HSI approach in the NIR range have also attracted attention in the recycling sector, where systems have been designed for the high-throughput detection of brominated flame retardants (BFRs) in plastics. Bonifazi *et al.* and Amigo *et al.* can show that high and low concentrations of these chemical additives in different plastic polymers within the 1000–2500 nm wavelength range was achievable.

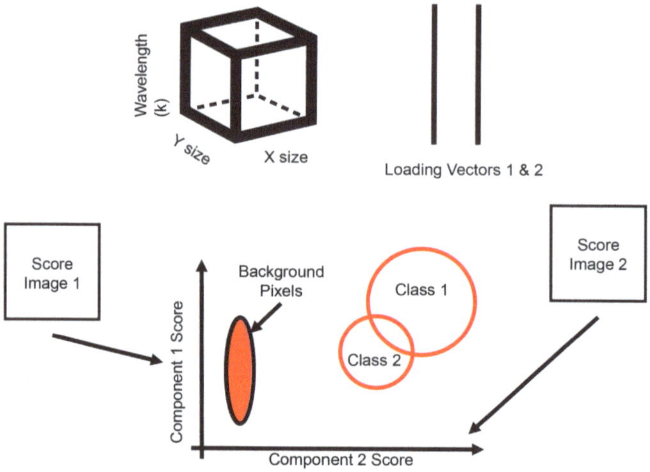

Figure 10.8 Hyperspectral images will have at least three dimensions. Two are physical usually (*X*) and (*Y*) the width and length. The third dimension is given by (*k*) the wavelength. PCA of mean centred data produces score images and vectors. The loadings vector can be used for spectral interpretation. This example shows three clusters: background, class 1, and class 2. These could filter out the background for plastic 1 and plastic 2.

10.9 Conclusions and Future Outlook

- Incorporating chemometrics into the basic research methods will assist in the reduction of errors in experimentation and increase the research's validity.
- Chemometrics will assist in pollutant remediation by assisting the researcher to focus on degradation of pollutants with maximum precision.
- Non-target and non-destructive methods such as NIR spectroscopy can be used as an advanced conceptual site modelling tool, acting as a reservoir of data that can be re-examined later.
- Field deployable instrumentation coupled with chemometrics will change the way environmental forensics will be conducted in the future.

References

1. J. Chapman, V. K. Truong, A. Elbourne, S. Gangadoo, S. Cheeseman, P. Rajapaksha, K. Latham, R. J. Crawford and D. Cozzolino, Combining Chemometrics and Sensors: Toward New Applications in Monitoring and Environmental Analysis, *Chem. Rev.*, 2020, **120**(13), 6048–6069.
2. V. Spikmans, The evolution of environmental forensics: from laboratory to field analysis, *WIREs Forensic Sci.*, 2019, **1**(3), e1334.

3. J. Lu, Use of chemical fingerprint data to enhance remedial site investigations at petroleum impacted sites, *Environ. Forensics*, 2016, **17**(1), 19–26.
4. I. G. Petrisor, *Environmental forensics fundamentals: A practical guide*, CRC Press, 2014.
5. R. J. Johnson, B. E. Smith, P. A. Sutton, T. J. McGenity, S. J. Rowland and C. Whitby, Microbial biodegradation of aromatic alkanoic naphthenic acids is affected by the degree of alkyl side chain branching, *ISME J.*, 2011, **5**(3), 486–496.
6. S. Shankar, Management and remediation of problem soils, solid waste and soil pollution, in *Principles and applications of environmental biotechnology for a sustainable future*, Springer, 2017, pp. 143–171.
7. C. Ohoro, A. Adeniji, A. Okoh and O. Okoh, Distribution and chemical analysis of pharmaceuticals and personal care products (PPCPs) in the environmental systems: A review, *Int. J. Environ. Res. Public Health*, 2019, **16**(17), 3026.
8. A. L. Van Nuijs, F. Y. Lai, F. Been, M. J. Andres-Costa, L. Barron, J. A. Baz-Lomba, J.-D. Berset, L. Benaglia, L. Bijlsma and D. Burgard, Multi-year inter-laboratory exercises for the analysis of illicit drugs and metabolites in wastewater: development of a quality control system, *TrAC, Trends Anal. Chem.*, 2018, **103**, 34–43.
9. D. Megson, E. J. Reiner, K. J. Jobst, F. L. Dorman, M. Robson and J.-F. Focant, A review of the determination of persistent organic pollutants for environmental forensics investigations, *Anal. Chim. Acta*, 2016, **941**, 10–25.
10. D. Cozzolino, S. Chandra, J. Roberts, A. Power, P. Rajapaksha, N. Ball, R. Gordon and J. Chapman, There is gold in them hills: Predicting potential acid mine drainage events through the use of chemometrics, *Sci. Total Environ.*, 2018, **619**, 1464–1472.
11. W. Lacour, F. De Alencastro and C. Weyermann, in *On-field chemical profiling of volatile and odorous compounds in water pollution cases: toward a forensic in situ approach*, 16th International Symposium on Olfaction and Electronic Noses (ISOEN 2015), 2015.
12. A. Horta, B. Malone, U. Stockmann, B. Minasny, T. Bishop, A. McBratney, R. Pallasser and L. Pozza, Potential of integrated field spectroscopy and spatial analysis for enhanced assessment of soil contamination: A prospective review, *Geoderma*, 2015, **241**, 180–209.
13. J. Chapman, R. Orrell-Trigg, K. Y. Kwoon, V. K. Truong and D. Cozzolino, A high-throughput and machine learning resistance monitoring system to determine the point of resistance for *Escherichia coli* with tetracycline: Combining UV-visible spectrophotometry with principal component analysis, *Biotechnol. Bioeng.*, 2021, **118**(4), 1511–1519.
14. A. Shaw, M. Winson, A. Woodward, A. McGovern, H. Davey, N. Kaderbhai, D. Broadhurst, R. Gilbert, J. Taylor and E. Timmins, Rapid analysis of high-dimensional bioprocesses using multivariate spectroscopies and advanced chemometrics. *Bioanalysis and Biosensors for Bioprocess Monitoring*, 1999, pp. 83–113.

15. A. de Juan and R. Tauler, Chemometrics applied to unravel multicomponent processes and mixtures: Revisiting latest trends in multivariate resolution, *Anal. Chim. Acta*, 2003, **500**(1–2), 195–210.
16. J. Li, Y. Tong, L. Guan, S. Wu and D. Li, Optimization of COD determination by UV-vis spectroscopy using PLS chemometrics algorithms, *Optik*, 2018, **174**, 591–599.
17. D. Cozzolino, M. F. Dupont, A. Elbourne, V. K. Truong, A. Power and J. Chapman, Role of sensors in fruit nutrition, in *Fruit Crops*, Elsevier, 2020, pp. 111–119.
18. D. Cozzolino, A. Power and J. Chapman, Interpreting and Reporting Principal Component Analysis in Food Science Analysis and Beyond, *Food Anal. Methods*, 2019, **12**(11), 2469–2473.
19. D. Cozzolino and J. Roberts, Applications and developments on the use of vibrational spectroscopy imaging for the analysis, monitoring and characterisation of crops and plants, *Molecules*, 2016, **21**(6), 755.
20. H. K. Rajendran, M. A. D. Fakrudeen, R. Chandrasekar, S. Silvestri, M. Sillanpää and V. C. Padmanaban, A comprehensive review on analytical and equation derived multivariate chemometrics for the accurate interpretation of the degradation of aqueous contaminants, *Environ. Technol. Innovation*, 2022, 102827.
21. V. D. Hoang, Chemometrics-assisted Spectrophotometric Determination of Ciprofloxacin and Naphazoline in Eye Drops, *Asian J. Res. Chem.*, 2014, 7(5), 461–465.
22. S. M. Sajjadi, Z. Asadollah-pour, S. H. Sajjadi, S. N. Nabavi, Z. Abed, F. Farzin, A. Emadi and B. Abdous, A thorough investigation of photocatalytic degradation of ortho and para-nitro phenols in binary mixtures: new insights into evaluating degradation progress using chemometrics approaches, *New J. Chem.*, 2021, **45**(29), 12974–12985.
23. A. Gałuszka, Z. M. Migaszewski and J. Namieśnik, Moving your laboratories to the field–Advantages and limitations of the use of field portable instruments in environmental sample analysis, *Environ. Res.*, 2015, **140**, 593–603.
24. S. K. Abid, N. Sulaiman, S. W. Chan, U. Nazir, M. Abid, H. Han, A. Ariza-Montes and A. Vega-Muñoz, Toward an integrated disaster management approach: how artificial intelligence can boost disaster management, *Sustainability*, 2021, **13**(22), 12560.
25. S. Cárdenas and M. Valcárcel, Analytical features in qualitative analysis, *TrAC, Trends Anal. Chem.*, 2005, **24**(6), 477–487.
26. D. Cozzolino, S. Chandra, J. Roberts, A. Power, P. Rajapaksha, N. Ball, R. Gordon and J. Chapman, There is gold in them hills: Predicting potential acid mine drainage events through the use of chemometrics, *Sci. Total Environ.*, 2018, **619–620**, 1464–1472.
27. S. J. Wenger and J. D. Olden, Assessing transferability of ecological models: an underappreciated aspect of statistical validation, *Methods in Ecol. Evol.*, 2012, **3**(2), 260–267.
28. A. S. Mikheyev and M. M. Tin, A first look at the Oxford Nanopore MinION sequencer, *Mol. Ecol. Resour.*, 2014, **14**(6), 1097–1102.

29. S. S. Johnson, E. Zaikova, D. S. Goerlitz, Y. Bai and S. W. Tighe, Real-Time DNA Sequencing in the Antarctic Dry Valleys Using the Oxford Nanopore Sequencer, *J. Biomol. Tech.*, 2017, **28**(1), 2–7.
30. K. Norman, P. Roffey and D. Royds, Application of a mobile laboratory for real time response to incidents, *Aust. J. Forensic Sci.*, 2009, **41**(1), 73–80.
31. B. Eckenrode, Environmental and forensic applications of field-portable GC–MS: an overview, *J. Am. Soc. Mass Spectrom.*, 2001, **12**(6), 683–693.
32. J. Chapman, L. Le Nor, R. Brown, E. Kitteringham, S. Russell, T. Sullivan and F. Regan, Antifouling performances of macro- to micro- to nano-copper materials for the inhibition of biofouling in its early stages, *J. Mater. Chem. B*, 2013, **1**(45), 6194–6200.
33. P. D. Taylor, M. H. Ramsey and P. J. Potts, Balancing measurement uncertainty against financial benefits: comparison of *in situ* and *ex situ* analysis of contaminated land, *Environ. Sci. Technol.*, 2004, **38**(24), 6824–6831.
34. V. Kumar, A. Sharma, R. Kumar, R. Bhardwaj, A. Kumar Thukral and J. Rodrigo-Comino, Assessment of heavy-metal pollution in three different Indian water bodies by combination of multivariate analysis and water pollution indices, *Hum. Ecol. Risk Assess.: Int. J.*, 2020, **26**(1), 1–16.
35. W. J. Shields, T. Saba, P. D. Boehm and J. Pietari, Chapter 10 – Congeners: A Forensics Analysis, ed. B. L. Murphy and R. D. Morrison, in *Introduction to Environmental Forensics*, Academic Press: San Diego, 3rd edn, 2015, pp. 347–393.
36. J. Liu, H. Kang, W. Tao, H. Li, D. He, L. Ma, H. Tang, S. Wu, K. Yang and X. Li, A spatial distribution - Principal component analysis (SD-PCA) model to assess pollution of heavy metals in soil, *Sci. Total Environ.*, 2023, **859**(Pt 1), 160112.
37. D. D. Suhr, in *Principal component analysis vs. exploratory factor analysis (paper 203–230)*, Proceedings of the thirtieth annual SAS® users group international conference, 2005, p. 30.
38. D. Baragaño, G. Ratié, C. Sierra, V. Chrastný, M. Komárek and J. R. Gallego, Multiple pollution sources unravelled by environmental forensics techniques and multivariate statistics, *J. Hazard. Mater.*, 2022, **424**, 127413.
39. C.-M. Wu, H.-T. Tsai, K.-H. Yang and J.-C. Wen, How reliable is X-ray fluorescence (XRF) measurement for different metals in soil contamination?, *Environ. Forensics*, 2012, **13**(2), 110–121.
40. P. Nowinski, V. F. Hodge and S. Gerstenberger, Application of field portable X-ray fluorescence to the analysis of desert varnish samples in areas affected by coal-fired power plants, *Environm. Chem.*, 2012, **9**(4), 379–388.
41. P. K. Davy, G. Gunchin, A. Markwitz, W. J. Trompetter, B. J. Barry, D. Shagjjamba and S. Lodoysamba, Air particulate matter pollution in Ulaanbaatar, Mongolia: determination of composition, source contributions and source locations, *Atmos. Pollut. Res.*, 2011, **2**(2), 126–137.

42. K. Goff, R. J. Schaetzl, S. Chakraborty, D. C. Weindorf, C. Kasmerchak and E. A. Bettis III, Impact of sample preparation methods for characterizing the geochemistry of soils and sediments by portable X-ray fluorescence, *Soil Sci. Soc. Am. J.*, 2020, **84**(1), 131–143.
43. G. L. Bosco, Development and application of portable, hand-held X-ray fluorescence spectrometers, *TrAC, Trends Anal. Chem.*, 2013, **45**, 121–134.
44. K. Hagiwara, T. Inui, Y. Koike, M. Aizawa and T. Nakamura, Speciation of inorganic arsenic in drinking water by wavelength-dispersive X-ray fluorescence spectrometry after *in situ* preconcentration with miniature solid-phase extraction disks, *Talanta*, 2015, **134**, 739–744.
45. K. Hagiwara, Y. Koike, M. Aizawa and T. Nakamura, On-site quantitation of arsenic in drinking water by disk solid-phase extraction/mobile X-ray fluorescence spectrometry, *Talanta*, 2015, **144**, 788–792.
46. J. Chou, G. Clement, B. Bursavich, D. Elbers, B. Cao and W. Zhou, Rapid detection of toxic metals in non-crushed oyster shells by portable X-ray fluorescence spectrometry, *Environm. Pollut.*, 2010, **158**(6), 2230–2234.
47. X. Hou, Y. He and B. T. Jones, Recent advances in portable X-ray fluorescence spectrometry, *Appl. Spectrosc. Rev.*, 2004, **39**(1), 1–25.
48. V. Panchuk, Y. Petrov, V. Semenov and D. Kirsanov, Quantification of elements in spent nuclear fuel using intrinsic radioactivity for sample excitation and chemometric data processing, *Anal. Chim. Acta*, 2023, **1239**, 340694.
49. F. F. Simán, Assessment of Machine Learning Applied to X-Ray Fluorescence Core Scan Data from the Zinkgruvan Zn-Pb-Ag Deposit, Bergslagen, Sweden, 2020.
50. S. Kogou, L. Lee, G. Shahtahmassebi and H. Liang, A new approach to the interpretation of XRF spectral imaging data using neural networks, *X-Ray Spectrom.*, 2021, **50**(4), 310–319.
51. S. Aidene, M. Khaydukova, G. Pashkova, V. Chubarov, S. Savinov, V. Semenov, D. Kirsanov and V. Panchuk, Does chemometrics work for matrix effects correction in X-ray fluorescence analysis?, *Spectrochim. Acta, Part B*, 2021, **185**, 106310.
52. A. Turner and K. R. Solman, Analysis of the elemental composition of marine litter by field-portable-XRF, *Talanta*, 2016, **159**, 262–271.
53. A. Turner, In situ elemental characterisation of marine microplastics by portable XRF, *Mar. Pollut. Bull.*, 2017, **124**(1), 286–291.
54. M. Cooper, P. Mijnarends, N. Shiotani, N. Sakai and A. Bansil, *X-ray Compton scattering*, OUP Oxford, 2004, vol. 5.
55. M. J. Farquharson and K. Geraki, The use of combined trace element XRF and EDXRD data as a histopathology tool using a multivariate analysis approach in characterizing breast tissue, *X-Ray Spectrom.*, 2004, **33**(4), 240–245.
56. R. S. Robin, R. Karthik, R. Purvaja, D. Ganguly, I. Anandavelu, M. Mugilarasan and R. Ramesh, Holistic assessment of microplastics in various coastal environmental matrices, southwest coast of India, *Sci. Total Environ.*, 2020, **703**, 134947.

57. A. Askarzadeh, A novel metaheuristic method for solving constrained engineering optimization problems: crow search algorithm, *Comput. Struct.*, 2016, **169**, 1–12.
58. W. Wu and Y. Chen, Application of isolation forest to extract multivariate anomalies from geochemical exploration data, *Global Geol.*, 2018, **21**(1), 36–47.
59. E. A. Silva, D. C. Weindorf, S. H. Silva, B. T. Ribeiro, G. C. Poggere, T. S. Carvalho, M. G. Goncalves, L. R. Guilherme and C. Nilton, Advances in tropical soil characterization *via* portable X-ray fluorescence spectrometry, *Pedosphere*, 2019, **29**(4), 468–482.
60. H. W. Borchers, Pracma: Practical Numerical Math Functions, R-package version 2.3.3, 2021, accessed 2022.
61. M. M. Breunig, H.-P. Kriegel, R. T. Ng and J. Sander, in *LOF: identifying density-based local outliers*, Proceedings of the 2000 ACM SIGMOD international conference on Management of data, 2000, pp. 93–104.
62. E. M. Höppener, S. Shahmohammadi, L. A. Parker, S. Henke and J. H. Urbanus, Classification of (Micro) Plastics Using Cathodoluminescence and Machine Learning, *Talanta*, 2023, **253**, 123985.
63. M. B. B. Guerra, B. T. Teaney, B. J. Mount, D. J. Asunskis, B. T. Jordan, R. J. Barker, E. E. Santos and C. E. G. Schaefer, Post-catastrophe analysis of the Fundão Tailings dam failure in the Doce River system, Southeast Brazil: Potentially toxic elements in affected soils, *Water, Air, Soil Pollut.*, 2017, **228**(7), 1–12.
64. S. Serranti, G. Capobianco, S. Malinconico and G. Bonifazi, Micro X-ray fluorescence imaging coupled with chemometrics to detect and classify asbestos fibers in demolition waste, *Detritus*, 2020, **12**, 150–159.
65. J. A. Contreras, J. A. Murray, S. E. Tolley, J. L. Oliphant, H. D. Tolley, S. A. Lammert, E. D. Lee, D. W. Later and M. L. Lee, Hand-portable gas chromatograph-toroidal ion trap mass spectrometer (GC-TMS) for detection of hazardous compounds, *J. Am. Soc. Mass Spectrom.*, 2008, **19**(10), 1425–1434.
66. P. E. Leary, G. S. Dobson and J. A. Reffner, Development and applications of portable gas chromatography–mass spectrometry for emergency responders, the military, and law-enforcement organizations, *Appl. Spectrosc.*, 2016, **70**(5), 888–896.
67. R. F. Kranenburg, D. Peroni, S. Affourtit, J. A. Westerhuis, A. K. Smilde and A. C. van Asten, Revealing hidden information in GC–MS spectra from isomeric drugs: Chemometrics based identification from 15 eV and 70 eV EI mass spectra, *Forensic Chem.*, 2020, **18**, 100225.
68. D. T. Snyder, C. J. Pulliam, Z. Ouyang and R. G. Cooks, Miniature and Fieldable Mass Spectrometers: Recent Advances, *Anal. Chem.*, 2016, **88**(1), 2–29.
69. G. L. Alexandrino, G. Tomasi, P. G. M. Kienhuis, F. Augusto and J. H. Christensen, Forensic Investigations of Diesel Oil Spills in the Environment Using Comprehensive Two-Dimensional Gas Chromatography–High Resolution Mass Spectrometry and Chemometrics: New Perspectives

in the Absence of Recalcitrant Biomarkers, *Environ. Sci. Technol.*, 2019, **53**(1), 550–559.
70. G. L. Alexandrino, J. Malmborg, F. Augusto and J. H. Christensen, Investigating weathering in light diesel oils using comprehensive two-dimensional gas chromatography–High resolution mass spectrometry and pixel-based analysis: Possibilities and limitations, *J. Chromatogr. A*, 2019, **1591**, 155–161.
71. D. P. Oliver, D. A. Navarro, J. Baldock, S. L. Simpson and R. S. Kookana, Sorption behaviour of per- and polyfluoroalkyl substances (PFASs) as affected by the properties of coastal estuarine sediments, *Sci. Total Environ.*, 2020, **720**, 137263.
72. M. Hemida, A. Ghiasvand, V. Gupta, L. J. Coates, A. A. Gooley, H.-J. Wirth, P. R. Haddad and B. Paull, Small-Footprint, Field-Deployable LC/MS System for On-Site Analysis of Per- and Polyfluoroalkyl Substances in Soil, *Anal. Chem.*, 2021, **93**(35), 12032–12040.
73. D. Bylund, A. Bergens and S. P. Jacobsson, Optimisation of chromatographic separations by use of a chromatographic response function, empirical modelling and multivariate analysis, *Chromatographia*, 1997, **44**(1–2), 74–80.
74. J. C. Berridge, Unattended optimisation of reversed-phase high-performance liquid chromatographic separations using the modified simplex algorithm, *J. Chromatogr. A*, 1982, **244**(1), 1–14.
75. D. Cozzolino, Infrared methods for high throughput screening of metabolites: food and medical applications, *Comb. Chem. High Throughput Screening*, 2011, **14**(2), 125–131.
76. M. Didham, V. K. Truong, J. Chapman and D. Cozzolino, Sensing the Addition of Vegetable Oils to Olive Oil: The Ability of UV-vis and MIR Spectroscopy Coupled with Chemometric Analysis, *Food Anal. Methods*, 2019, 1–7.
77. O. Abbas, A. Pissard and V. Baeten, Near-infrared, mid-infrared, and Raman spectroscopy, in *Chemical Analysis of Food*, Elsevier, 2020, pp. 77–134.
78. J. M. Chalmers, H. G. Edwards and M. D. Hargreaves, *Infrared and Raman Spectroscopy in Forensic Science*, 2012.
79. K. Ashley, Field-portable spectroscopy, *Appl. Occup. Environ. Hyg.*, 2003, **18**(1), 10–15.
80. D. Levy and E. G. Diken, Field identification of unknown gases and vapors *via* IR spectroscopy for homeland security and defense, *IEEE Sens. J.*, 2010, **10**(3), 564–571.
81. L. J. Robinson, A. H. Jean Robertson, L. A. Dawson and A. M. Main, *In Situ* FTIR Analysis of Soils for Forensic Applications, *Spectroscopy*, **2015**, **30**, 22–30.
82. D. Sorak, L. Herberholz, S. Iwascek, S. Altinpinar, F. Pfeifer and H. W. Siesler, New developments and applications of handheld Raman, mid-infrared, and near-infrared spectrometers, *Appl. Spectrosc. Rev.*, 2012, **47**(2), 83–115.

83. R. Mukhopadhyay, *Product review: portable FTIR spectrometers get moving*, ACS Publications, 2004.
84. K. Carron and R. Cox, *Qualitative analysis and the answer box: a perspective on portable Raman spectroscopy*, ACS Publications, 2010.
85. S. K. Sharma, A. K. Misra and B. Sharma, Portable remote Raman system for monitoring hydrocarbon, gas hydrates and explosives in the environment, *Spectrochim. Acta, Part A*, 2005, **61**(10), 2404–2412.
86. W. Windig, Spectral data files for self-modeling curve resolution with examples using the Simplisma approach, *Chemom. Intell. Lab. Syst.*, 1997, **36**(1), 3–16.
87. J. Moreau and E. Rinnert, Fast identification and quantification of BTEX coupling by Raman spectrometry and chemometrics, *Analyst*, 2015, **140**(10), 3535–3542.
88. C. Duan, J. Li, Y. Zhang, K. Ding, X. Geng and Y. Guan, Portable instruments for on-site analysis of environmental samples, *TrAC, Trends Anal. Chem.*, 2022, **154**, 116653.
89. S. Pakhomova, I. Zhdanov and B. van Bavel, Polymer type identification of marine plastic litter using a miniature near-infrared spectrometer (MicroNIR), *Appl. Sci.*, 2020, **10**(23), 8707.
90. F. Sales, M. Callao and F. Rius, Multivariate standardization techniques using UV-vis data, *Chemom. Intell. Lab. Syst.*, 1997, **38**(1), 63–73.
91. F. Antonucci, P. Menesatti, N. M. Holden, E. Canali, S. Giorgi, A. Maienza and S. R. Stazi, Hyperspectral Visible and Near-Infrared Determination of Copper Concentration in Agricultural Polluted Soils, *Commun. Soil Sci. Plant Anal.*, 2012, **43**(10), 1401–1411.
92. T. M. Karlsson, H. Grahn, B. van Bavel and P. Geladi, Hyperspectral Imaging and Data Analysis for Detecting and Determining Plastic Contamination in Seawater Filtrates, *J. Near Infrared Spectrosc.*, 2016, **24**(2), 141–149.

CHAPTER 11

Practitioner Aspects

K. PITTS

Forensic Science Laboratory, ChemCentre, Australia
Email: KPitts@chemcentre.wa.gov.au

11.1 Introduction

Forensic science is the application of multidisciplinary scientific principles, techniques, and methods to questions within the justice system. *Forensic practitioner* is the typical name for those within the area that specialise in the processing of casework, in contrast to those within academic or purely research-focussed roles. This 'real-world' focus means that forensic practitioners must meet the exacting standards of the justice system, as well as the scientific one. In the justice system, facts and statements are often phrased in a definitive manner; guilty or not guilty, yes or no. In some cases, the scientific system is able to meet these expectations – but there are complications in science, such as inherent uncertainties due to measurements. This means that a scientist can never truly say that something is known with 100% certainty, as although a result or potential sequence of events may be highly improbable, it remains in the realms of the possible. Within this and other scientific constraints, it is the forensic practitioner's job to assist the court in analysing evidence and interpreting the results in a robust, objective, and reproducible manner.

The field of forensic science has evolved over the years to include various sub-disciplines, many of which include the collection of chemical data. Chemometrics is consequently applicable to many areas of forensic science, including data for body fluid determination within forensic biology, species identification of skeletal remains within forensic anthropology, and data collection within forensic chemistry. Forensic chemistry involves the analysis of chemical evidence (such as fibres, paints, soils, glass) with the typical

Chemometric Methods in Forensic Science
Edited by Vishal Sharma, Georgina Sauzier and Simon W. Lewis

Published by the Royal Society of Chemistry, www.rsc.org

purpose of determining its composition and origin. The analysis of chemical evidence is a complex process that involves the use of various analytical techniques including modes of chromatography, mass spectrometry, and spectroscopy. Chemometrics is one tool in a forensic practitioner's arsenal to manage the complex datasets generated through these techniques. The application of mathematical and statistical methods to chemical data can identify hidden patterns and trends, extract meaningful information, or highlight relevant areas warranting further focus.

Previous chapters have illustrated that chemometrics can be used to analyse a variety of different types of evidence, including biological samples, illicit drugs, ignitable liquids and explosives, lubricants, and many others. The use of chemometrics by forensic practitioners can prove invaluable to support or refute hypotheses and assist the practitioner in the interpretation of scientific results for the justice system. As with any tool, chemometrics also has limitations that must be considered. This chapter will focus on the aspects and challenges of chemometrics in forensic science from a practitioner perspective, including a review of available software, the integration of chemometrics with casework (including the presentation of chemometric-related evidence in court), and the place of chemometrics within existing frequentist and Bayesian frameworks of evidence evaluation.

11.2 Chemometric Use by Practitioners

There are a wide range of different techniques used in chemometrics, many of which have been well discussed in previous chapters. Practitioners must balance the capabilities of research with the demands of the justice system. As such, any techniques used must be rigorously validated prior to use on evidential items – a process that requires considerable time and labour. This means that forensic practitioners may be slower to incorporate new techniques or applications; not because they are dismissive, but because they are oftentimes working against case backlogs and timeframe demands from the courts. However, practitioners always seek the truth and are oftentimes keen to develop capabilities that further enhance the evidential analysis they have undertaken. Once such area of growth in recent times is the interpretation of evidential data using chemometric techniques. After (or during) appropriate pre-processing of data, this can be thought of as being applicable to three areas of data analysis in a forensic practitioner's evaluation: visualisation, classification, and comparison.[1]

For data visualisation, one of the most commonly used chemometric techniques by forensic practitioners is principal component analysis (PCA). PCA is an unsupervised technique used to identify patterns in large data sets by identifying the variables that contribute the most to the variation in the data set. PCA is commonly used by practitioners to explore structure or trends in databases or reference collections to gain an understanding of the variability within the population being examined. This information may be particularly relevant, for example, when determining if an association between two items can be considered compelling evidence based on the 'distinctiveness' of the items in question. Hierarchical cluster analysis (HCA) may be used as an alternative or complementary visualisation tool, depending on the application.

Visualisation is typically followed by more 'application ready' classification techniques, such as linear discriminant analysis (LDA). Discriminant analysis (DA) is a class of supervised techniques used to identify and model the variables that best discriminate between two or more groups of samples. Regression analysis is also a common chemometric application used by practitioners, most often in the form of partial least squares regression (PLSR). Regression analysis is used to relate a dependent variable to one or more independent variables. In forensic science, regression analysis is typically used to quantify the relationship between instrumental signals and chemical concentrations. Along with discriminant analysis, regression can assist in the classification of data to workable groups, ready for comparison.

Finally, the forensic practitioner may use chemometrics for comparison applications. After all, in many trace evidence cases, the ultimate question is one of a questioned exhibit compared to a known source or activity. For this purpose, methods that generate distance measures (such as Canberra, Pearson or Euclidean) from exhibit data will best allow for the evaluation of 'shared source' propositions and aid in the interpretation of the significance of the results.

Overall, regardless of which chemometric technique(s) are chosen for use, datasets and interpretational protocols must undergo rigorous evaluation before implementation by a practitioner. In many cases, these protocols and evaluations have resulted from the focus of academic research. The main benefit of this is the scientific evaluation of the chemometric methods as part of a peer-review publication process, allowing for strong academic evaluation, followed by casework specific practitioner evaluations.

11.3 Review of Available Software

There are several software programs available for chemometric analysis in forensic science. Each of these programs has its strengths and weaknesses, and the choice of software will depend on the specific needs of the user as well as considerations such as ease of use, transparency of data analysis and also financial considerations. Many practitioners are familiar with the simple data analyses tools available within Microsoft Excel but move on from the slightly more advanced XLSTAT to more advanced statistical programs. Whichever program chosen, the software used is designed to help forensic scientists analyse complex datasets in a more efficient and accurate manner. Some of the most commonly used software packages available for use within forensic science include MATLAB, JMP, R, Python, and Unscrambler. Table 11.1 summarises the key points of comparison between packages, but each are discussed in detail below.[2]

Table 11.1 Summary of available software for chemometrics for forensic practitioners. Adapted from ref. 2 with permission from the Royal Society of Chemistry.

Software name (and supplier) *OS compatibility*	Licensing options and indicative pricing (single year in USD)	Key features and user-relevant notes
MATLAB (Mathworks, USA) *Windows, macOS, Linux*	Annual or perpetual licenses MATLAB: \$815 Mathworks Toolboxes: \$405–\$470 PLS Toolbox (Eigenvalue Research): \$2095	Most chemometric functionalities with MATLAB require a separate Toolbox. Most pattern recognition, regression and experimental design methods area available within the Statistics and Machine Learning Toolbox. PLS Toolbox allows expanded options including pre-processing.
JMP (SAS, Cary USA) *Windows, macOS*	Single, multi-user and corporate plans (\$1080)	Strong visual and data analysis features, including data pre-processing, pattern recognition, experimental design and regression. Routine work can be simplified but software shows flexibility for more complex requirements.
R (R Core Team) *Windows, macOS, Linux*	Open source (no cost other than time)	Essentially unlimited applications as new functionalities can be coded and shared within community. However, novel applications require coding and new applications and code packages must be rigorously tested prior to use within casework. No official support provided.
Python *Windows, macOS, Linux*	Open source (no cost other than time)	Similar to R in features, not as widespread so available packages not as extensive. Additionally, coding experience is valuable.
The Unscrambler X (AspenTech, USA) *Windows only*	Single and multi-users, but limited options (\$1500+)	Base software offers a broad range of data pre-processing, pattern recognition and regression methods. Versions 10.4+ include Design-Expert (Stat-Ease, USA) add-in allowing experimental design.
SIMCA (MKS/Umetrics, Sweden) *Windows only*	Single-user license with annual maintenance support fee: \$1720	Basic pattern recognition, regression and design of experiments. Novel methods are scriptable using Python.
SPSS Statistics (IBM Corporation, USA) *Windows, macOS*	Single- and multi-user. \$1020 annual, but monthly and perpetual licensing available	Basic pattern recognition and regression, not typically suited for chemometrics.

Table 11.1 (*Continued*)

Software name (and supplier) *OS compatibility*	Licensing options and indicative pricing (single year in USD)	Key features and user-relevant notes
XLSTAT (Addinsoft, Paris) *Windows, macOS*	Single- and multi-user. XLSTAT Basic: $295 XLSTAT Basic Plus: $495 (minimum 12 months) XLSTAT Premium: $1495	XLSTAT Basic version allows basic pattern recognition and regression analysis, Basic Plus expands features within these and XLSTAT Premium enables experimental design. All are limited in functionality, flexibility and maximum data size. Not recommended for chemometrics beyond simplistic applications.
Minitab (Minitab LLC, USA) *Windows, macOS*	Single- and multi-user plans (single user $1600) perpetual licensing available	Basic pattern recognition and regression, as well as a broad range of experimental design options.

MATLAB is a popular software package for chemometric analysis that is widely used in science applications. The core package includes a variety of different tools and functions for data analysis, including PLSR, PCA, and DA. Chemometrics Toolbox for MATLAB is a powerful add-on that allows users to perform a wider range of chemometric analyses, incorporating 70 specialised functions. It includes modules for data pre-processing, exploratory data analysis, additional modes of regression analysis (such as principal component regression and multiple linear regression), classification, and clustering. The program also has a user-friendly interface and can be customized to suit individual needs. However, it does represent a steep learning curve and requires a good understanding of the techniques being applied. PLS_Toolbox is another optional addition to MATLAB that is specifically designed to support PLSR. It includes modules for data pre-processing, exploratory data analysis (notably PCA), regression analysis, and validation. The program again has a user-interface with customisable outputs but is restricted to PLSR and related algorithms.

JMP (pronounced as jump) Statistical Discovery LLC (a subsidiary of SAS) is a commercial statistical software program gaining popularity with chemists and associated forensic applications. With a focus on engineering and science data, it includes an extensive range of statistical and visualisation tools. These can be customised to suit individual needs and JMP is particularly noted for the ability to create good-quality, customised visual representations. One strong point in support of JMP is the publication of a white paper with each version (currently JMP 17) detailing the comparison with the National Institute of Standards and Technology Statistical Reference Data sets, with reference to numerical accuracy. This publication aids in the quality control management requirements of software validation for use within forensic laboratories and enhances assurance in the product.

R is an open-source programming language that is gaining strong popularity for data analysis in a variety of different fields, including forensic science. In its infancy, R required advanced coding skills, making it out of reach for many practitioners. However, with the development of an extensive online community who create and share 'packages', the need for coding experience is diminished. The Comprehensive R Archive Network (CRAN) open-source forums now include a variety of different packages for chemometric analysis, including several that enable high-quality visual representation of generated models. A typical workspace within R has a few different windows (see Figure 11.1). Coded commands can be written and saved in the code editor before being subsequently run to the R console that will execute the commands and display the results. Variables and objects making up the workspace are recorded in the environment, whilst visual representations of results will be shown in the files/plots/packages window.

Despite the requirement for some basic coding knowledge, R has a number of advantages for forensic scientists over commercial packages. Because the R console shows the evaluation and execution of all input commands, its analyses are relatively transparent. This allows practitioners to demonstrate the exact mathematical processes that have been carried out, and to more readily examine loadings, variable influences, and impacts of particular decisions on outcomes. R is also highly extensible due to new packages being frequently added, and has several packages for data visualisation such as ggplot2 and lattice that give the user greater control over how results are represented. This is valuable when optimising visuals for jury presentations or scientific publications. However, given the shared and unaccredited nature of these packages, extensive validation is required of new and novel applications of the software prior to use within forensic casework.

Python is a general-purpose open-source programming language that is also gaining popularity in scientific computing and data analysis. There are a range of examples of its use in geotechnical data analysis, geology, physics and astronomy, although not all include chemometric techniques. It includes a variety of different libraries and packages for chemometric analysis, including Scikit-learn, pandas, NumPy (also compatible with R), and SciPy. Scikit-learn is a machine learning library that provides functions for various clustering, classification and regression methods including support vector machines (SVMs) and random forests (RFs). It is also designed to interface with NumPy, which allows rapid computation of numerical matrices, and SciPy, intended for scientific computing. Pandas meanwhile provides a library for data manipulation and analysis, particularly for tabular data such as the outputs from chemometric methods. Like R, Python also allows customisation of visual modelling and diagrams. Compared to R, Python has an easier to read syntax that users may find more straightforward to learn, while also producing more 'robust' code that

Figure 11.1 A typical RStudio workspace for undertaking chemometric analysis. This workspace consists of four windows: the code editor (upper left), R console (lower left), environment (upper right), and files/plots/packages (lower right). Note that the order and shapes of these windows may be customised by the user. Image credit T. Newland.

requires less maintenance to keep functional. On the other hand, it is a general-purpose software that is not optimised for data visualisation and statistical analysis in the same way that R is.

Camo (now AspenTech) Unscrambler X is a commercial software package specifically designed for multivariate data analysis. It was first oriented towards users of synchrotron data, though many of these users have now moved away to other packages. It includes a variety of different tools and functions for data pre-processing and chemometric analysis, including clustering, partial least squares (PLS), and soft independent modelling of class analogy (SIMCA). The program has a fairly user-friendly interface with comprehensive in-built tutorials, and Version 10.4 was co-bundled with the Stat-Ease Design Expert software for experimental design. Graphics in Unscrambler X are customisable but can be limited in visual output options compared with JMP, R or Python. It also represents a far higher cost compared to the open-source alternatives (or indeed many other commercial software packages), with single-user licenses no longer available for academics. Additionally, technical support has been found to be very limited with the shift to AspenTech.

Other software programs that are may be familiar to scientists but aren't commonly used within forensic chemometric applications include SIMCA (not to be confused with the chemometric technique), SPSS, Minitab and the previously mentioned XLSTAT. SIMCA is a multivariate data analysis software package that includes a variety of different tools and functions for chemometric analysis, including PCA, PLS, and discriminant analysis. It is capable of running R or Python scripts, although its range of techniques is limited in comparison to some other software packages. SPSS is another commercial statistical software program designed to rapidly extract insights from data. It can be customised and has a range of univariate statistical tools but is not as extensive in its chemometric tools and seems to be more suited to social sciences than forensic chemistry applications. Minitab is similar to SIMCA in that its features can be integrated with open-source languages R or Python, and unlike SIMCA it also enables design of experiments. However, its applications also tend to be more manufacturing and business-focussed rather than academic or chemical.

Overall, the choice of software is a matter of personal and organisational preference. No matter the choice, the software must be rigorously tested to ensure user capability and sound performance prior to use in casework. Subtle differences caused by the order of processes or factors chosen within the software should be explored. Many software packages also offer machine learning, an emerging field showing promise to enhance many statistical analyses. Like with many things though, caution must remain at the forefront due to the potential to inherited bias from training data sets and those 'teaching' the machinations. No software should be accepted as a 'black-box' or 'thing that goes ping' to be trusted implicitly, but rather should be used as a transparent and well-understood tool to facilitate clear, reproducible, and reliable decision-making.

11.4 Integrating Chemometrics Within Casework

Previous chapters have demonstrated that the use of chemometrics in forensic science can provide valuable insights and increase the objectivity of analyses. However, successful application of these methods in research does not automatically translate to operationalisation in casework. To meet the exacting standards of the justice system, the use of chemometrics within real cases needs to be rigorously evaluated. This necessitates using appropriate statistical methods, validating the results obtained, and ensuring that the sample preparation and analysis procedures are standardised and reproducible. In general terms, the following aspects of chemometrics are applicable to practitioners: data acquisition and processing, feature extraction, classification, visualisation and interpretation.

Data acquisition is the first step in any analytical process, including chemometric analyses. Forensic scientists must ensure that suitable sample selection and preparation occurs, producing data that is representative, reliable, and of sufficient quality for chemometric analysis. The choice of instrumentation and the specific parameters of those analytical methods are critical to ensuring suitable data acquisition. In this manner, chemometric analysis, particularly experimental design, can be used to ensure the optimal methodology for reliable, robust and reproducible acquisition.

After data acquisition comes data processing. This may involve several steps, including data cleaning, outlier detection and normalisation. Data cleaning involves the removal of incomplete or inconsistent data; a process that can be assisted through chemometric-supported algorithms. Outlier detection involves the identification and removal of data points that deviate significantly from the mean. Pattern recognition methods such as PCA can assist in identifying outliers by visually representing the overall structure of the data, and several chemometric software packages also have built-in outlier detection for generated models. Normalisation involves scaling the data to eliminate differences in measurement units or acquisition effects such as instrument drift. This capability is standard in many chemometric software programs, whilst shared packages exist for use within open-source platforms such as R or Python.

Once the data has been processed appropriately, extracting relevant features from the data is possible. This involves using chemometric techniques to first identify patterns and trends in data relevant to the case at hand (unsupervised analysis), followed by classifying the data (supervised analysis). This classification especially involves examining how a case exhibit fits (or does not fit) within a specific group. This may be in order to undertake source determination, quantitate a purity, or aid in evaluating evidential support for a proposed scenario. As part of this, it is important to present the results of the chemometric analysis in a format that is useful both for forensic scientists to interpret and for law enforcement or justice officials to understand. This may involve creating visualisations of the data, such as scatter plots or heatmaps, that help highlight patterns or interpret trends in the data.

In addition to the need for rigorous evaluation, casework differs from research in that the analyst may not have time and/or resources to devote to long, large data acquisition schemes. Similarly, they often have little control over the condition of exhibits that are potentially non-representative of the true population. With this in mind, practitioners may rely on previously acquired datasets, such as reference collections or databases for populations that have been well studied before implementation, to aid in interpretation of case results. Ensuring an adequate overlap between the exhibits and the database or reference collection then becomes paramount to justifying its use.

One example of the overlap between databases and casework is the Paint Data Query (PDQ) library maintained by the Royal Canadian Mounted Police.[3,4] This represents a vast collection of automotive samples, estimated at over 85 000 layers, collected over decades. They have strict sample submission criteria and practitioners contribute by submitting samples collected from automotives with known histories from within their jurisdictions. PDQ is very useful for identifying potential candidate vehicles in cases such as hit-and-runs but is dominated by North American vehicles due to the distribution of contributors. Additionally, it predominantly uses a text-based search system, although recent research efforts have sought to undertake a more statistical comparison by utilising algorithms to assist in determining associations between questioned samples and submitted references.[5,6] The library of spectra is still a valuable resource to jurisdictions outside of North America, but it does not represent the true population of vehicles in those jurisdictions. By potentially combining this library with data from vehicle licensing centres and locally focussed collections using chemometric assessment tools, a more representative population can be established.[7–10] This can then be used for case interpretation, contributing more realistic frequencies to evaluative reporting, including the more advanced evaluation of activity level reporting.

11.4.1 Example of Chemometric Application Within Casework

The following casework example illustrates how chemometrics may be applied by forensic practitioners, with consideration given to the challenges outlined above. In this scenario, the use of an already established database and chemometric model was used to aid in the interpretation of evidence. The focus of the case was on soil; specifically soil from a precise site of a potential drug concealment plot. The forensic analysis involved the comparison of soil from recovered illicit material to the proposed burial site. It should be noted that this investigation took place in Western Australia, where the majority of soils are heavily leached quartz sands with minimal deviations in their mineral compositions.

11.4.1.1 Case Circumstances

As part of an ongoing investigation, detectives from the local police force conducted a vehicle stop, locating approximately 1.5 kg of alleged illicit

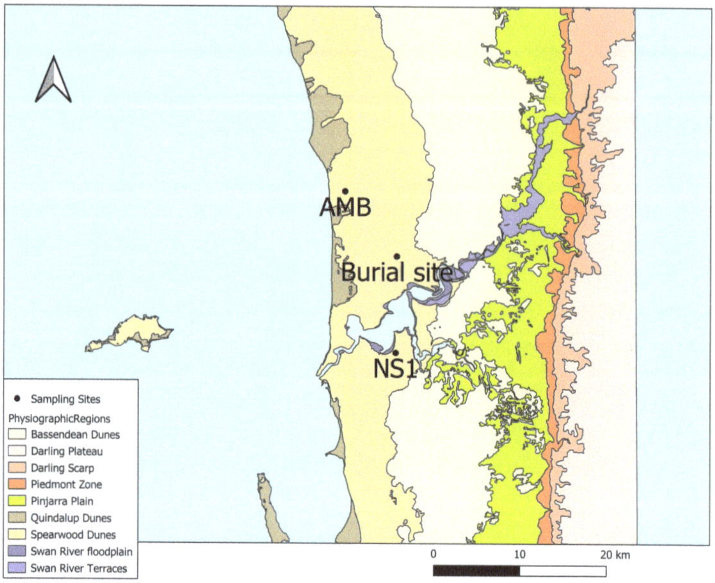

Figure 11.2 Map showing location of alleged burial site as well as the two closest sites from the subsequent PCA (see Figure 11.3). The coloured areas represent physiographic regions, *i.e.*, areas with different general soil characteristics. Image prepared by authors using physiographic regions data from Gozzard.[14]

drugs in the vehicle. Within the outer plastic bag containing the drugs was an amount of soil. Additionally, in a second vehicle stop in the same operation, another plastic bag containing illicit drugs with adhering soil was seized. After investigation, it was alleged that the offenders were burying the material at the grounds of a local soccer club, denoted as 'burial site' (see Figure 11.2). A total of three samples were taken from the garden bed of the soccer club and submitted, along with the two samples seized from the plastic bags containing the illicit material for comparison.

Upon receipt, a representative sub-sample of the quartz recovered fine fraction (<20 μm) was obtained and analysed using X-ray diffraction (XRD). This allowed the determination of the relative abundance of the mineral species in the soils. This was then compared with already modelled data from previous research, consisting of soils of the surrounding area that had been examined using the same XRD preparation to evaluate variations in the quartz grain coatings.[11–14]

11.4.1.2 Chemometric Analysis

The net intensities of reflections for characteristic minerals identified in the quartz fine fractions were normalised as a percentage of the sum of their intensities.[13] The resulting processed data for the case samples were then projected onto a previously constructed PCA plot representing lab-based reference samples from the surrounding area (see Figure 11.3).

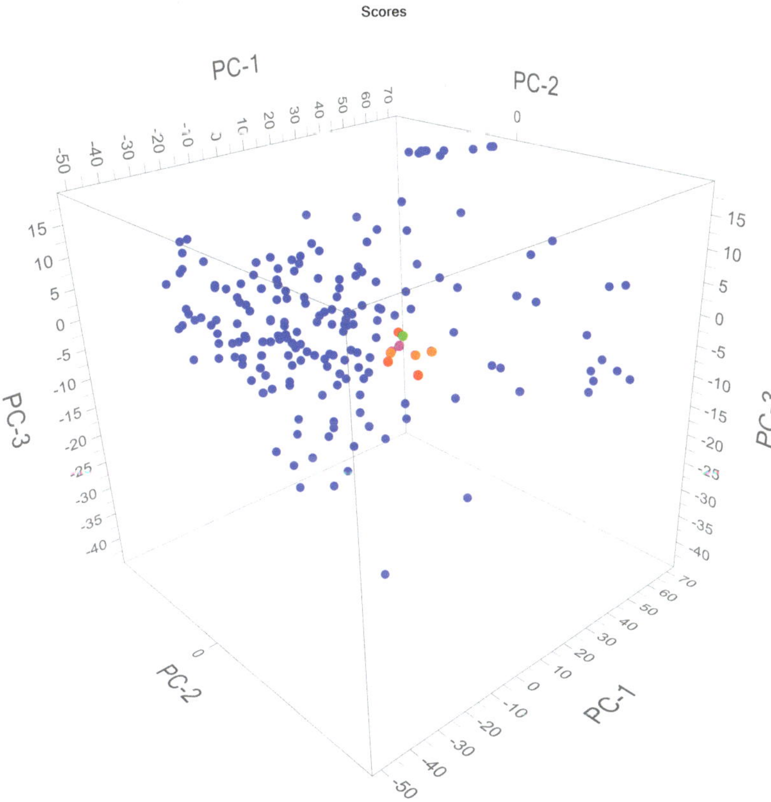

Figure 11.3 Projection of the five case samples, consisting of two recovered from seized illicit drug packaging (pink) and three control samples collected from the suspected concealment location on the grounds of a local soccer club (red), onto a previously generated PCA scores plot of known soils from the surrounding metropolitan area, relevant to the case at hand (blue). The two closest known soils within the PCA are NS1 (green) and AMB (orange).

There were several immediate conclusions made from the chemometric output. First, the three control samples were clustered closely together in the model space, indicating shared characteristics and that soil composition throughout the garden bed is relatively homogenous. Additionally, the two samples seized from the plastic bags containing illicit drug material were also close in the overall sample space to the control samples taken from the garden bed, where the illicit material was suspected to have been concealed. There were other samples in the existing reference collection that were also identified as being similar to the case samples; the first being NS1. This reference soil was from a site in a similar longitudinal line to that of the control site in the present case but residing on the other side of the river. The next closest in the PCA projection were two samples of yellow-brown sand from a northern suburb of the city (AMB), but these were slightly separated along PC2 to the group from this case.

11.4.1.3 Case Conclusions

The chemometric results suggested that the two soil samples recovered from the plastic bags originated from a site with quartz fine fraction characteristics corresponding to that of the garden bed samples. This in turn supported the proposition that the soil found in the bags containing illicit materials originated from the garden bed of the soccer club. It is worth noting that the chemometric assessment of this case was undertaken as a supporting exercise after the defendants had pleaded guilty, but prior to this information being released to the analyst to avoid unduly influencing interpretation of the output.

11.5 Presentation in Court

It should go without saying that the presentation of results (such as those obtained in the case above) to the court must be reliable, robust, and accurate. The justice system is the ultimate trier of fact but cannot be expected to be all-knowing and can be led astray by poor science or communication thereof. Any laboratory working within the forensic science space must have a strong, well-designed quality management system that recognises inherent limitations and works to minimise the impact of uncertainties on their results. Maintaining equipment calibrations, ensuring the traceability and accuracy of standards, controls, methods and reference materials, and an overall culture of accepting nothing less than the best is imperative. The same applies to the use of chemometrics in forensic cases and their presentation in court. The presentation of chemometrics results in court can be challenging, as the concepts and terminology used in chemometrics may be unfamiliar to non-experts, including judges and jury members. Hence, it is essential to present the results in a clear and concise manner that can be easily understood and not overstate or understate the value of the evidence.

Probably the most common approach to presenting chemometrics results in court is to use visual aids, such as scores plots from a PCA, to illustrate the key findings. However, these plots may allow jury members to infer erroneous conclusions and should always be strongly evaluated prior to use. Breaking down the process by first using simpler analogies to demonstrate how the technique works, as shown in a recent presentation using a cookie example to describe PCA,[15] means that juries can be led through the process, highlighting decision points along the way and then expanding the application to the matter at hand. It is therefore essential the scientist understands, and is able to provide a clear explanation of, the statistical methods used and the assumptions underlying the analysis, as well as any limitations. It is for this reason that many practitioners undertake training in science communication, which is a valuable but underrated skill in any expert witness testifying in court. The ability to clearly explain, in simple terms, advanced scientific and statistical processes is vital to ensure evidential value is optimised.

By having a strong understanding of the techniques used, reflecting not just the instrumentation but the data analysis too, the expert witness can guard against errors of interpretation, such as the well documented prosecutor's fallacy where conditional probabilities are erroneously reversed. There have been many documented cases where a misinterpretation of statistics has led to miscarriages of justice and chemometrics is no different in this risk.[16–18] The use of chemometrics in casework enhances objectivity in interpretation but only when it is used appropriately and explained by an expert with sufficient knowledge as to its limitations.

11.6 The Place of Chemometrics Within Frequentist and Bayesian Frameworks

In many areas of forensic science, interpretational frameworks are used to aid in evaluation of evidential results and chemometrics can contribute to these frameworks in a number of ways. In general terms, frameworks tend to be frequentist or Bayesian in focus. Chemometrics can be used within both of these frameworks for statistical analysis. The choice of approach and whether the use of chemometrics is valuable will depend on the specific needs of the practitioner, as well as the nature of the evidence type, collected data and the case-specific circumstances being investigated.

Frequentist statistics is a classical approach to statistical inference that involves the use of probability theory to make inferences about the population based on the sample data. Frequentist statistical analysis is based on the notion of repeated sampling of a random event, where the probability of observing a particular result is based on the frequency of that result in repeated experiments. A probability of 0 means that the event never occurs and a probability of 1 means the event always occurs. Hypothesis testing to allow inferences about the population based on the collected data is the most common frequentist approach.

A good example of this is the use of the Welsh-modified Student *t*-test to evaluate the probability that recovered fragment(s) of glass, with a measured average refractive index (RI), originated from the same population as glass from a broken window at a crime scene.[19–22] By evaluating a resulting *t*-value against tables of calculated *t*-critical numbers, a decision to accept or reject the null hypothesis (that the difference in the means of RI between recovered and control glass groups is statistically insignificant) can be made. However, frequentist glass evidence evaluations are often criticised by Bayesian approach proponents for only considering the probability under one proposition. Additionally, this approach only allows for an 'accept' or 'reject' outcome, although the use of probability density curve area can help to mitigate this. This binary choice can be seen as too arbitrary, restrictive, or even incorrect, as a failure to reject does not equate to an accept – that is, one cannot prove a null. Additionally, RI is well known to have inherent variation that is not captured in this evaluation. This may be where the glass characteristics measured are affected by heat, where recovered fragment quality is poor due to dirt or toughening, or even where the subjective grouping of the fragments as individual source(s) for comparison is performed differently between analysts.

This is even further complicated when trace elemental profiling is used in glass comparison cases. More than 18 elements, many with correlations between them, are typically analysed for relative or semi-quantitative concentrations, and this multivariate data rules out simplistic population statistical tests.[23] Chemometrics can be used to estimate the parameters of a model and test whether the model fits the data. By analysing the population relevant to the case, the frequentist requirement to assume random generation of datasets can be examined, as biases or inherent gaps in data may be highlighted. Chemometrics can also be used to identify outliers and assess the validity of the assumptions underlying the analysis. Frequentist methods are commonly used in forensic science, as they provide a clear and objective framework for statistical analysis, but the issue arises in the question of how to repeat experiments where the event cannot be repeated or are not in fact random in nature – inherent in case specific circumstances – and this is where Bayes theorem proves useful. In fact, the use of chemometrics has been proposed as a method to analyse glass populations and allow for the calculation of realistic likelihood ratios, evaluating whether a questioned and known sample could share a common origin using a Bayesian approach.[24]

Bayesian statistical analysis is based on the notion of probability as a measure of belief, where the probability of observing a particular result is based on the prior probability of that result and the evidence provided by data. The Bayesian statistical approach hence involves updating conditional probabilities based on new evidence (data) for each scenario. Bayesian methods have become almost the norm in forensic science over the last 20 years, as they provide a flexible framework for statistical analysis that can incorporate prior knowledge and uncertainty, something that is inherent in forensic circumstances but not reflected in frequentist approaches such as hypothesis testing. It allows quantitative evaluation of the strength or weight of the evidence, most often represented by a verbal scale of reporting such as no/weak/strong support for a particular pair of propositions.

This chapter will not detail the finer aspects of Bayesian statistics, as there are multiple sources readily available on the matter.[19,22,25–27] However, there are a few points that should be reinforced for the reader to assist in the implementation of chemometrics in a Bayesian interpretation. The first is that Bayesian testing relates to the evaluation of two alternative propositions in the light of conditional and related prior knowledge. In forensic science, the alternate propositions are most often referred to as the prosecution hypothesis (H_p) and the defence hypothesis (H_d) and as such the Bayesian odds-based derivation can be written as shown in Figure 11.4.

There are three terms in this scheme, simplified as the posterior odds, prior odds, and likelihood ratio. It is important to highlight at this stage the

$$\frac{Pr(H_p|E,I)}{Pr(H_d|E,I)} = \frac{Pr(H_p|\,I)}{Pr(H_d|\,I)} \times \frac{Pr(E|H_p,I)}{Pr(E|H_d,I)}$$

Posterior Odds Prior Odds Likelihood Ratio

Figure 11.4 Odds-based derivation of the evaluation of two hypotheses, in accordance with a Bayesian approach.

distinction between probability and likelihood: probability is attached to possible results; likelihood is attached to the hypotheses. The posterior odds is the ratio of the probability of hypothesis H_p, given the evidence and information (E, I); and the probability of hypothesis H_d, given the evidence and information (E, I). Overall, this represents the evaluation of the two propositions given all evidence and prior information relating to the case at hand. The second term represents the 'prior odds', the ratio of the probability of each hypothesis given purely the information. This term acts almost a measure of credibility of the propositions, before considering any supporting data. The likelihood ratio (LR) is the ratio of the probability of the observed evidence (E) given that the prosecution hypothesis (H_p) and the information provided are true, and the probability of the evidence (E) given that the defence hypothesis (H_d) and the information provided are true. Care must be taken to avoid transposing the conditionals for the LR calculations, *i.e.*, it is always evaluating the probability of the observed evidence given propositions, not the probability the propositions are true given the evidence. This is summarised as the probability of obtaining a particular set of analytical results, within the model used to produce it. In most forensic cases, the prior odds are not evaluated or known, given the tenet of presumption of innocence of the accused but may be assumed to be 1/1 in the most conservative approach. It is therefore often only the likelihood ratio that is used by practitioners to interpret the results of analysis.

In any Bayesian likelihood calculation or evaluative reporting framework, the development of mutually exclusive hypotheses can sometimes be challenging but is essential.[25] In simple source-association cases, the prosecution hypothesis may be heavily summarised as 'the questioned material originated from the same source as the known sample' whilst the defence proposition is 'there is an unrelated overlap between questioned and known samples'. The posterior odds in this instance represent the probability of a source-association between the questioned and known samples given their overlapping characteristics. Chemometrics of large-scale datasets can assist in evaluating these propositions, as it can assist in first determining the appropriate characteristics for discrimination, followed by evaluating the variability and overlap between unrelated populations.

In more complex cases, or in the application to evaluations at the activity level of interpretation, hypothesis creation can become more and more challenging, especially as a lot of questions or unknowns arise. Let's examine a theoretical example in simplistic terms and then more complex terms: a physical assault outside, where the offender violently attacked the victim from behind. We will assume for this case that the victim and the offender are unrelated, unknown to each other and have had no disclosable contact prior to the alleged incident. The evidence (or analysis findings) is that 100+ fibres, indistinguishable in characteristics from that of the offender's shirt, were recovered from the victim's shirt, with the majority specifically recovered from the back outer surface.

In source-level associations, the H_p proposition would be that all the fibres recovered originated from the shirt of the offender, whilst the defence proposition (H_d) would likely be that some other item was the source of the fibres recovered. The LR calculation would therefore be evaluated using knowledge of the frequency (occurrence) of fibres with the determined characteristics in a relevant population and the degree to which it can be differentiated from other fibres of 'unrelated sources'. Chemometric pattern recognition is useful for the second part as it can generate quantitative measures of similarity between fibres amongst the population. Both factors contribute to the probability of observing the evidence (100+ corresponding fibres) by chance.

In a true 'source-only' assessment only the background occurrence is considered, however a practitioner may also include aspects of transfer and persistence. Therefore potentially included, depending on the practitioner, are data contributing to the background probability of the 100+ specific fibres being present due to circumstances unrelated to the alleged offence. This background data may be approximated using studies that sampled foreign fibres on garments, or at a stretch, on a proxy surface such as cinema seats.[28,29] Finally, data that assists in approximating the probability of the fibres being transferred, persisting on and then being recovered by the analyst, is needed. Data related to the initial transfer of this type of fibre from the shirt and its persistence to detection (simulations such as shedding ability and persistence studies will assist here) may be included, dependent on the analyst's consideration. By calculating the likelihood ratio, one can quantify the strength of the evidence provided by the fibres in favour of $E|H_p$ (association to the known source) *versus* $E|H_d$ (association to an unknown or unrelated source). A high LR indicates that the fibres are more likely to have come from the known source (the suspect's shirt), while a low LR suggests that alternative explanations are more plausible. It is worth noting that the location of the recovered fibres may be incorporated into this assessment, provided that background data publications are detailed enough about recovery locations to allow for specificity to the 'back' of a shirt.

In comparison, for an evaluation at the activity level, the propositions would likely be H_p: the offender had violent, extended contact with the victim and H_d: the offender was not present when the victim was attacked. It is worth highlighting here the need to ensure clarity around the meaning of "contact" in a case *i.e.* a violent attack *versus* a brush-past of a stranger. The defence proposition is not often provided to a practitioner but must always be exclusionary to the prosecution proposition. This is especially difficult for complicated cases involving co-habitation or previous contact between parties or objects. In the evaluation of a proposed activity, the rarity of the activity being evaluated plays a crucial role. If the proposed activity is common, such as what occurs in everyday contact, the LR and subsequent evidential value of the examination will be lower. The evaluation of competing activities that could explain the evidence also must be undertaken. If multiple activities could result in the same type of evidence, it becomes more challenging to attribute the evidence to a specific activity. In fact, a pre-assessment to evaluate the overall benefit of undertaking an activity level evaluation is often undertaken for this reason. It is also noted here that propositions need to be carefully crafted to ensure that they do not evaluate the offence but rather the activity, ensuring not to overstep into the jurisdiction of the 'ultimate question'.

However, assuming the two propositions outlined above, the interpretation of the evidence would include the previously outlined information regarding occurrence, discrimination, background, transfer, persistence and recovery but would also require data regarding the action. Firstly, what information is available to determine what levels of fibre transfer may be estimated during extended violent physical contact, such as during a tussle.[30,31] In comparison, what information is available to determine what level of fibre transfer may result from everyday activities such as sharing an elevator, or from secondary transfer such as successive chair use[28]? Is there data that can be used to evaluate the background of the fibres? In this case, considering the context of the crime scene and the available evidence, the pool of individuals or items from which the fibres could plausibly originate would be expected to be relatively small and limited to those present during, or just prior to, the reported incident.

Regardless of the evaluation undertaken, reliable data is the key resource needed for robust evaluations of analytical results. From this perspective, chemometrics offers significant benefits to both frequentist and Bayesian interpretations of forensic evidence. In a frequentist framework, chemometrics allows for the exploration and interpretation of complex datasets, enabling forensic scientists to identify relevant variables and extract meaningful information. Chemometrics assists in quantifying the significance and reliability of evidence, enhancing the objectivity and rigour of frequentist interpretations. On the other hand, chemometrics also complements Bayesian interpretations by providing tools for modelling and understanding uncertainty in forensic evidence. Bayesian approaches incorporate prior knowledge and update it using observed data to calculate posterior probabilities. Chemometrics techniques facilitate the estimation of prior distributions, the selection of appropriate likelihood ratio evaluation parameters, and the assessment of evidence strength. By incorporating chemometrics, Bayesian interpretations gain a more comprehensive understanding of the underlying data and offer a systematic and robust approach to forensic inference.

Overall, chemometrics empowers forensic scientists in both frequentist and Bayesian interpretations by enhancing data analysis, pattern recognition, uncertainty quantification, and evidence evaluation. Its application leads to more accurate and informed interpretations of forensic evidence, thereby contributing to the advancement of forensic science and the fair administration of justice.

11.7 Barriers to Acceptance of Chemometrics Within Forensic Casework

Despite the various applications of chemometrics in forensic science, several challenges exist in its application. One of the main challenges is the need for extensive training and expertise. Chemometrics requires a high level of statistical and mathematical knowledge, and it is essential to have a thorough understanding of the chemical and physical properties of the samples being analysed. Additionally, the interpretation of results can be challenging, and it is essential to have expertise in the field of forensic science to apply chemometric methods effectively.

Another challenge is the quality and variability of the data. Forensic samples can be highly variable in their quality, and it can be challenging to obtain consistent and reliable data. However, any forensic practitioner knows that you must do your best with the sample you are given, you cannot ask a suspect to redo their action to provide a better sample! The quality of the data can also be affected by sample preparation, instrumentation, and analytical methods, which can introduce bias and errors into the analysis. A further challenge is the issue of data interpretation and presentation. Chemometric methods can generate large amounts of data, and it can be challenging to extract meaningful information from the results. It is essential to have appropriate data visualisation tools and techniques to facilitate data interpretation and presentation, as well as well-trained, skilled staff to evaluate these results. It is therefore crucial to have well-defined, established protocols for sample collection, preparation, analysis and interpretation to ensure that the best quality data arises from the limited samples submitted. This in turn allows for the best outcome possible, but, in some cases, a practitioner must be willing to indicate that the exhibit, quality or resulting data is too compromised to be reliable and that the results are inconclusive.

Chemometric methods can be time-consuming and resource intensive. The development and validation of chemometric models require significant time and effort, and the availability of appropriate software and hardware can be a limiting factor. Therefore, it is essential to have adequate resources and support to apply chemometric methods effectively in forensic science.

11.8 Future Directions

The use of chemometrics in forensic science is a rapidly evolving field. New techniques and methods are constantly being developed and their implementation into practitioner methodologies is likely. Nonetheless, there are several areas of focus where further research is needed to fully realise the potential of chemometrics in forensic science.

One area is the development and implementation of chemometric approaches for the analysis of complex samples by practitioners. Many types of forensic evidence are complex mixtures, such as illicit drugs, paint, explosive residues, or environmental pollutants. The development of chemometric approaches that can accurately identify and quantify these is a major challenge (see previous chapters in this book). Efforts in the ignitable liquid residues space is one such example of this, with complex hydrocarbon profiles being interrogated using chemometrics to identify potential ignitable liquid residues indicating arson.[32–34] Coupled to this is machine learning and artificial intelligence programs, which are emerging as important tools for interpretation.[35,36] These techniques can be used to develop more sophisticated models for data analysis and interpretation, and they may provide insights into complex data sets that are difficult to analyse using traditional statistical methods, although questions remain concerning their transparency.

Moreover, the integration of chemometric methods with other forensic techniques, such as DNA analysis and imaging, can provide a more comprehensive approach to forensic analysis. For example, chemometric methods can be used to analyse the chemical composition of fingerprints, which can be used in combination with DNA analysis to provide a more robust identification of an individual, possibly even leading to the potential for aging or sexing prints.[37–39] In another example, the combination of chemometrics with high-resolution mass spectrometry can provide even more detailed information about a sample. This may help to improve the analysis for emerging, data-heavy focusses such as proteomics.[40]

Finally, there is a need to ensure research continues into the validation and reliability of chemometric methods. Validation studies and ground-truth testing such as blind validation or pseudo-operational trials are essential to ensure that the results obtained are accurate and reliable, and that there is empirical evidence that establishes standardised validation protocols applicable across different applications and laboratories.

11.9 Conclusion

Chemometrics is a powerful tool in a forensic practitioner's competency. It can provide valuable insights into evidence and can be used in a wide range of applications, including drug analysis, toxicology, arson investigation and environmental forensics. By applying chemometrics, forensic scientists can analyse complex and diverse datasets with greater efficiency and accuracy. The ability to distinguish between different types of evidence and to identify the source of samples can improve the accuracy and reliability of forensic investigations. Furthermore, the use of chemometric methods can provide valuable information to investigators that may not be accessible through other forensic techniques.

The integration of chemometrics with casework can be challenging, and it is essential to ensure that the results obtained are reliable and accurate. This can be achieved by using appropriate sample preparation and analysis procedures, validating the results obtained, and ensuring that the interpretation approach is standardised and reproducible. The presentation of chemometrics results in court can also be challenging, and it is important to ensure that the results are presented by a skilled, knowledgeable scientist in a clear and concise manner that can be understood by non-experts.

It is also worth noting that chemometrics can play an essential role in quality control and assurance in forensic laboratories. By monitoring the performance of analytical instruments and procedures using chemometric methods, forensic laboratories can ensure that the data produced is consistent, reliable and accurate. Within the courts, the use of chemometric methods can also offer increased transparency for how an evidential assessment was determined.

The application of chemometric methods in forensic science also presents several challenges, including the need for extensive training and expertise, the quality and variability of the data, data interpretation and presentation, and time and resource requirements. Despite these challenges, chemometrics has made a significant contribution to the field of forensic science and is expected to continue to grow in the future as analytical instrumentation and computing power continue to improve. In conclusion, the use of chemometrics in forensic science has the potential to revolutionise the field by providing valuable insights into the analysis and interpretation of a wide range of evidence types. It is therefore vital to continue the exploration and expansion of applications and its potential in the realms of the practitioner.

Acknowledgements

The author would like to acknowledge Dr Georgina Sauzier (Curtin University, Perth, Australia) and Professor Simon Lewis (Curtin University, Perth, Australia) for assistance in preparing this chapter.

References

1. M. Bovens, B. Ahrens, I. Alberink, A. Nordgaard, T. Salonen and S. Huhtala, *Forensic Sci. Int.*, 2019, **301**, 82–90.
2. G. Sauzier, W. van Bronswijk and S. W. Lewis, *Analyst*, 2021, **146**, 2415–2448.
3. J. L. Buckle, D. A. Macdougall and R. R. Grant, *J. – Can. Soc. Forensic Sci.*, 1997, **30**, 199–212.
4. T. Hodgins, A. Ho and M. Sandercock, *J. Am. Soc. Trace Evidence Exam.*, 2015, **6**, 46–63.
5. B. K. Lavine, N. Mirjankar, S. Ryland and M. Sandercock, *Talanta*, 2011, **87**, 46–52.

6. B. Lavine, J. Almirall, C. Muehlethaler, C. Neumann and J. Workman, *Forensic Chem.*, 2020, **18**, 100224.
7. M. Maric, W. van Bronswijk, K. Pitts and S. W. Lewis, *J. Raman Spectrosc.*, 2016, **47**, 948–955.
8. K. J. van der Pal, G. Sauzier, M. Maric, W. van Bronswijk, K. Pitts and S. W. Lewis, *Talanta*, 2016, **148**, 715–720.
9. M. Maric, W. van Bronswijk, S. W. Lewis and K. Pitts, *Talanta*, 2014, **118**, 156–161.
10. M. Maric, W. van Bronswijk, S. W. Lewis, K. Pitts and D. E. Martin, *Forensic Sci. Int.*, 2013, **228**, 165–169.
11. K. M. Pitts, MPhil thesis, Curtin University, 2022.
12. T. G. Newland, K. Pitts and S. W. Lewis, *Forensic Chem.*, 2022, **28**, 100412.
13. K. M. Pitts and R. M. Clarke, *Forensic Sci. Int. Rep.*, 2020, **2**, 100130.
14. J. R. Gozzard, *Geological Survey of Western Australia*, 2010, p. 72.
15. J. D'Uva, R. Fillingham, S. W. Lewis, D. DeTata and R. Dunsmore, in *25th International Australian and New Zealand Forensic Science Society (ANZFSS) Symposium*, Brisbane, Queensland, Australia, 2022.
16. P. Gill, *Forensic Sci. Int.: Genet.*, 2016, **23**, 9–18.
17. J. M. Chin and C. M. Ibaviosa, *Sci. Justice*, 2022, **62**, 272–283.
18. A. P. A. Broeders, *Forensic Sci. Int.*, 2006, **159**, 148–157.
19. S. Coulson, in *Encyclopedia of Forensic Sciences*, ed. M. M. Houck, Elsevier, Oxford, 3rd edn, 2023, pp. 315–324.
20. J. M. Curran, in *Encyclopedia of Forensic Sciences*, ed. M. M. Houck, Elsevier, Oxford, 3rd edn, 2023, pp. 766–773.
21. J. M. Curran, *Sci. Justice*, 2013, **53**, 251–252.
22. J. M. Curran, C. M. Triggs, J. Buckleton and S. Coulson, *Forensic Sci. Int.*, 1998, **91**, 181–196.
23. ASTM Standard E2927-16e1, in ASTM Volume 14.02, West Conshohocken, PA, 2016.
24. D. Ramos, J. Maroñas and J. Almirall, *Chemom. Intell. Lab. Syst.*, 2021, **217**, 104399.
25. A. Biedermann, S. Bozza and F. Taroni, *Sci. Justice*, 2018, **58**, 159–165.
26. G. S. Morrison and N. Poh, *Sci. Justice*, 2018, **58**, 200–218.
27. S. Gittelson, T. Kalafut, S. Myers, D. Taylor, T. Hicks, F. Taroni, I. W. Evett, J. Bright and J. Buckleton, *J. Forensic Sci.*, 2016, **61**, 186–195.
28. R. Palmer, K. Sheridan, J. Puckett, N. Richardson and W. Lo, *Forensic Sci. Int.*, 2017, **278**, 334–337.
29. S. Cantrell, C. Roux, P. Maynard and J. Robertson, *Forensic Sci. Int.*, 2001, **123**, 48–53.
30. K. J. Sheridan, R. Palmer, D. A. Chalton, J. N. Bacar, J. Beckett, K. Bellerby, L. Brown, E. Donaghy, A. Finlayson, C. Graham, B. Robertson, L. Taylor and M. D. Gallidabino, *Sci. Justice*, 2023, **63**, 509–516.
31. K. J. Sheridan, E. Saltupyte, R. Palmer and M. D. Gallidabino, *Forensic Sci. Int.*, 2020, **315**, 110432.
32. M. E. Sigman, M. R. Williams, N. Thurn and T. Wood, *Forensic Chem.*, 2021, **26**, 100358.

33. A. Akmeemana, M. R. Williams and M. E. Sigman, *Forensic Chem.*, 2017, **5**, 91–108.
34. M. Lopatka, M. E. Sigman, M. J. Sjerps, M. R. Williams and G. Vivó-Truyols, *Forensic Sci. Int.*, 2015, **252**, 177–186.
35. Z. Liu, W. Wang and X. Liu, *TrAC, Trends Anal. Chem.*, 2023, **160**, 116956.
36. H.-P. Wang, P. Chen, J.-W. Dai, D. Liu, J.-Y. Li, Y.-P. Xu and X.-L. Chu, *TrAC, Trends Anal. Chem.*, 2022, **153**, 116648.
37. M. González, R. P. Gorziza, K. de Cássia Mariotti and R. Pereira Limberger, *J. Forensic Sci.*, 2020, **65**, 1040–1048.
38. M. A. Souza, A. S. Santos, S. W. da Silva, J. W. B. Braga and M. H. Sousa, *Forensic Chem.*, 2022, **27**, 100395.
39. C. Manis, C. Malegori, E. Alladio, M. Vincenti, P. Garofano, F. Barni, A. Berti and P. Oliveri, *Talanta*, 2022, **245**, 123472.
40. G. J. Parker, H. E. McKiernan, K. M. Legg and Z. C. Goecker, *Forensic Sci. Int.: Genet.*, 2021, **54**, 102529.

Subject Index

Locators in *italic* refer to figures; those in **bold** to tables

www.ingramcontent.com/pod-product-compliance
Lightning Source LLC
LaVergne TN
LVHW051155210926
848486LV00005B/282